CONSTRUCTIONAL ENGINEERING AND ECOLOGICAL ENVIRONMENT

Constructional Engineering and Ecological Environment contains papers presented at the 4th International Symposium on Architecture Research Frontiers and Ecological Environment (ARFEE 2022, Guilin, China, 23–25 December, 2022). With a focus on hot research topics and difficulties in construction technology and ecological environment, this book provides the latest research results on a variety of topics:

- building structures
- civil engineering
- seismic technology
- ecological environment repair

The book is aimed at engineers, scholars and researchers in construction, structural engineering and environmental sciences.

PROCEEDINGS OF THE 4TH INTERNATIONAL SYMPOSIUM ON ARCHITECTURE RESEARCH FRONTIERS AND ECOLOGICAL ENVIRONMENT (ARFEE 2022), GUILIN, CHINA, 23–25 DECEMBER 2022

Constructional Engineering and Ecological Environment

Edited by

Chih-Huang Weng
I-Shou University, Taiwan

CRC Press
Taylor & Francis Group
Boca Raton London New York Leiden

CRC Press is an imprint of the
Taylor & Francis Group, an **informa** business
A BALKEMA BOOK

First published 2024
by CRC Press/Balkema
4 Park Square, Milton Park, Abingdon, Oxon, OX14 4RN

and by CRC Press/Balkema
2385 NW Executive Center Drive, Suite 320, Boca Raton FL 33431

CRC Press/Balkema is an imprint of the Taylor & Francis Group, an informa business

ISBN: 978-1-032-53198-4 (hbk)
ISBN: 978-1-032-53200-4 (pbk)
ISBN: 978-1-003-41084-3 (ebk)

DOI: 10.1201/9781003410843

Typeset in Times New Roman
by MPS Limited, Chennai, India

Table of Contents

Preface

The 2022 4th International Symposium on Architecture Research Frontiers and Ecological Environment (ARFEE 2022) was successfully held in Guilin, China from December 23rd to 25th, 2022 (virtual event). This Seminar was the fourth in a series of topical meetings, which began with the Wuhan meeting in 2018 and continued with the Guilin meeting in 2019, and the Zhangjiajie meeting in 2020. ARFEE 2022 was attended by about 60 participants from different countries.

For this fourth meeting we returned to Guilin and met with our old or new friends again. While the location of the Conference has been changed sometimes, what remained invariant is the aim of these meetings, which is to discuss recent advances and new perspectives in architecture research frontiers and ecological environment in a pleasant and friendly atmosphere. At this meeting we had a comprehensive overview of this fascinating field and of future scenarios thanks to the participation of leaders of the most important projects.

The Conference presented an outstanding program of papers covering the most recent advances in architecture research frontiers and ecological environment, including 3D Printing Architecture, Architectural Design and Theory, Green Building, Building Energy Saving Technology, Sustainable Development, Wetlands and Aquatic Ecosystems, Waste Treatment and Recycling, etc. The papers in this Proceedings published by CRC Press / Balkema – Taylor & Francis Group represent a collection of the invited talks.

The conference program consisted of five invited speeches presented in plenary sessions, including five "hot topic" speeches highlighting the most recent advances in the field, and various oral presentations. The program included speeches by Prof. Baoyu Gao who honored as the Highly-cited Scholar of China (environmental science) by Elsevier from 2014 to 2022, a speech title with "Influence of Coagulation Behavior and Floc Properties on Membrane Fouling: The Role of Different Al Species", by Prof. Zheng Chen on "Improvement in Strength and Chloride Resistance of Concrete Blended with SCBA", etc. In the latter one, SCBA was variously treated by three different processing protocols to improve the pozzolanic activity of the resulting SCBA.

We would like to thank the participants, especially those who contributed speeches, posters and manuscripts, for making ARFEE 2022 such an exciting and memorable conference. We thank the Program Committee for putting together an outstanding program and the ARFEE International Advisory Committee for their professional advice and suggestions. Finally, we thank the Organizing Committee for their tireless and expert efforts in the organization of ARFEE 2022, and all of our colleagues whose friendly and efficient service contributed much to the success of the Conference.

The Committee of ARFEE 2022

© 2024 The Editor, ISBN 978-1-032-53198-4

Committee Members

Conference General Chairs
Prof. Chih-Huang Weng, *Taiwan I-Shou University, China*
Prof. Baoyu Gao, *Shandong University, China*

Technical Program Committee Chair
Prof. Zheng Chen, *Guangxi University, China*

Publication Chair
Prof. Chih-Huang Weng, *Taiwan I-Shou University, China*

Technical Program Committee
Prof. Xueying Liu, *Zhejiang University of Water Resources and Electric Power, China*
Prof. Hua Wang, *China Coal Research Institute, China*
Prof. Rui Neves, *Polytechnic Institute of Setubal, Portugal*
Prof. Darve, *University Grenoble Alpes, France*
Prof. Antonio Caggiano, *Darmstadt University of Technology/CONICET and University of Buenos Aires, Germany and Argentina*
Assoc. Prof. Jun Hu, *Hainan University, China*
Prof. Feng Lan, *Xi'an University of Architecture and Technology, China*
Prof. Xianguo Wu, *Huazhong University of Science and Technology, China*
Assoc. Prof. Shengqi Guo, *Hebei University of Technology, China*
Prof. Tao Zhu, *China University of Mining and Technology (Beijing), China*
Prof. Li Di, *Gansu Agricultural University, China*
Prof. Yong Zhang, *CMA Centre for Atmosphere Watch and Services, China*
Prof. Yongxiu He, *North China Electric Power University, China*
Prof. Zhibao Huo, *Shanghai Ocean University, China*
Prof. Yi Wang, *China University of Geosciences, China*
Prof. Qiuxin Liu, *Wuhan University of Science and Technology City College, China*
Dr. Congmi Cheng, *Guangzhou University, China*
Prof. Changji Shan, *Zhaotong University, China*
Prof. Caiyun Qian, *Nanjing Tech University, China*
Dr. Tingguang Ma, *Shanghai Institute of Technology, China*
Prof. Wei Gao, *Northeast Forestry University, China*
Assoc. Prof. Chunming Wang, *South China Agricultural University, China*
Prof. Shiyuan Huang, *University of South China, China*
Prof. Zheng Chen, *Guangxi University, China*
Prof. Zhenfu Chen, *University of South China, China*
Assoc. Prof. Yanning Wang, *Shantou University, China*
Prof. Lei Wang, *Changsha University of Science & Technology, China*
Dr. Ling Qi, *Beijing University of Technology, China*
Prof. Shaosen Wang, *Xiamen University, China*
Assoc. Prof. Xiaojun Dai, *Southwest Petroleum University, China*
Prof. Yunzhang Li, *Sichuan University, China*
Prof. Xiaoshuang Shi, *Sichuan University, China*
Prof. Luoping Zhang, *Xiamen University, China*
Prof. Lunwei Zhang, *Tongji University, China*
Prof. Shuqing Hao, *China University of Mining and Technology, China*
Prof. Shuili Yu, *Tongji University, China*

Dr. Shibin Yuan, *Southwest University of Science and Technology, China*
Prof. Jorge de Brito, *University of Lisbon, Portugal*
Prof. Yanhua Wan, *Huazhong University of Science and Technology, China*
Dr. Hachemi Samya, *University of Biskra, Algeria*
Assistant Prof. Behrouz Behnam, *Amirkabir University of Technology, Iran*
Prof. James Yang, *Coventry University, UK*
Dr. F. Pacheco Torgal, *University of Minho, Portugal*
Prof. Xia Huo, *Jinan University, China*
Prof. Yuning Miao, *Guanxi Minzu University, China*
Prof. Ping Xiang, *Central South University, China*
Prof. Ming Li, *Dalian University of Technology, China*
Assoc. Prof. Hongxu Peng, *Fuzhou University, China*
Dr. Thota Sivasankar, *NIIT University, India*
Prof. Guoyun Lu, *Taiyuan University of Technology, China*
Prof. Guoping Xiong, *Southeast University, China*
Assoc. Prof. Xin Wu, *Fuzhou University, China*
Prof. Ying Fu, *University of Jinan, China*
Prof. Bo Wang, *Chang'an University, China*
Assoc. Prof. Rui Guo, *Southwest Jiaotong University, China*
Assoc. Prof. Pingping Zhang, *South China Agricultural University, China*

Building structure and civil engineering seismic technology

Constructional Engineering and Ecological Environment – Chih-Huang Weng (Ed)

Analysis on lateral behavior of viscoelastic damping horizontal corrugated steel shear wall

Fengchen Li & Xiaotong Peng*
School of Civil Engineering and Architecture University of Jinan, China

Chen Lin
School of Architecture and Landscape Design Shandong University of Art & Design Jinan, China

ABSTRACT: In order to make up for the shortage of lateral stiffness of traditional lateral force resisting systems and reduce the seismic requirements of steel frames for joints, we proposed a new lateral force resisting system called Viscoelastic Damping Horizontal Corrugated steel shear wall System (VDHCS), which was considered by sliding long hole viscoelastic dampers, corrugated steel plates, and peripheral steel frames. The nonlinear finite element model of the new lateral force-resisting system was established by ABAQUS finite element software. The lateral force resisting capacity of the system was analyzed by displacement loading method to study the influence of viscoelastic damper, thickness, and wavelength of corrugated steel plate on the bearing capacity. The results showed that the system which added a viscoelastic damper possesses better lateral resistance. With the increase of plate thickness and wavelength, the amplification of each lateral resistance index of the system shows a downward trend. Compared with the wavelength of corrugated steel plate, the thickness has a greater impact on the bearing capacity of the system.

1 INTRODUCTION

The traditional steel frame structure has small lateral stiffness and large lateral displacement; the steel plate shear wall structure is prone to local buckling. If the thickened steel plate or ribbed structure is used to prevent premature buckling, the economic cost will be increased. For this reason, we proposed a Viscoelastic Damping Horizontal Corrugated steel shear wall System (VDHCS). The corrugated steel plate with greater out of plane stiffness is selected to replace the flat steel plate, and the sliding long-hole viscoelastic damper is introduced to combine with the peripheral steel frame. Anwar Hossain and Wright (2004) proposed to combine corrugated steel plates as shear walls with steel frames and proved the feasibility of this scheme through relevant experimental studies; W. Berman and Bruneau (2005) found that corrugated steel plate shear wall has better ductility, buckling resistance, and bearing capacity through comparative tests; Li Feng (2005) found that the steel plate shear wall with honeycomb section has reasonable stress and clear force transmission through finite element variable parameter analysis; Fan Jiaqi *et al.* (Fan 2020, 2022) conducted a comprehensive study on the lateral performance of corrugated steel plate shear walls, and obtained the influence of geometric parameters on the lateral performance; Zhao Qiuhong *et al.* (Zhao 2016, 2018) studied the seismic performance of structures related to corrugated steel plate shear walls and summarized the influence of trapezoidal corrugated steel plates on the

*Corresponding Author: pengxito@163.com

DOI: 10.1201/9781003410843-1

seismic performance of shear walls; Zhang Tingting (2017) used ABAQUS to carry out finite element analysis and research on the new viscoelastic lateral force resisting system. The results show that the new viscoelastic lateral force-resisting system can achieve better seismic performance. Two systems and six variable parameter models are established by ABAQUS finite element software to explore the influence of dampers and the main dimensions of corrugated steel plates on the lateral resistance of the structural system.

2 FINITE ELEMENT SIMULATION OVERVIEW

According to the relevant standard (GB50009-2012), VDHCS is formed by combining a corrugated steel plate shear wall and sliding long-hole viscoelastic damper with the peripheral steel frame, as shown in Figure 1. When VDHCS is subjected to frequently occurring earthquakes or strong wind loads, the long hole sliding viscoelastic damper is only used as a force transfer component to transfer the lateral force to the lower corrugated steel plate, and the corrugated steel plate is mainly used for lateral resistance; When VDHCS suffer from a rarely occurred earthquake, the damper helps the structure absorb some of the vibration energy, and the damper will eventually enter the plastic state before the beam-column frame and other components. This elastic-plastic deformation is one of the most effective mechanisms for consuming seismic energy. According to the above mechanism, the VDHCS nonlinear Finite Element Model (FEM) is established using ABAQUS finite element software. In this paper, a shell element is adopted. Two trapezoidal corrugated steel plates are selected to form a honeycomb combined trapezoidal plate, which is deformed and dissipates energy together by combining commands. The steel frame of the structure is Q345 grade steel, and the embedded corrugated steel plate is Q235 grade steel. Due to the complex deformation and energy dissipation of the viscoelastic damper with long-hole bolts, it is necessary to consider how to replace the preload imposed by high-strength bolts. Therefore, the viscoelastic damper with long hole bolts is simplified as a spring element, symmetrically arranged on both sides of the corrugated steel plate. The reference point is set at the centroid of the top of the column to couple with the section at the end of the column, and the displacement load is applied in the Z-axis direction of the reference point. The displacement corresponding to the reduction of the bearing capacity to 85% of the peak value is taken as the ultimate displacement. If the bearing capacity of the model does not decrease to 0.85Vmax after reaching the peak value, the ultimate displacement is taken as the maximum displacement.

3 EFFECT OF VISCOELASTIC DAMPER

In order to study the influence of viscoelastic dampers on the bearing capacity and ductility of the new lateral force-resisting system, a nonlinear finite element model of VDHCS and Horizontal Corrugated steel plate Shear wall System (HCSS) is established. Except for whether dampers are added, the dimensions of other members are completely the same. Displacement control loading is simulated to 100mm, and the ductility and bearing capacity of the structure are compared and analyzed. As shown in Figure 2, with the increase of displacement load, the lateral deformation of both structures increases gradually. When the displacement is 100mm, the ultimate bearing capacity of VDHCS is 2673.29KN, which is 19.6% higher than that of HCSS 2149.33KN, which is 19.6% higher than that of HCSS. The ultimate bearing capacity of the new viscoelastic lateral force-resisting system is significantly improved. Through the analysis of the results of the finite element simulation, it can be seen that the difference in the initial stiffness of the structure is small, indicating that the viscoelastic damper has an impact on the initial stiffness; After entering the plastic stage, the shear wall generates a tension band in the diagonal direction, and the bearing capacity of the

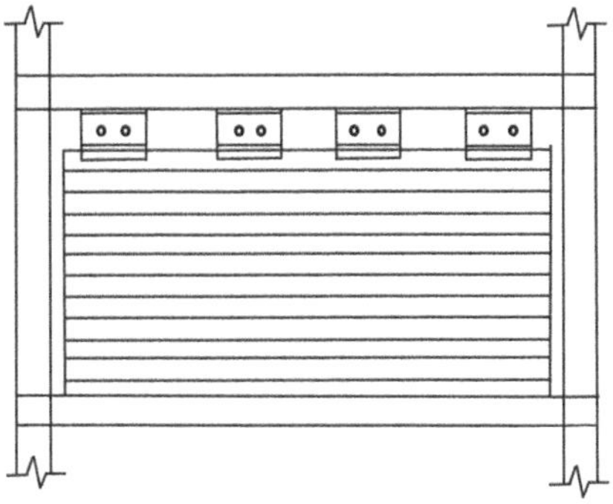

Figure 1. VDHCS model (Self-drawn).

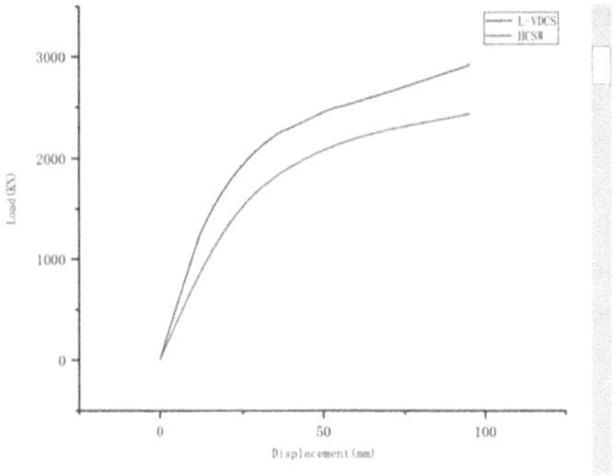

Figure 2. Load displacement curve of VDHCS and HCSS(Self-drawn).

structure slowly rises to a stable state; after loading, the ductility coefficient is also improved compared with that of HCSS. Therefore, VDHCS shows a high bearing capacity and ductility performance. At the same time, due to the introduction of viscoelastic dampers, the trend of structural curve changes has not changed significantly, and some energy is consumed in the process of structural lateral resistance so that the ultimate bearing capacity and ductility of the structural system are improved.

4 PARAMETER ANALYSIS

On the basis of the VDHCS model, considering the different parameters of corrugated steel plate that affect the lateral resistance of the structure: the thickness and wavelength of corrugated steel plate, 6 models of 2 series in total are designed for monotonic loading analysis. The selection of relevant model parameters is shown in Table 1. t_w is the thickness of the plate, and q is the wavelength of the plate.

Table 1. VDHCS model dimensions with different plate thicknesses (Self-drawn).

Model	M-1	M-2	M-3	M-4	M-5	M-6
t_w / (mm)	4	2	3	5	4	4
q / (mm)	240	240	240	240	180	300

4.1 *Influence of plate thickness*

To analyze the influence of the thickness of the corrugated steel plate on the lateral resistance of VDHCS, four VDHCS finite element analysis models were established. The thickness of a single corrugated steel plate was 4 mm, 2 mm, 3 mm, and 5 mm, and the wavelength was the same. As shown in Figure 3(a), The thickness of the corrugated steel plate will not affect the development trend of the load-displacement curve of the structure, and the curves of each model have obvious rising and yielding segments. The structure shows lateral stiffness, which is proportional to the plate thickness. The displacement of each model when entering the yield section and reaching the ultimate bearing capacity is different, indicating that the plate thickness greatly influences the lateral stiffness and bearing capacity of VDHCS. M-1 dissipates energy after reaching the plastic stage. Then there is a second rising stage, gradually rising steadily to the peak point, which means that the damper is excited. The auxiliary system dissipates energy after reaching the yield point. M-2 will be damaged when it is loaded to 91 mm, so the thickness of the single-layer plate in this system should be at least greater than 2 mm. In order to more accurately analyze the influence of plate thickness on the lateral resistance performance of the structure, the lateral resistance indicators of each model are extracted for analysis: With the increase of the thickness of corrugated steel plate, the initial stiffness of VDHCS increases continuously, but the increased amplitude shows a downward trend.

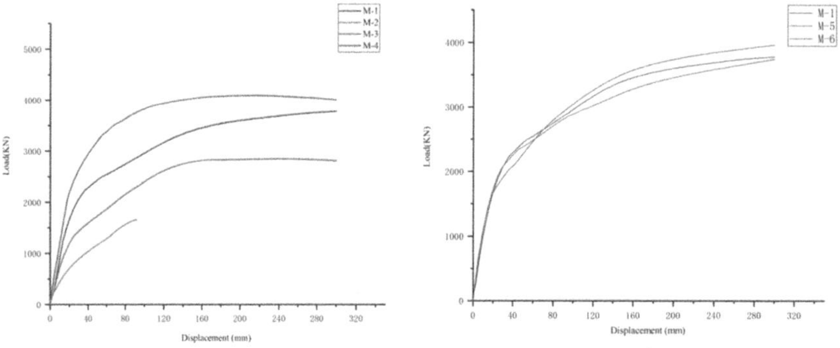

Figure 3. Load displacement curve (Self-drawn). (a) Influence of thickness (b) Influence of wavelength.

For each 1 mm increase in the thickness of each model, the increased amplitude is 41.5%, 24.1%, and 17.4%, respectively. The ultimate bearing capacities of the structure increase by 72.2%, 32.4%, and 10.3% with each 1 mm increase in the plate thickness. Except for M-2, there is little difference in the change of the ductility coefficient of each model. Starting from 3 mm, the ductility coefficient increases by 11.4% and 7.4% for every 1 mm increase. The ductility of M-1 and M-4 structures is good. It can be seen from the above that the corrugated steel plate has a great impact on the initial stiffness, ultimate bearing capacity, and ductility of the structure. Increasing the thickness of the corrugated steel plate will improve the initial stiffness, ultimate bearing capacity, and ductility of the structure. However, with the increase in the thickness of the plate, the increase of the lateral resistance index of the structure begins to decrease. At this time, continuing to increase the thickness of the plate will cause poor economic efficiency. At the same time, the thickness of the single-layer plate should not be less than 2 mm. Therefore, it is reasonable to comprehensively consider the single-deck corrugated steel plate thickness of 4 mm.

4.2 *Influence of wavelength*

In order to analyze the influence of the wavelength of corrugated steel plate on the lateral resistance of VDHCS, three finite element analysis models of VDHCS were established, with wavelength q of 180mm, 240mm, and 300mm, respectively. As can be seen from the load-displacement curve in Figure 3(b), the lateral bearing capacity, lateral stiffness, and ductility coefficient of each model are nearly the same, and the influence of wavelength is small. The curves of M-1, M-5, and M-6 all rise steadily to the top. We extract anti-lateral indicators of each model for more detailed analysis: The difference between the initial stiffness and ductility coefficient of VDHCS with different wavelengths is small, and each model shows higher lateral stiffness and ductility. The ultimate bearing capacity changes obviously and decreases with the increase of wavelength. The ultimate bearing capacity of M-1 and M-5 is 3780kN and 3739kN, respectively, which is about 200kN lower than that of M-6. This is because the wavelength is shorter, and the ripple of the corrugated plate is more compact. Therefore, when the steel plate is subjected to shear yielding, its area of action is larger, resulting in its out-of-plane stiffness increasing, thus improving its lateral bearing capacity. The gap between M-1 and M-5 is small, and the improvement caused by the change in wavelength is not obvious. Therefore, the lateral resistance of the structure can be improved by appropriately shortening the wavelength, but the wavelength is not easy to be too small to avoid increasing the steel consumption of steel plates.

5 CONCLUSION

By comparing the results of VDHCS and HCSS under the displacement loading method, the ultimate bearing capacity and ductility coefficient of VDHCS are improved, which indicates that the introduction of viscoelastic dampers can effectively dissipate some energy in the process of lateral resistance of the structure so that the ultimate bearing capacity and ductility of the structure are improved. The purpose of energy dissipation is achieved. The thickness of the corrugated steel plate has a great impact on the initial stiffness, ultimate bearing capacity, and ductility coefficient of the structure. Properly increasing the thickness of the steel plate can improve the lateral resistance of the structure, but the thickness of a single layer of steel plate should not be less than 2mm. When the thickness of the plate is 3mm, continue to increase the thickness of the plate, and the increase of the lateral resistance of the structure starts to decrease, reducing the economy. Therefore, it is reasonable to take 4mm as the simulation result of a single layer of corrugated steel plate. The wavelength has little influence on the ductility and initial stiffness of the structure, but the shorter the wavelength is, the higher the ultimate bearing capacity of the structure is, and the greater the loading displacement for bearing capacity degradation (overall buckling) occurs. This is because the smaller the wavelength is, the higher the ripple density is, and the greater the shear buckling capacity of the steel plate is. The buckling time of the steel plate is delayed, so the bearing capacity of the structure is also increased.

ACKNOWLEDGMENTS

The work was sponsored by the Natural Science Foundation of Shandong Province (ZR2019MEE009), Ministry of Education University-Industry Collaborative Education Program (201802276002; 201902204001; 201902204002), Shandong Postgraduate Education and Teaching Reform Research Project (SDYJG19039), Science and Technology Project of Housing and Urban-Rural Development of Shandong Province (2020-K2-3). The writers gratefully acknowledge all the support provided.

REFERENCES

Anwar Hossain K.M. and Wright H.D. Experimental and Theoretical Behaviour of Composite Walling Under In-Plane Shear. *Journal of Constructional Steel Research*, 60(1):59–83(2004).

Berman J.W. and Bruneau M. Experimental Investigation of Light-Gauge Steel Plate Shear Walls. *Journal of Structural Engineering*, 131(2):259–267(2005).

Li Feng and Zhao Liang. Elastic Buckling Analysis of Transverse Honeycomb Shaped Section Steel Plate Shear Wall. *Steel Construction*, 29(2):182(2014).
Fan Jiaqi. *Research on Lateral Performance of Corrugated Steel Shear Wall Structure[D]*. Harbin Institute of Technology (2020).
Fan Jiaqi, Guo Ming, Li Wentao, Zhao Qunchang and Cha Xiaoxiong. Simplified Analysis Model and Seismic Analysis of Corrugated Steel Plate Shear Wall. *Building Structure*, 52(S1):1207–1211(2022).
Fan Jiaqi, Guo Ming, Li Wentao, Zhao Qunchang and Cha Xiaoxiong. Shear Buckling Behaviour of Corrugated Steel Plate Shear Wall. *Building Structure*, 52(S1):1212–1216(2022).
Zhao Qiuhong, Li Nan and Sun Junhao. Analysis on Lateral Performance of Sinusoidally Corrugated Steel Plate Shear Walls. *Journal of Tianjin University: Science and Technology*, 49(S1):152–160(2016).
Zhao Qiuhong, Qiu Jing, Li Nan and Li Zhongxian Experimental Study on Seismic Performance of Trapezoidally Corrugated Steel Plate Shear Walls. *Journal of Building Structures*, 39(S2):112–120(2018).
Zhang Tingting. *Seismic Analysis of New Viscoelastic Lateral Resisting System[D]*. University of Jinan. (2017).
(GB50009-2012). *Load Code for the Design of Building Structures[S]*.

Constructional Engineering and Ecological Environment – Chih-Huang Weng (Ed)

Research on detection and treatment of old road in upgradation and reconstruction of trunk highway

Zhanhong Hu*
Henan Yanhuang Expressway Co., Ltd, Zhengzhou, China

ABSTRACT: As the traffic volume of trunk highways grows fast with the rapid development of China's economy, various distresses appear inevitably in the asphalt pavement, and upgradation and reconstruction become a necessity to the pavement with the increase in the service time of trunk highways. At present, it is unfounded and subjective to propose a detection scheme for old pavement in the upgrading and reconstructing of trunk highways. No specifications or guidelines especially designed for major maintenance or upgradation and reconstruction of trunk highways are available. By summarizing the previous experience of old road detection and in view of the features of trunk highway upgrading, the paper proposes formulating the detection scheme by determining typical road sections through investigation of typical distresses, performing general detection of the whole line and key detection of typical road sections and combining field detection and indoor material test. In addition, the paper provides a case study that shows the detection scheme of the old road, proposed based on features of upgradation and reconstruction of trunk highways, can be well applied to practical projects.

1 INTRODUCTION

Trunk highways are the important highways that occupy the main position and play a leading role. By the end of 2020, the total length of highways opened to traffic nationwide reached 4.7735 million km. With the increase in service time of trunk highways, most asphalt pavements show various distress due to the influence of the environment, climatic conditions and traffic volume growth. In view of serious damage to trunk highways, the traffic management departments generally adopt major maintenance or upgrading and reconstruction technology to deal with the original roads by investigating the traffic composition of road sections and surveying and evaluating pavement conditions and distresses. At present, the detection scheme for old pavements in the major maintenance or upgrading and reconstruction of trunk highways is mostly empirical, but there are some problems to be solved. The first reason for these problems is the particularity of major and intermediate maintenance or upgrading and reconstruction. As major and intermediate maintenance or upgrading and reconstruction works generally take a short time, there is not enough time to arrange detection, which will inevitably affect the quality of distress treatment and subsequent construction. The second reason is the unavailability of specifications or guidelines especially designed for old pavement detection in major maintenance or upgradation and reconstruction of trunk highways. In practice, pavement condition assessment specifications or maintenance specifications are generally referred to. But after all, the detection for upgradation and reconstruction works is different from routine detection and often

*Corresponding Author: 627224349@qq.com

DOI: 10.1201/9781003410843-2

encounters thorny problems that are not described in routine detection specifications (Ma 2010; Wang 2009). Therefore, the prerequisite and guarantee for major and intermediate maintenance or upgrading and reconstruction of trunk highways are studying the old pavement detection scheme to reflect pavement distress conditions and put forward targeted maintenance measures.

2 PROJECT DESCRIPTION

The studied project in the paper is the section (K828 + 720 to K843 + 720) of the G312 National Highway in Xinyang, Henan Province, with a total length of 15.0 km. The original pavement structure includes an upper surface course (4 cm thick asphalt concrete), middle surface course (5cm thick asphalt concrete), lower surface course (7 cm thick asphalt concrete), base course (18 cm thick cement stabilized crushed aggregate) and subbase course (20 cm thick cement and lime soil stabilized sand).

3 ANALYSIS AND EXAMPLE OF OLD PAVEMENT DETECTION

3.1 *Analysis of features of upgrading and reconstruction of trunk highways*

Typical distresses have existed in the pavement before the upgradation and reconstruction of the trunk highway, which differs in degree and mainly includes a transverse crack, longitudinal crack, alligator cracking, pumping, depression and patching (Cui *et al.* 2005; Yang 2016).

At present, there are three main schemes for upgrading and reconstruction of trunk highways: (i) direct overlaying scheme, which mainly involves the overlay of seal coat and structure layer depending on traffic volume; (ii) mill-and-fill scheme, which mainly includes removal of existing surface course with a milling machine and replacement of the milled location with a new surface layer, and removal of base course with a milling machine and replacement of the milled located with new base course and surface course; and (iii) recycling scheme for milled materials from the pavement.

Different schemes need to be designed according to the investigation results of road conditions and the features of pavement distresses, so it is necessary to investigate the typical pavement distresses before old pavement detection and put forward targeted detection schemes according to the distribution of distresses.

Based on the studied project, the specific detection scheme, as shown in Figure 1, is put forward by determining typical road sections through investigation of typical distresses, performing general detection of the whole line and key detection of typical road sections and combining field detection and indoor material test.

First, in view of actual conditions of trunk highway reconstruction, important and necessary items are selected to be detected along the whole line of a highway that is to be reconstructed, with the intention of analyzing typical distresses of the highway and understanding the severity and distribution of diseases. Second, the road sections with concentrated typical distresses are selected from the highway to be reconstructed. The road sections with serious typical distress and the general road sections are provided with contrast detection. The detection items and depth are more detailed than the whole line detection to have a final understanding of the features and development of distresses and provide a scientific basis for reconstruction and treatment scheme. As the detection of the whole highway in the detection scheme reduces the items specified in Highway Performance Assessment Standard (JTG H20-2007), detection time can be shortened, and unnecessary expenses can be cut. Meanwhile, the detection company, the design institute and the

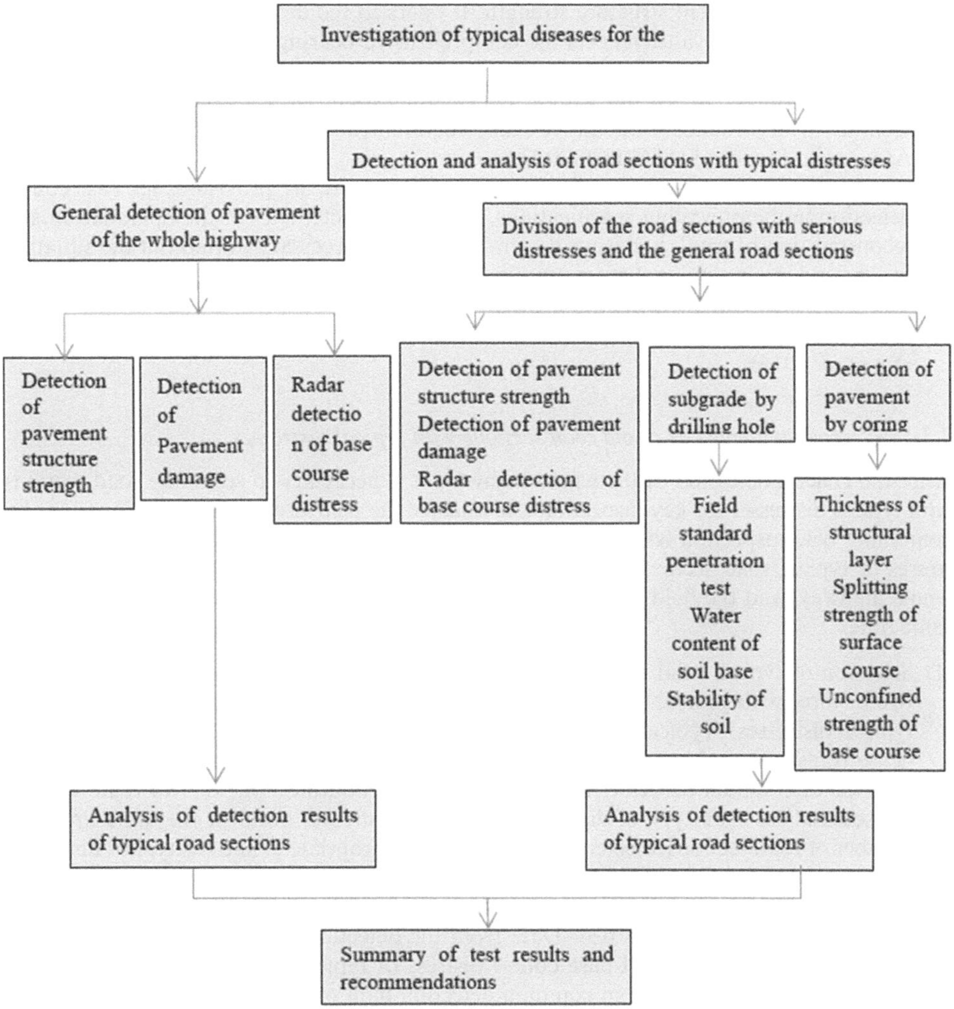

Figure 1. Flow chart for detection of old pavement in upgradation and reconstruction of trunk highway [Self-drawn].

construction contractor can work together to carry out detailed detection of the road sections with typical distresses, which will provide an accurate basis for the distress treatment scheme.

3.2 *General detection of old pavement of the whole highway*

According to the purpose of detection for upgrading and reconstruction, general detection is carried out on the pavement of the whole highway to understand the overall situation of the whole line, such as the severity and distribution of pavement distress. The selected items are mainly for rapid and qualitative detection (Chen 2010; Jiang 2022; Zhang 2014).

(1) Detection of pavement damage. Continuous detection shall be carried out on the detected road sections in accordance with the classification standard of pavement distresses in Highway Performance Assessment Standard (JTG H20-2007).

(2) Detection of pavement structure strength. It refers to the detection of pavement deflection value, and the value reflects the comprehensive bearing capacity of subgrade and pavement and is an important index of pavement performance evaluation. Continuous detection shall be carried out with a falling weight deflectometer on the detected road sections in accordance with the relevant requirements of the Highway Performance Assessment Standard (JTG H20-2007).
(3) Radar detection of base course distress. At present, there are provisions for base course detection in the prevailing specifications, but the characteristics of the upgradation and reconstruction of trunk highways determine that it is necessary to find out the situation of the old base course before putting forward a targeted reconstruction scheme. Therefore, the ground penetrating radar shall be used in the upgradation and reconstruction of trunk highways to detect the internal defects and raveling of the base course (Yu *et al.* 2014).

3.3 *Detection and analysis of old road sections with typical distresses*

After the general detection of the whole highway, it is necessary to select the road sections with typical distresses for key inspection and analyze the distresses of these road sections by combining field inspection with the indoor material test. The analysis and detection of distresses of typical road sections shall be refined on the basis of the detection items for the whole highway, and the field coring verification test shall be additionally performed at the same time.

(1) Selection of typical road sections. One or two road sections with good conditions and two or three with general conditions are selected to be compared with road sections with typical distresses. Typical distresses, such as depression, rutting, alligator cracking, dense transverse and longitudinal cracks and pumping, are selected in consideration of the type of common distresses and actual project conditions. One or two road sections are selected for each type of distress. If multiple distresses occur at the same time, the number of road sections selected can be adjusted appropriately, and the typical distresses of the detected items must be widely covered.
(2) Detection of pavement structure strength, detection of pavement damage, and radar detection of base course distress. Data from the detection of pavement damage, pavement structure strength and base course distress of typical road sections are extracted and evaluated separately from common detection data of the whole highway.
(3) Detection of pavement by coring. They take core samples from the road sections with typical shall be carried out according to features of each distress. Core drilling is carried out for cracks, core taking is performed within the range of alligator cracking, and core samples are taken at the peaks and valleys of rutting, with the depth of core samples reaching the bottom of the pavement structure layer. The detection contents include a description of the core sample condition, the actual thickness of each structural layer of asphalt pavement, and the development depth of distress.
(4) Detection of subgrade by drilling hole. Detection of subgrade by drilling holes must be carried out for road sections with depression and those with high fill and high groundwater levels. The main purpose of drilling holes in the subgrade is to detect soil properties, water content and bearing capacity of the subgrade.
(5) Mechanical strength test of core sample. All drilled core samples shall be cut according to the horizon of the pavement structure layer. The splitting strength test is carried out for the surface course, and the unconfined compressive strength test is for the base course. Ten representative road sections with typical distress were selected from the studied project, including five sections with serious distresses, 3 sections with general conditions and two sections with good conditions, as shown in Table 1.

Table 1. Conditions of road sections with typical distresses [self-drawn].

Number of typical road sections	Starting and ending stake number of the road section	Pavement condition	Description of typical distress condition
1	K829 + 060-K829 + 260	Good	Pavement with good condition (with a small amount of slight alligator cracking)
2	K830 + 620-K830 + 720	Inferior	Large depression, accompanied by pumping
3	K833 + 700-K833 + 900	General	Slight transverse cracks (with a small amount of pumping, white mud) and a small quantity of cracks (without pumping)
4	K834 + 700-K834 + 800	Inferior	Large depression and alligator cracking (located at the end of the bridge)
5	K835 + 640-K835 + 840	Inferior	High fill with large depression and alligator cracking
6	K836 + 720-K836 + 920	General	Continuous longitudinal cracks, accompanied by pumping and partial alligator cracking
7	K838 + 620-K838 + 820	General	Longitudinal and transverse cracks are typical and have been sealed
8	K840 + 520-K840 + 720	Good	Pavement with good conditions
9	K841 + 920-K842 + 020	Inferior	It is an excavated section where depression, alligator cracking and pumping are concentrated
10	K842 + 600-K842 + 800	Inferior	Pavement depression with serious alligator cracking and large deflection

3.4 *Analysis of detection results of road sections with typical distresses*

Detection results of road sections with typical distresses are shown in Table 2.

As can be seen in Table 2:

(1) The minimum representative deflection of the typical road section is 22 (0.01 mm), which corresponds to typical road section 8 with good pavement condition and a PCI of 86.3.

Table 2. Detection results of road sections with typical distresses [self-drawn].

Number of typical road sections	Pavement condition index (PCI)	Representative deflection/ (0.01 mm)	Length of distress of base course/ m	Thickness of the pavement structure layer/cm	Strength of pavement structure layer/Mpa				Standard penetration test/times	
					Upper surface course	Lower surface course	Base course	Subbase course	1.05–1.35m	1.65–1.95m
1	80.1	37.4	0	33.3	0.40	–	14.00	11.10	10	7
2	56.5	91.8	30	42.9	0.44	0.38	17.20	6.80	9	3
3	70.2	72.5	10	40.1	0.56	0.41	11.30	8.00	5	8
4	50.3	90.4	55	39.9	0.46	0.37	20.10	9.40	9	–
5	58.3	78.7	70	35.2	0.51	–	13.50	7.00	7	5
6	71.0	52.9	20	38.2	0.64	0.32	14.50	8.30	8	11
7	72.2	55.7	0	41.8	0.84	0.69	15.80	8.50	11	13
8	86.3	22.0	0	40.3	0.76	0.58	16.40	10.30	–	–
9	66.0	94.6	15	40.1	0.65	0.42	14.70	11.20	4	11
10	58.0	133.2	23	36.3	0.58	–	9.70	10.20	4	8

The maximum representative deflection of the typical road section is 133.2 (0.01 mm), which corresponds to typical road section 10, which has poor pavement conditions and a PCI of 58. There is a good corresponding relationship between deflection and PCI of each typical road section.

(2) The thickness of the pavement structure layer of typical road section 1 is the smallest, which is 33 cm. The thickness of the pavement structure layer of the typical road section 2 is the largest, which is 42.9 cm. The average thickness of the pavement structural layer is 38.8 cm, and the thickness of the structural layer of each typical road section has no obvious change.
(3) No damage to the base course was detected in typical road sections 1 and 8, which have good pavement. Different damages were found in the base course of other typical road sections.
(4) The splitting strength of the asphalt surface course is small, and the strength of the base and subbase course s meets the requirements.
(5) For sections (typical road sections 2, 4, 5, 9 and 10) with serious pavement depression, alligator cracking and pumping, and large deflection, standard penetration tests were carried out at two horizons with depths of 1.05–1.35 and 1.65–1.95 m. Generally, the blow count of the standard penetration test of subgrade in one horizon is 5 or less (the compactness of subgrade is evaluated as loose or very loose). For road sections with good pavement and those with general pavement, in two different standard penetration depths, there is generally a depth in which the soil subgrade compactness is evaluated as medium density, and the blow count of a standard penetration test is generally more than 5 times.

3.5 *Summary of detection results and recommendations*

Summary of Detection Results. In view of the detection results of old roads of the section of G312 highway in Xinyang, the analysis and summary are as follows:

(1) The main typical distresses of pavement are longitudinal and transverse cracks, depression, alligator cracks and pumping. Some sections have structural distresses, and PCI value changes obviously with road sections.
(2) The pavement deflection test data changes greatly and obviously with road sections.
(3) There are some distresses such as raveling, fragmentation and depression in the base course, subbase course and subgrade of some road sections. The conclusions of core drilling investigation and deflection detection show that in the road sections with good pavement, the core samples taken out from the upper base course are complete and dense, and the strength of the base course and subbase course meets the specification requirements.
(4) The core drilling test results indicate that (i) some cracks are temperature shrinkage cracks, with no cracks in the base course; (ii) some cracks are reflection cracks, with cracks in the base course; (iii) the road sections with alligator cracking and depression is generally subjected to damages to the base course; (iv) the base course of road sections pumping white mud is basically intact; (v) the base course of road sections pumping yellow mud is obviously damaged; and (vi) the base course of the road sections with serious pavement distresses is seriously broken.
(5) The average thickness of the pavement structure layer is 38.8 cm, and the coefficient of variation is 34.7%. The variability of thickness of the base course of old asphalt pavement is large.
(6) The subgrade soil is silty clay, and the water content of the subgrade at different depths is between 21.6% and 27.8%. The soil consistency is generally stiff, and some road sections with serious distress are plastic.

Recommendations for Detection and Assessment of Old Road. (1) PCI value, deflection and blow counts of standard penetration tests of subgrade soil have a good corresponding relationship. The blow counts of a standard penetration test are small for road sections with serious pavement damage and large deflection. Therefore, improvement of the bearing capacity of pavement structure shall be considered for road sections with serious distresses in the later treatment of distresses (Zeng & Li 2011).

(2) It is recommended that according to PCI and pavement deflection, the pavement is roughly classified into three types for targeted treatment. The pavement classification is as follows: (i) road sections with good pavement condition, whose PCI is generally greater than 80 and deflection is 15 – 30 (0.01 mm); (ii) road sections with fair pavement condition, whose PCI is generally 70 -80 and deflection is 30 -50 (0.01 mm); and (iii) road sections with poor pavement condition, whose PCI is generally less than 70, and deflection is 60 -120 (0.01 mm).

Recommendations for Treatment Measures. (1) For road sections with poor pavement condition, it is recommended that: (i) the pavement structure layer is removed, and the original soil subgrade is treated to ensure the modulus of soil subgrade meets design requirements; (ii) cement stabilized macadam is laid and overlaying for reinforcement is carried out; and (iii) cold in-place recycling with cement can be considered for old materials excavated from the base course.

(2) For road sections with fair pavement condition, the treatment scheme shall be designed for pavement distresses (including cracks, alligator cracking, pothole and depression) in accordance with their damage rate. It is recommended that mill-and-fill or cold in-place recycling technology is applied to road sections with a damage rate greater than 10% for continuous distress treatment, and targeted local treatment measures are adopted for road sections with a damage rate less than 10%.

(3) For road sections with good pavement conditions but isolated distress (cracks, alligator cracking, pumping and local depression), it is recommended to adopt targeted local treatment measures and carry out overlaying for reinforcement (Gao 2012; Zhao 2014).

4 CONCLUSION

Targeted detection schemes are proposed on the basis of the features of old road detection in upgrading and reconstruction of trunk highways, pavement classification is provided according to old road detection results, and different treatment measures are given depending on pavement condition.

REFERENCES

Chen Zhaojun. Cause Analysis and Maintenance Strategy Study of Pavement Distresses of Leiyang-Yizhang Highway. *Journal of China and Foreign Highway*, 2010 (5).

Cui Yingchao, Sun Mingang & Pan Guoqiang. Detection and Evaluation of Asphalt Concrete Pavement Diseases in Highway Reconstruction. *Chinese Journal of Underground Space and Engineering.*, 2005 (7):1163–1167.

Gao Fei. Research on Key Technologies of Major Maintenance of Asphalt Pavement of Heavy Traffic Trunk Highway [D]. *Master's Degree Thesis from Hebei University of Technology*, 2012.

Jiang Guoshuai. Research on Common Diseases and Treatment Measures in Highway Asphalt Pavement Inspection. *Western China Communication Science & Technology*, 2022(9):68–70.

Ma Jin. Research on Distress Features and Major Maintenance Decisions of Asphalt Pavement of Trunk Highway in Shaanxi Province [D]. *Master's Degree Thesis from Chang'an University*, 2010.

Wang Lei. Research on Cause Analysis and Maintenance Technology of Typical Distresses of Asphalt Pavement of Trunk Highways in Shaanxi province [D]. *Master's Degree Thesis from Changan University*, 2009.

Yang Changgui. Detection and Evaluation of Old Highway. *Transpoworld*, 2016(26): 46–47.

Yu Xianjiang, Wang Chuntao & Gu Zhangchuan. Application of Comprehensive Evaluation System in Detecting Internal Quality of Pavement Structure by Ground Penetrating Radar. *Journal of China and Foreign Highway*, 2014 (2).

Zeng Sheng & Li Zhencun. Discussion on Several Issues in Highway Performance Assessment Standard. *Journal of China and Foreign Highway*, 2011 (2).

Zhang Dongjin. Application of Pavement Inspection in Old Road Reconstruction Project. *Journal of Henan Science and Technology*, 2014(2):166.

Zhao Peng. Study on Treatment Scheme of Old Road Pavement in Tongxi Reconstruction and Extension Project. *Journal of Highway and Transportation Research and Development*, 2014(4):31–33.

Constructional Engineering and Ecological Environment – Chih-Huang Weng (Ed)
© 2024 The Author(s), ISBN 978-1-032-53198-4

Restaurant architecture design of reuse of railway sleeper materials: Wumi Congee (Congee Shop) in Panyu Guangzhou

Qiang Tang
Shunde Polytechnic. Shunde District, Foshan City, Guangdong Province, China

Yile Chen & Liang Zheng
Faculty of Humanities and Arts, Macau University of Science and Technology. Avenida Wai Long, Taipa, Macau, China

Junzhang Chen*
Faculty of Innovation and Design, City University of Macau. Avenida Padre Tomás Pereira Taipa, Macau, China

ABSTRACT: This paper takes the reuse of railway sleeper materials as the core analysis and discusses it in conjunction with the Guangzhou Panyu Old Town Renewal Project Wumi Congee (Congee Shop). The study found that the use of sleeper materials for recycling and application in the restaurant's building facade and interior decorations have the best effect of integrating into the environment.

1 INSTRUCTIONS

1.1 *Origin of railway sleeper*

A railway sleeper is a rectangular support for the rails in railroad tracks. Generally laid perpendicular to the rails, ties transfer loads to the track ballast and subgrade, hold the rails upright and keep them spaced to the correct gauge. Railway Sleeper Materials are traditionally made of wood, but prestressed concrete is now also widely used, especially in Europe and Asia. Steel ties are common on secondary lines in the UK (Tie 2022); plastic composite ties are also employed, although far less than wood or concrete. As of January 2008, the approximate market share in North America for traditional and wood ties was 91.5%, the remainder being concrete, steel, azobé (red ironwood) and plastic composite (Remennikov 2008).

Railways in mainland China have long used wood as sleepers. Timber prices in China have been rising due to low forest cover and the huge population's demand for furniture and building materials. However, China's cement production has increased significantly after the 1980s, and the price of cement has dropped relatively. The railway department has replaced wooden sleepers with cement sleepers. Most of the high-speed railway systems built after 2000 have abandoned the traditional railway mode of sleeper plus ballast and switched to ballastless railways.

1.2 *Literature review*

Over the past 140 years, the shape of railway sleepers has remained largely unchanged. At present, more than 2.5 billion railway sleepers have been laid around the world. Due to the

*Corresponding Author: U19091105192@cityu.mo

DOI: 10.1201/9781003410843-3

technical and economic advantages of wooden sleepers in tunnels, bridges, turnouts, vibration reduction sections, small curve radii and industrial lines, it is still an important type of railway sleeper in North America, Europe and the Commonwealth of Independent States. For example, railway sleepers account for more than 85% of rail sleepers in the United States and Canada. In the maintenance and repair of the railway sleeper, in order to avoid the sudden change of the mechanical characteristics of the track line, a sleeper with similar mechanical properties should be used to replace the failed railway sleeper, and the prestressed reinforced concrete sleeper should not be used to replace the wooden sleeper. Therefore, although the share of wooden sleepers in railway sleepers worldwide has stagnated, it is still a major sleeper type.

Before the founding of the People's Republic of China and in the early days of the People's Republic of China, wooden sleepers were generally used on railways in my country. Since the 1950s, with the development of seamless lines and concrete technology, my country has begun to develop and use concrete sleepers on a large scale. Although the proportion of wooden sleepers in my country's railway sleepers is only 3.6%, the number in service still exceeds 11.7 million (Monson 2009). The average service life of wooden sleepers is 20 years, and 5% of them need to be replaced every year, that is, about 585,000 wooden sleepers. Considering that the volume of each wooden sleeper is about 0.1 m^3, it is equivalent to consuming 58,500 m^3 of high-quality wood every year. In addition, wasted wooden sleepers contain heavy metals and dangerous carcinogens, which will pose a threat to the ecological environment (soil, water, etc.) And human health if not handled properly. Therefore, extending the service life of wooden sleepers and properly disposing of discarded wooden sleepers have positive significance for the sustainable development of railways and the construction of ecological civilization in my country (McNeill 2004).

1.3 *Research methods*

This study focuses on an old city renovation project in Canton, Guangdong. Firstly, through on-the-spot investigation and analysis, the direction of architectural renovation design is determined. Secondly, our research team visited the building material market, compared the performance of the materials, and determined the railway sleeper as the main recycling design material. Finally, a series of drawings were drawn, and construction was carried out.

2 VENUE SITUATION

2.1 *Location*

The project site is near No. 41, Shiqiao Qiao South Road, Panyu District, Guangzhou City, Guangdong Province, China. It is at a crossroads and has a central position in the life of the residents of this area. However, this place used to be an old market and has been deserted for almost a decade (Figure 1).

2.2 *Consideration of regional culture in architectural design*

The design of the restaurant building is inseparable from the regional food culture. Going back to history, there is Wumi Congee, a pot of seemingly ordinary porridge water, but it condenses the Shunde people's pursuit of food and also carries Shunde catering enterprises' adherence to and inheritance of Lingnan food culture. In addition to its birthplace Shunde, there are also branches in Foshan Chancheng, Guangzhou, Zhuhai, Shenzhen, and even Macau and Beijing outside the province. However, how to reflect the characteristics of Lingnan food culture in architectural design?

Figure 1. Location of Wumi Congee (Congee Shop) in Panyu District, Guangzhou City.

The so-called "Wumi Congee" is made from chicken and herbs. The fragrant rice was boiled in broth for several hours. It is necessary to stir repeatedly to prevent the rice grains from sticking to the bottom during the process. It produces porridge water that is delicate and smooth, milky white and translucent, and water and rice are one. Afterward, porridge water is used as the soup base to scald seafood, meat, and seasonal vegetables.

The porridge water contains a large amount of minerals and various trace elements. Its high temperature not only locks in the moisture of the ingredients but also retains the original flavor of the ingredients. Therefore, "Wumi Congee" makes the food cooked with porridge water fresh and refreshing, accompanied by a refreshing rice fragrance. What's even more surprising is that this pot of porridge water will not stick to the bottom, no matter how you cook it. This is also the magic and skill of "Wumi Congee". This kind of special cooking skill often amazes foreign tourists: "Hotpot can still be eaten like this. Shunde's food culture is really broad and profound". In 2011, the porridge water-making technique of "Wumi Congee" was selected for the "Intangible cultural heritage" list of Shunde.

2.3 *Scheme design concept*

With the influx of migrant workers and new immigrants, the Lingnan cultural city with Guangzhou as its core has gradually become culturally diverse. In order to retain this local catering culture, we maximize the use of local building materials for recycling. Therefore, the "principle of subtraction" was utilized when conceiving the conceptual scheme (Figures 2

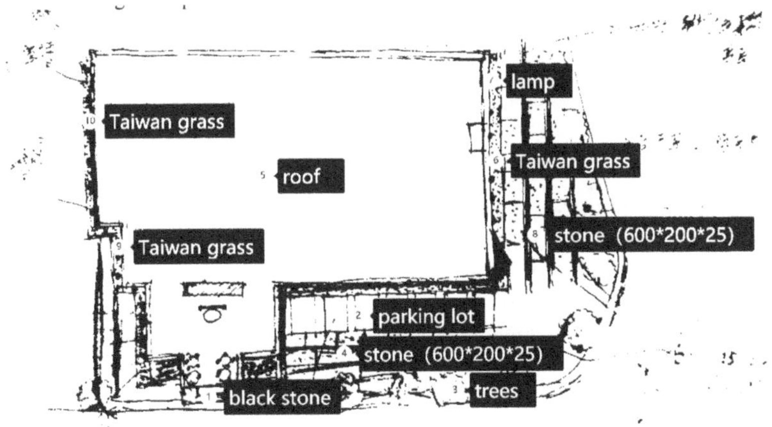

Figure 2. The design concept of Wumi Congee (Congee Shop) in Panyu District, Guangzhou City.

to 3). Take a rectangular building plan, cut the surrounding space, and expand the landscape design around this Congee Shop.

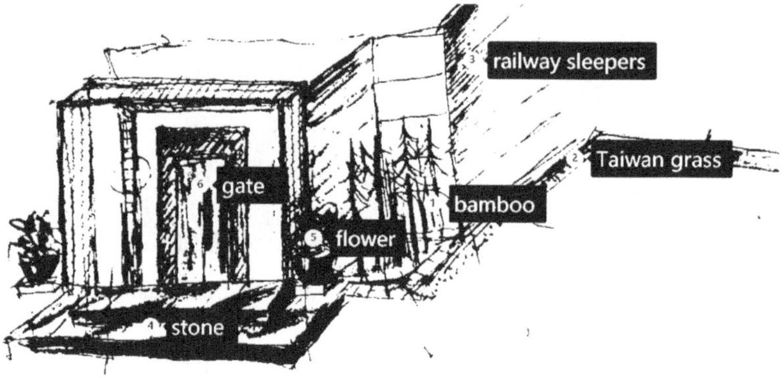

Figure 3. The design concept of Wumi Congee (Congee Shop) in Panyu District, Guangzhou City.

Since ancient times, Guangdong has been rich in bamboo. Guangning, Zhaoqing, and Guangdong have a history of bamboo cultivation and utilization for more than 2000 years. Beginning in the late Ming Dynasty, with the development of Guangdong's commerce and trade, bamboo was widely known as a bulk export commodity in Guangning. Among all the varieties, the most prominent is Bambusa textiles. After the Bambusa textiles are cut, as long as the roots are not damaged, shoot tips will emerge from the roots again. Bamboo grows faster, and the shoot buds will rise to Hsinchu within a year, and the more you cut, the lusher it becomes. As long as it is thinned every three years, the old bamboo is chopped down, the new ones are left, and the green bamboo in Guangning can be used sustainably. Therefore, the exterior facade of the restaurant is designed to plant bamboo, which is not only a continuation of the regional context but also a reflection of the environmental protection of the building materials.

3 APPLICATION OF RAILWAY SLEEPER MATERIALS

3.1 *Building facade*

On the building facade, a large area of railway sleeper is used. The horizontal laying appears to have a more extended feeling (Figure 4). However, for the consideration of the restaurant's lighting, a part of the floor-to-ceiling glass windows are reserved for the building's facade. The strip railway sleeper splicing treatment is more flexible.

3.2 *Entrance gate*

In the past, residential buildings in Guangdong had triple doors. The first heavy foot door is a four-fold small folding door, about half the height of the door. It separates the interior and exterior areas of the house and reduces outside disturbances. At the same time, it can be ventilated, lighted, flexible, and artistic. The second gate is a sliding gate similar to a fence, called the "Tanglong Gate". It consists of 13, 15 or 17 hard round wooden bars, which open and close horizontally as a whole, but do not close. The door is solid and heavy, and there are hardwood matches and chains behind the door, which are generally closed when no one is in the house, which can effectively prevent theft.

Figure 4. Building facade of Wumi Congee (Congee Shop) in Panyu District, Guangzhou City.

Therefore, the entrance door of the restaurant building also imitates the door frame of the triple door, and the new design is carried out in a fusion style. At the decorative position on the edge of the door frame, sleepers are also used (Figure 5).

Figure 5. Entrance gate of Wumi Congee (Congee Shop) in Panyu District, Guangzhou City.

3.3 *Landscape walk outside the building*

The outdoor landscape trails are also paved with sleepers. But the fly in the ointment is that there is a certain degree of damage after the railway sleeper is recycled. It is difficult to achieve seamless butt joints in laying joints. Therefore, small stones were incorporated into the design (Figure 6).

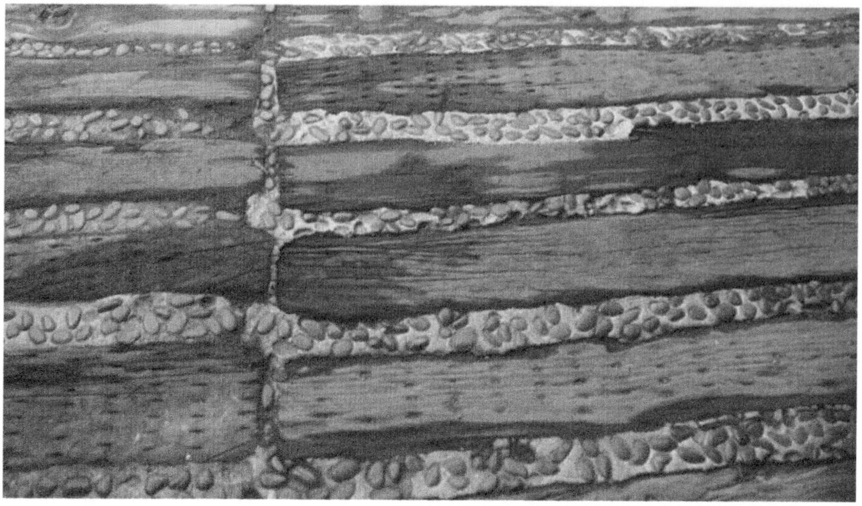

Figure 6. Landscape walk of Wumi Congee (Congee Shop) in Panyu District, Guangzhou City.

4 CONCLUSIONS

A study in Croatia showed that the concentrations of copper and zinc in railway sleepers treated with creosote were 66 times and 21 times higher than the maximum allowable concentrations in agricultural land soil, respectively. The allowable upper limit value has also been reached. Therefore, wasted railway sleepers must be properly disposed of. In the recycling of non-load-bearing or decorative components such as outdoor furniture and fences, the decoration design of the restaurant is currently a successful attempt.

ACKNOWLEDGMENT

This article is based on the research staff from 2019 to 2020 from the 2020 Guangdong Province Ordinary Universities Characteristic Innovation Project (No. 2020WTSCX286); and the phased achievements of the Tang Qiang Guangfu Cultural Heritage Protection Skills Master Studio of Shunde Polytechnic.

REFERENCES

McNeill J. R. (2004). Woods and Warfare in World History. *Environmental History*, 9(3), 388–410.

Monson J. (2009). *Africa's Freedom Railway: How a Chinese Development Project Changed Lives and Livelihoods in Tanzania.* Indiana University Press.

Remennikov A.M. & Kaewunruen S. (2008). A Review of Loading Conditions for Railway Track Structures Due to Train and Track Vertical Interaction. *Structural Control and Health Monitoring: The Official Journal of the International Association for Structural Control and Monitoring and of the European Association for the Control of Structures.* 15(2), 207–234.

Tie K.H.P.R. (2022). *Wikipedia's Railroad tie as Translated by GramTrans.*

Constructional Engineering and Ecological Environment – Chih-Huang Weng (Ed)
© 2024 The Author(s), ISBN 978-1-032-53198-4

Research on seismic performance of new assembled semi-rigid beam-column joints

Qianwen Han
Key Laboratory of Building Structure of Anhui Higher Education Institutes, Anhui Xinhua University, Hefei, China

Ziming Fan*, Chaohao Wang & Xiangran Ren
School of Urban Construction, Anhui Xinhua University, Hefei, China

ABSTRACT: A new type of assembled semi-rigid beam-column joint is studied, in which 10.9-grade high-strength bolts and 4.6-grade common bolts are used to embed the beam-column respectively, and Q345 steel plates and L-shaped connectors are used for welding between the prefabricated corbel column and the prefabricated rectangular beam. Through numerical simulation of ABAQUS, its hysteretic curve, skeleton curve, energy consumption and other seismic performance are analyzed. The results show that although the basic mechanical properties of high-strength bolts are slightly higher than those of common bolts, their hysteretic curves, skeleton curves and energy dissipation curves are basically consistent, indicating that the seismic performance of the two types of bolts is close. As the bolts damaged in the earthquake can be replaced, using common bolts in the construction process can save construction costs. The installation process discards the sand and cement used in traditional projects while reducing construction waste.

1 INTRODUCTION

Assembled construction means that some or all building components are processed and manufactured in a factory, transported to the site, and installed through reliable connections. Since the assembled components can be reused to achieve resource regeneration, it avoids the waste of raw materials and also reduces construction waste. At present, the connection types of assembled beam-column joints are mainly divided into dry connection and wet connection. The nodal connection form is the main problem that needs to be solved for the current implementation of modular buildings, which has been studied by scholars at home and abroad.

Congxiao Wu *et al.* carried out a numerical analysis on prefabricated assembled concrete beam-column joints with sector lead viscoelastic dampers and compared it with the test results. It was found that sector leads viscoelastic dampers changed the stress mode of prefabricated concrete frame joints, effectively reduced and controlled the development of cracks in the core area of the joints and reduced the strain values of load-bearing bars and stirrups in the post-cast area of the beam end (Wu *et al.* 2017). Dongfeng Zhu *et al.* proposed a reinforcement sleeve extrusion and steel pipe positioning connection beam-column joint. This joint can effectively solve the reinforcement connection problems of fabricated structures and components. It is concluded that the prefabricated concrete structure with

*Corresponding Author: 253760722@qq.com

DOI: 10.1201/9781003410843-4

this joint system can achieve similar or even the same seismic performance as the full cast-in-place concrete structure (Zhu *et al.* 2020). José F *et al.* proposed a beam-column joint, which connects the reinforcement at the end of the column together and conducts tests under cyclic load. The results of the connection between precast columns and columns are compared with those of similar cast-in-place elements without joints. It is concluded that the performance of the two elements is equivalent, with only slight differences in cracking, damage distribution and hysteretic performance. Therefore, it can be considered appropriate to use the proposed precast column-to-column connection (José *et al.* 2018). R. A. Hawileh *et al.* established a three-dimensional nonlinear finite element model of precast composite beam-column joints. The model considers the pretension effect of post-tensioned steel strands and the nonlinear material behaviour of concrete. The model response is compared with the test results, and good consistency is achieved in each stage of loading (Hawileh *et al.* 2009).

In this paper, the joints are mainly welded by bull-leg columns and rectangular beams through connectors and steel plates. However, through the previous earthquake damage found that the semi-rigid node of the bull leg has the disadvantage that the bull leg is easily pulled. Therefore, this new node adopts the combination of bolt connection and bull leg column to play the role of supporting and transferring the load. This node abandons the connection method of traditional joints and improves the seismic performance of the overall members. And ABAQUS numerical analysis using grade high strength bolts and common bolts, and the corresponding conclusions were drawn.

2 DESIGN OF ASSEMBLED SEMI-RIGID BEAMS-COLUMNS

2.1 *Joints design and production*

The joints in this paper are made of prefabricated concrete members, of which beams are made of C40 rectangular beams and columns are made of C80 bull-leg columns. The junction of the rectangular beam and bull leg column on both sides is welded with a Q345 steel plate with a thickness of 20 mm and pre-buried bolt holes with a radius of 30 mm. The upper beam and column are welded with L-shaped connectors with a thickness of 20 mm and pre-buried bolt holes with a radius of 30 mm. The beams and columns used in this experiment were all reinforced with HRB500, where the beam section is 750 mm × 750 mm. Both the upper and lower parts of the beam are 4 bars of 22 mm diameter. The hoop reinforcement is made of 8 mm diameter steel bars. The encrypted area is spaced 100 mm apart, and the non-encrypted area is 200 mm apart. Lateral reinforcement is 3 pieces of 22 mm diameter reinforcement on each side. The column section is 750 mm × 750 mm. The longitudinal reinforcement is 8 pieces of 25 mm diameter steel bars. The hoop reinforcement is made of 10 mm diameter steel bars. The encrypted area is spaced 100 mm apart, and the non-encrypted area is 200 mm apart. The bull-leg column is equipped with 4 bars of 20 mm diameter. The degree of the concrete protection layer is set at 25 mm (Zhao 2021). The design dimensions of the main components and reinforcement diagram of the joints are shown in Figure 1. The concrete material and its steel components parameters are shown in Tables 1 and 2.

2.2 *Realization of the design concept of "strong column and weak beam"*

Strong column and weak beam refer to the actual bending bearing capacity of the column end being greater than the actual bending bearing capacity of the beam ends at the beam-column joint. That is, the plastic hinge appears at the beam end first. Among them, the bending bearing capacity of the column end and beam end can be calculated according to the code for the seismic design of buildings (China Construction Industry Press 2010), and

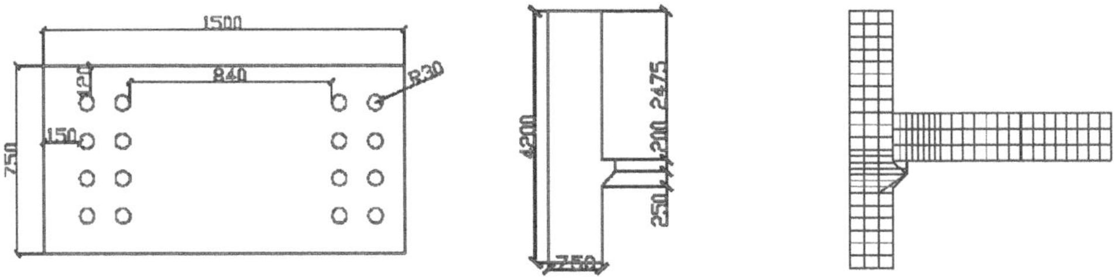

(a) (b) (c)

Figure 1. Dimensions of main components and reinforcement diagram. (a) Q345 steel plate (b) C80 prefabricated column (c) Reinforcement diagram.

Table 1. Mechanics performance of concrete.

Concrete strength	Cube compressive strength (MPa)	Axis compressive strength (MPa)	Elastic modulus (GPa)	Poisson ratio
C40	40	26.8	32.5	0.2
C80	80	50.2	38.0	0.2

Table 2. Mechanical properties of steel bars and steel members.

Category/ bolt grade	Steel bar (bolt) diameter/steel plate thickness (mm)	Yield Strength (MPa)	Limit Strength (MPa)	Elastic modulus (GPa)	Poisson ratio
HRB500	8–25	500	630	200	0.28
Q345	20	345	550	200	0.28
L-type	20	345	550	200	0.28
10.9	60	940	1040	200	0.28
4.6	60	240	400	200	0.28

the calculation formula is as follows.

$$\sum M_{cu} \geq \sum M_{bu} \tag{1}$$

$$\sum V_{cu} \geq \sum V_{bu} \tag{2}$$

where, M_{cu} is the bending moment of the column, M_{bu} is the bending moment of the beam, V_{cu} is the shear resistance of the column, and V_{bu} is the shear resistance of the beam.

For bending moments,

$$h_0 = h - \alpha_s = 750 - 44 = 706\text{mm} \tag{3}$$

Where, h_0 is the effective height of the section, h is the height of the section, and α_s is the distance from the edge of the tensile zone to the acting point of the resultant force of the tensile reinforcement.

For the C40 concrete beam,

$$x = \frac{f_y A_s}{\alpha_1 f_c b} = \frac{435 \times 1520}{1.0 \times 19.1 \times 750} = 46.16\text{mm} \tag{4}$$

$$\xi = \frac{x}{h_0} = 0.065 < \xi_b = 0.518 \tag{5}$$

$$\rho_{min} = \max\left\{0.2\%, 45\frac{f_t}{f_y}\%\right\} = \max\{0.002, 0.0017\} = 0.002 \tag{6}$$

$$\rho = \frac{A_s}{bh_0} = \frac{1520}{750 \times 706} = 0.28\% > 0.2\% \tag{7}$$

$$M_{bu} = \alpha_1 f_c bx\left(h_0 - \frac{x}{2}\right) = 1.0 \times 19.1 \times 750 \times 46.16 \times \left(706 - \frac{46.16}{2}\right) = 751.6\text{kN} \cdot \text{m} \tag{8}$$

Where, x is the compression zone height, f_y is the design value of the tensile strength of the reinforcement, A_s is the sectional area of the longitudinal load-bearing reinforcement in the tension zone, $\alpha_1 = 1.0$ is the influence coefficient of concrete strength, f_c is the design value of concrete axial compressive strength, b is the width of rectangular section, ξ is the relative compression zone height of the section, ξ_b is the height of the compression zone relative to the boundary, ρ is the reinforcement ratio of the section, and f_t is the design value of the axial tensile strength of the concrete.

For the C80 concrete columns,

$$x = \frac{f_y A_s}{\alpha_1 f_c b} = \frac{435 \times 3927}{35.9 \times 750} = 63.45\text{mm} \tag{9}$$

$$\xi = \frac{x}{h_0} = 0.09 < \xi_b = 0.518 \tag{10}$$

$$\rho_{min} = \max\left\{0.2\%, 45\frac{f_t}{f_y}\%\right\} = \max\{0.002, 0.0023\} = 0.002 \tag{11}$$

$$\rho = \frac{A_s}{bh_0} = \frac{3927}{750 \times 706} = 0.7\% > 0.2\% \tag{12}$$

$$M_{cu} = \alpha_1 f_c bx\left(h_0 - \frac{x}{2}\right) = 1.0 \times 35.9 \times 750 \times 63.45 \times \left(706 - \frac{63.45}{2}\right) = 1151.9\text{k}N \cdot \text{m} > M_{bu} \tag{13}$$

For shear resistance,

$$V_{cu} = \frac{0.2\beta_c f_c bh_0}{r_{RE}} = 4472.7\text{k}N > V_{bu} = 2379.6\text{k}N \tag{14}$$

where, β_C is the influence coefficient of concrete strength, and r_{RE} is the seismic adjustment coefficient of bearing capacity, taken as 0.85.

The stability of the structure cannot be achieved without mechanical support, the essence of which is also the confrontation of forces between the columns and beams (Liu 2018). The results show that the ultimate bearing capacity of reinforced concrete precast columns is greater than that of precast beams, so this joint meets the idea of strong columns and weak beams.

3 ESTABLISHMENT OF FINITE ELEMENT MODEL

3.1 *Element type and mesh division*

In this paper, the boundary points are selected for research. The reinforcement is a double-line model. The concrete, steel plate, and connectors are solid elements, and the reinforcement is a truss element. In the analysis step, the dynamic implicit method is adopted for circulation, angle steel connectors and steel plates are welded with concrete beam-columns in constraint settings, and the action between high-strength bolts and common bolts and nuts is also welded. The concrete beam and column, the nut and the pressure-bearing steel plate, the bolt rod, and the hole wall adopt surface contact, the tangential behaviour is set as penalty contact, the friction coefficient is 0.45, and the normal behaviour is set as hard contact. The mesh division of prefabricated concrete is shown in Figure 2.

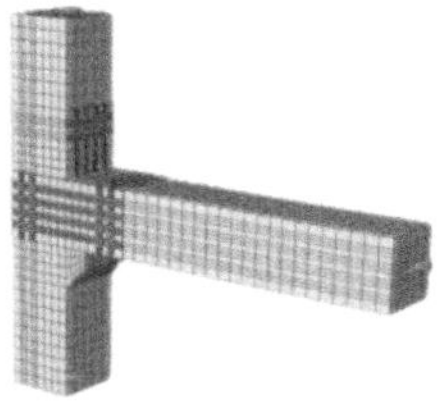

Figure 2. Mesh division.

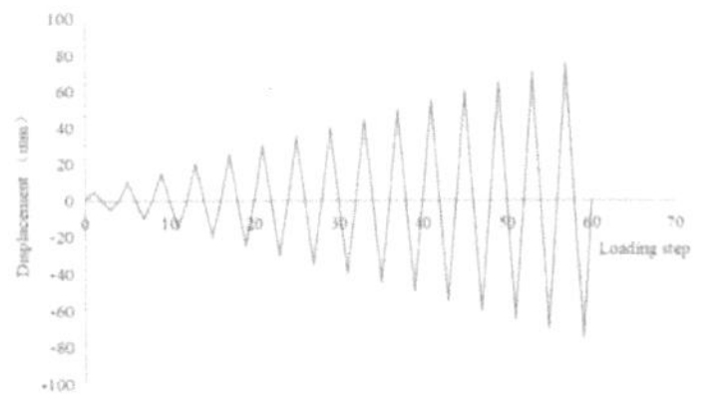

Figure 3. Displacement loading curve.

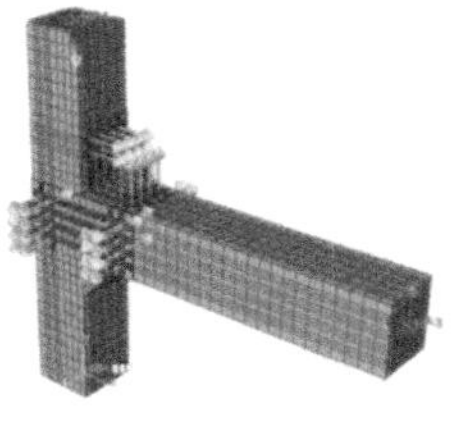

Figure 4. Loading mode.

3.2 *Boundary conditions and loading methods*

(1) At the top and bottom of the column, the degrees of freedom in three directions are fixed, and the lateral direction of the beam end is also fixed.
(2) Let the reference point RP1 at the centre of the column end section and coupled with the column surface, Applying displacement along the negative direction of the Y-axis. A vertical load with an axial pressure ratio of 0.15 is imposed on the top of the column along the minus direction of the Y-axis.
(3) Different preload forces are applied to the bolts in different analysis steps and applied to the bolt cross-section in the indicated direction. The beam-column joint displacement loading curve and the loading method are shown in Figures 3 and 4.

4 ANALYSIS OF FINITE ELEMENT SIMULATION RESULTS

4.1 *Hysteresis curve*

The hysteresis curve can well reflect the relationship between the force and deformation of the test member under low circumferential repeated loading. It can comprehensively reflect the seismic performance of the joints. The hysteresis curves of high-strength bolts and common bolts are shown in Figure 5.

By comparing the hysteretic curves in Figure 5(a) and (b), the following conclusions can be obtained.

(1) High-strength bolts and common bolts' hysteresis curve is similar and are shuttle shaped. The full hysteretic curve reflects the strong plastic deformation capacity of the whole structure or component.
(2) As the load increases, the two different strength bolts begin to enter the yielding phase, when the stiffness of the structure begins to degrade, and the hysteresis curve encircles an increasingly large area.
(3) Depending on the use of different strength bolts, the increase in bolt strength will also lead to an increase in the load capacity of the node and enhanced energy dissipation performance.

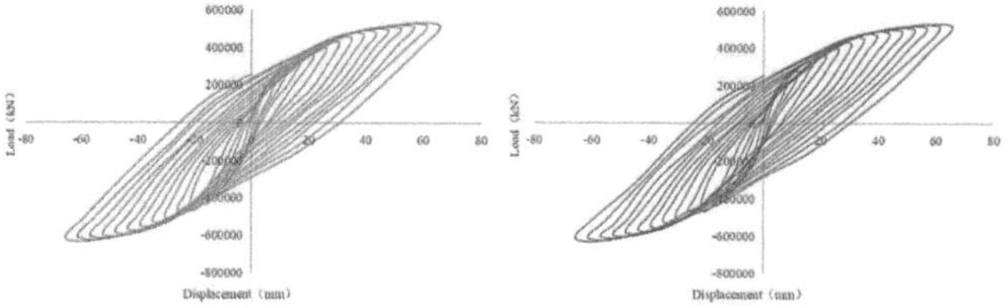

(a) (b)

Figure 5. Hysteresis curves for bolts of different strength. (a) High strength bolt lag curve (b) Common bolt lag curve.

4.2 *Skeleton curve*

The skeleton curve is a curve in which the peak points of hysteretic loops of the first cycle are connected when hysteretic curves are subjected to different levels of loads. Its slope indicates stiffness. The skeleton curve can better reflect the changes in force and displacement of the member during the loading process. The skeleton curves of high-strength bolts and common bolts are shown in Figure 6, and the following conclusions can be obtained.

(1) In the pre-loading period, the load and displacement of the two different strength bolts vary linearly in the elastic phase. As the load increases step by step, the slope of the skeleton curve for both joints begins to decrease, indicating that the stiffness is also decreasing. However, the slope of the skeleton curve of high-strength bolts is still greater than the slope of the skeleton curve of common bolts.

(2) By comparing the skeleton curves, it is found that the yield load of high-strength bolts is 455kN, and the yield load of common bolts is 440kN. The ultimate load of high-strength bolts is 552kN, and the ultimate load of common bolts is 532kN. In contrast, high-strength bolts yield load and ultimate load improve by 3.4% and 3.8% separately.

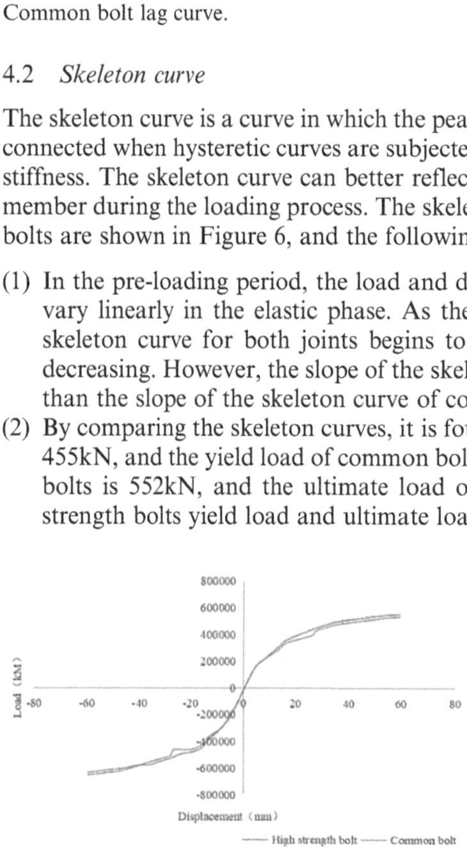

Figure 6. Skeleton curves.

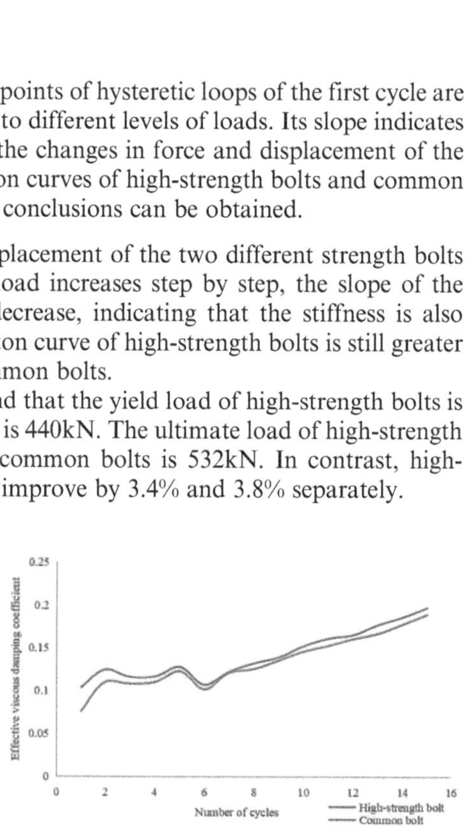

Figure 7. Effective viscous damping coefficients.

4.3 *Energy consumption analysis*

Energy dissipation is measured in terms of the area enclosed by the hysteresis curve and is usually expressed in terms of the equivalent viscous damping coefficient ζ_{eq}. Energy consumption is related to the shape of the hysteresis curve and ductility capacity. The larger ζ_{eq}, the greater the energy dissipation capacity and the greater the resistance to deformation. The effective viscous damping coefficients of bolts with different strengths are shown in Figure 7, and the following conclusions can be obtained.

(1) In the early stage of loading, the equivalent viscosity coefficient grows more slowly. With the increase of cyclic load, the equivalent viscous damping coefficient of two different strength bolts is still increasing. This means that they have a high energy consumption capacity.

(2) The equivalent viscous damping coefficient of high-strength bolts is close to the size of common bolts. The equivalent viscous damping coefficient of high-strength bolts increased by 3.7% compared with that of common bolts in the last loading. The equivalent viscous damping coefficient of high-strength bolts per cycle is between 0.1 and 0.3, which meets the requirements.

5 CONCLUSION

Through the combination of bolts with different strengths and bull-leg columns, and the study of semi-rigid connection, the same geometric dimensions, materials, loads, etc., are selected. Through the comparison of finite element analysis, the following conclusions can be drawn.

(1) The mechanical properties of all types of high-strength bolts are slightly greater than those of common bolts under repeated low circumferential loads.
(2) Since the hysteresis curve, skeleton curve, and energy dissipation curve shapes of the two different strength bolts basically coincide. The seismic performance of high-strength bolts does not increase much compared to common bolts. It means that the mechanical properties of high-strength bolts are approximately equal to those of common bolts.
(3) Under the standard of seismic performance, according to the joint performance and project cost of bolts with different strengths, common bolts can also meet the structural design requirements.

In this paper, the new joint discards the traditional connection mode and improves the seismic performance of the whole component. At the same time, it is also verified that the use of common bolts can also have good seismic performance in accordance with the specifications. Hence, the use of common bolts in building construction also has good application prospects.

ACKNOWLEDGMENTS

This research was supported by the Key Project of Anhui Key Laboratory at School Level (KLBSZD202006), the Key Project of Natural Science Research of Anhui Higher Education Institution (KJ2019A0882), and the Research Project of Anhui School Level (KLBSZD202103).

REFERENCES

Congxiao Wu, Yufeng Zhang, Xuesong Deng and Chao Zhang. Study on the Seismic Performance of Assembled Energy Dissipating and Vibration Damping Concrete Beam-column Joints. *Journal of Disaster Prevention and Mitigation Engineering*. 37 (01): 62–70(2017).

Dongfeng Zhu, Shuiquan Lian, Weisheng Zhang, Jinhui Xie, Jian Cai, Qingjun Chen and Jianfa Zeng. Research on the Application of New Reinforcement Sleeve Extrusion Connection Technology in Assembled Buildings. *Building Structure*. 50(22):51–56 + 76(2020).

GB 50011-2010, *Seismic Design Code for Buildings*. China Construction Industry Press, Beijing. (2010).

Hawileh R.A. Rahman A. and Tabatabai H. Nonlinear Finite Element Analysis and Modeling of a Precast Hybrid Beam–Column Connection Subjected to Cyclic Loads. *Applied Mathematical Modelling* 34(9): 2562–2583(2009).

Jianhua Liu. *Experimental Study on Seismic Performances of a New Type of Fabricated Semi-rigid Beam-to-column Connection*. Anhui Jianzhu University, Hefei. (2018).

José F. Rave-Arango et al. Seismic Performance of Precast Concrete Column-to-Column Lap-Splice Connections. *Engineering Structures*. 172: 687–699(2018).

Zibin Zhao and Liping Zhao. Research on the Seismic Performance of Beam-column Joints of Newly Assembled Frame Structures. *Residence*. (21): 176–178 + 1800(2021).

Constructional Engineering and Ecological Environment – Chih-Huang Weng (Ed)
© 2024 The Author(s), ISBN 978-1-032-53198-4

Research on load test of long-span continuous rigid frame bridge

Zailin Qiu
Luoyang Branch Offices of Henan Transport Investment Group Co., Ltd, Luoyang, China

Bingwei Liu & Xijian Liang*
Henan Provincial Communications Planning & Design Institue Co., Ltd, Zhengzhou, China

ABSTRACT: Taking a long-span continuous rigid frame bridge as an example, this paper systematically introduces an operation safety detection method, which mainly judges whether the technical condition of the bridge meets the operation safety requirements by systematically detecting the appearance of distress. After three years of continuous detection, it is found that an increase in the number of cracks in the beam body is obvious, and there is a downward deflection trend in the center of the main span, so the technical condition of the bridge operation is evaluated as class 4. The load test was carried out on the bridge before reinforcement to ensure the safety bridges and provide a basis for later maintenance and reinforcement. The test results show that the actual bearing capacity of the bridge can still meet the normal operation requirements under design load, but the existing distresses of the bridge need to be treated through maintenance and reinforcement to improve emergency capacity and slow down the development of the distresses.

1 INTRODUCTION

The total length of the bridge is 577.32 m, the span arrangement is 4 × 30 m + (65 + 110 + 65) m + 7 × 30 m, and the main girder has a section of the single box and single chamber. The design load level is: automobile-super 20 class and trailer-120, and the design speed is 100 km/h. The total width of the bridge deck is 26 m, including 0.5 m guardrail, 11.5 m lane, 2 m median divider, 11.5 m carriageway, and 0.5 m guardrail. The basic earthquake intensity is VI, and the design flood frequency is a 300-year return period. In view of the actual conditions of the bridge, the operation safety inspection mainly involves appearance distress, material condition, and geometric shape and is carried out once a year. If necessary, a load test is carried out to judge the actual bearing capacity of the bridge.

2 DETECTION OF APPEARANCE DISTRESS

A bridge detection vehicle was used to detect the superstructure, substructure, and bearings of the bridge. The detection focuses on the following points: (i) the web and pier at the consolidation position of piers outside the box girder; (ii) the web at the quarter position and the bottom plate at the middle position of the main span; (iii) the cast-in-place section of the internal side span of the box girder; (iv) the top and web of the box girder at the quarter position; (v) and the anchor plate at each closure section (Niu 2008; Lian 2015; Lu 2006).

*Corresponding Author: 275310524@qq.com

 DOI: 10.1201/9781003410843-5

The detection results show that: (i) a transverse long crack appears in the bottom plate near the middle of the main span of the left deck, which extends to the web by 0.7 m and has a width of 0.23 mm; (ii) a transverse long crack with a width of 0.20 mm exists in the bottom plate near the middle of the main span of the right deck, which extends to the web by 0.3 m; (iii) the above two transverse cracks have developed rapidly compared with the detection results obtained in 2014; (iv) there are 1093 longitudinal cracks in the roof of the left and right box girders, which is significantly more than the 895 cracks detected in 2013, especially number of the longitudinal cracks in each closure section and cast-in-place section of the box girder side span lies between 10 and15; (v) the length and width of longitudinal cracks at the same position of roof have no obvious development; (vi) there are 63 oblique web cracks in the quarter position inside the box girder, which has increased by 25 cracks compared with 38 cracks in 2014; and (vii) the cracks in the same position basically have no obvious change, except for the slight development of longitudinal crack length in a few positions compared with 2014. There are 6 prestressed anchor plates with longitudinal cracks on each of the left and right decks, among which the longitudinal cracks in 3 anchor plates are radial, and the local crack width is 3 mm. Except that the longitudinal crack width of one anchor plate increases slightly compared with that detected in 2014, there is no other obvious development. The external web of the left box girder has corrosion, and exposed steel bars at the consolidation position of the pier girder, with a corrosion area, is 0.5 m^2. The internal roof of the left and right box girders has breakage and exposed steel bars at the hoisting hole of the construction hanging basket, and the total damaged and exposed bar area of the whole bridge is 2.53 m^2.

For the prestressed concrete bridge, transverse cracks are not allowed, and the maximum width of a longitudinal crack is 0.20 mm. According to the Standards for Technical Condition Evaluation of Highway Bridges (JTG/T H21-2011), the technical condition of the bridge is rated as Class 4 and shall be treated with major maintenance (Ministry of Transport of the People's Republic of China 2011). The distresses found in bridge detection shall be repaired immediately according to their causes. For structural distresses that cannot be repaired in a short time, attention should be paid to their development during the monthly inspection so as to take corresponding management measures (Ministry of Transport of the People's Republic of China 2011).

3 LOAD TEST

A static load test is a field test that can test the mechanical effect of control sections under an equivalent static load (Ministry of Transport of the People's Republic of China 2004). Dynamic load test is a field test that can test vibration characteristics under dynamic load and environmental load. As the detection showed the rapid development of transverse cracks near the middle of the main span of the box girder, a load test was carried out on the bridge to ensure operation safety before major maintenance and reinforcement, determine whether the bearing capacity of the bridge can meet the requirements of normal use and provide reasonable management measures for the maintenance department.

3.1 *Static load test*

Determination of Test Sections. The special program was used for structural calculation and analysis of the bridge. Three lanes are designed for each side of the bridge. Automobile-super 20 class is adopted for calculation, and load is applied on the three lanes. According to the specification, the transverse reduction factor of 0.8 is compared with the calculation results of two loaded lanes, and the larger value is taken (Shi 2003, 2013). After calculation, the results concerning the three lanes are unfavorable. The internal force envelope diagram under the action of design live load is shown in Figure 1, and the load test section is determined according to the bending moment envelope diagram, as shown in Figure 2.

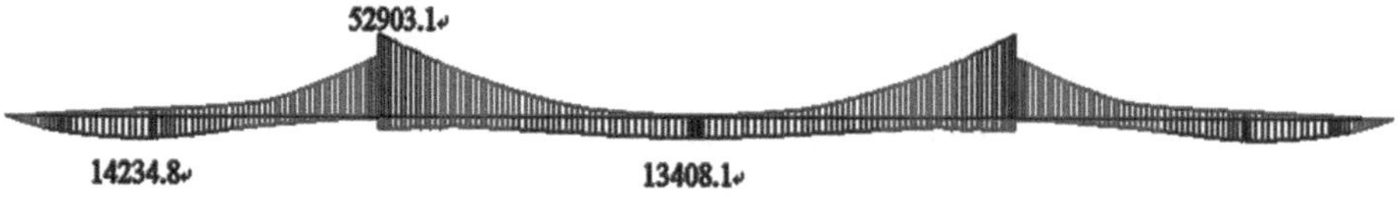

Figure 1. The moment envelope diagram of the main beam (kN·m).

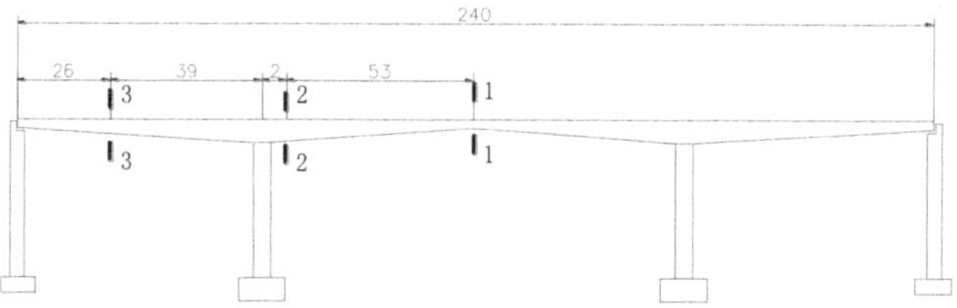

Figure 2. Schematic diagram of the control section(m).

Test Items and Contents. The right-side deck of the bridge was selected for load test in accordance with the Specification for Inspection and Evaluation of Load-bearing Capacity of Highway Bridges (JTG/T J21-2011) and the regular inspection results of the bridge. The test conditions and contents are as follows:

(i) Maximum positive bending moment effect of Section 1-1 under symmetrical loading and eccentric loading conditions; the deflection, stress, and changes of a transverse crack in the middle span of the section were tested.
(ii) Maximum negative bending moment effect of Section 2-2 under eccentric loading conditions; the stress of this section was tested.
(iii) Maximum positive bending moment effect of Section 3-3 under symmetrical loading and eccentric loading conditions; the deflection and stress of the section were tested.

According to the calculation, seven rear eight-wheeled trucks (capacity 41t) were used for equivalent loading in the static load test of the bridge, which can meet the requirement that the efficiency coefficient of the load test shall be between 0.95 and 1.05. The efficiency coefficient of the load test under each condition is shown in Table 1.

Table 1. Efficiency coefficient of bending moment of different conditions.

Condition Test item	Condition 1 Medium load on Section 1-1	Condition 2 Eccentric load on Section 1-1	Condition 3 Eccentric load on Section 2-2	Condition 4 Medium load on Section 3-3	Condition 5 Eccentric load on Section 3-3
Efficiency coefficient	0.99	0.99	0.95	0.99	0.99

Static Load Test Data. It can be seen from Tables 2 to 4 that: (i) the measured deflection and strain values under various conditions are less than the theoretical ones; (ii) the deflection calibration coefficient ranges from 0.75 to 0.83, and the calibration check coefficient is between 0.43 and 0.84, both of which are less than 1, indicating that the actual bearing

Table 2. The deflection data of different conditions.

Condition	Measuring point	Theoretical value (mm)	Measured value (mm)	Calibration Coefficient	Condition	Measuring point	Theoretical value (mm)	Measured value (mm)	Calibration Coefficient
Condition 1	1	−29.27	−23.44	0.80	Condition 2	1	−30.16	−24.71	0.82
	2	−29.27	−23.62	0.81		2	−29.78	−24.43	0.82
	3	−29.27	−23.73	0.81		3	−29.29	−22.73	0.78
	4	−29.27	−24.11	0.82		4	−28.80	−22.59	0.78
	5	−29.27	−24.18	0.83		5	−28.46	−22.64	0.80
Condition 4	1	−14.56	−11.07	0.76	Condition 5	1	−15.30	−11.83	0.77
	2	−14.56	−11.43	0.78		2	−14.99	−11.61	0.77
	3	−14.56	−11.35	0.78		3	−14.59	−10.95	0.75
	4	−14.56	−11.71	0.80		4	−14.19	−11.06	0.78
	5	−14.56	−11.51	0.79		5	−13.91	−10.67	0.77

Table 3. The strain data of different conditions.

Condition	Location	Theoretical value (με)	Measured value (με)	Calibration Coefficient	Condition	Location	Theoretical value (με)	Measured value (με)	Calibration coefficient
Condition 1	1	90.6	40.1	0.44	Condition 2	1	90.7	41.5	0.46
	2	90.6	38.9	0.43		2	90.6	40.2	0.44
	3	90.6	41.9	0.46		3	90.5	42.7	0.47
	4	90.6	39.9	0.44		4	90.4	39.2	0.43
Condition 4	1	82.6	69.1	0.84	Condition 5	1	83.0	65.4	0.79
	2	82.6	68.6	0.83		2	82.8	63.0	0.76
	3	82.6	66.1	0.80		3	82.6	61.8	0.75
	4	82.6	65.8	0.80		4	82.5	60.3	0.73
Condition 3	1	−51.0	−40.8	0.80	Condition 3	3	33.2	24.5	0.74
	2	−9.7	−4.4	0.45		4	43.2	30.6	0.71

Table 4. The variation of transverse crack width of conditions 1 and conditions 2.

Condition	Location	Extended width (mm)	Closed width (mm)
Condition 1	1	0.173	0.173
	2	0.155	0.150
	3	0.165	0.162
	4	0.163	0.161
Condition 2	1	0.173	0.173
	2	0.165	0.160
	3	0.162	0.160
	4	0.158	0.158

capacity of the bridge meets the normal use requirements under design load. The strain calibration coefficient of Section 1-1 is between 0.43 and 0.47 under conditions 1 and 2, which is smaller than the strain calibration coefficient of other sections and that of the same section, indicating that the strain of the id-span section is greatly affected by transverse cracks in mid-span. The maximum extension width of the original transverse cracks in Section 1-1 is 0.173 mm under conditions 1 and 2. The original width of the cracks is basically restored after unloading, indicating that the bridge is substantially in an elastic working state.

3.2 *Dynamic load test*

Analysis of Natural Vibration Characteristics. The dynamic test system was used to collect and analyze the modal and natural frequency of the bridge in the open driving state. The vertical vibration pickup was arranged at the quarter point of each span of the bridge, and 9 measuring points in total were provided. It can be seen from Table 5 that the first four orders measured vertical vibration frequency values are close to the theoretical values, indicating that the overall stiffness of the bridge is close to the theoretical value. The measured damping ratio ranges from 0.838% to 1.180%, which lies within the normal range of conventional bridges. The first four orders measured vertical vibration modes of the bridge are basically consistent with the theoretical modes calculated with the solid bridge model. The first four orders measured vertical vibration modes are shown in Figure 3 through Figure 6.

Table 5. The 4-order vertical frequency values of the long-span continuous rigid-frame bridge.

Order	First order	Second order	Third order	Fourth order
Theoretical frequency (Hz)	1.383	2.435	2.616	3.705
Measured frequency (Hz)	1.409	2.478	2.727	3.875
Frequency deviation (%)	1.88	1.77	4.24	4.59
Damping ratio (%)	1.083	1.130	1.180	0.838

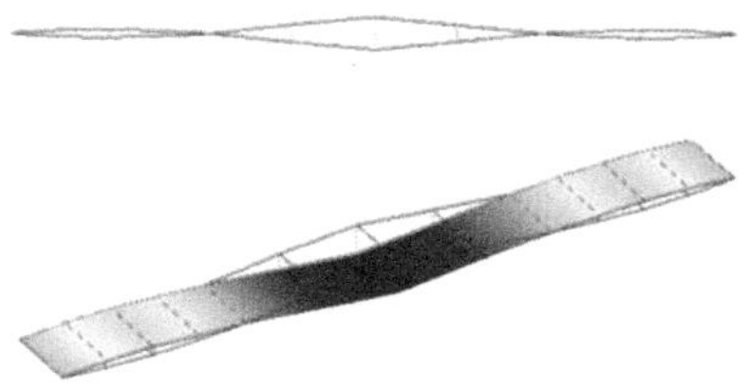

Figure 3. 1-order measured mode.

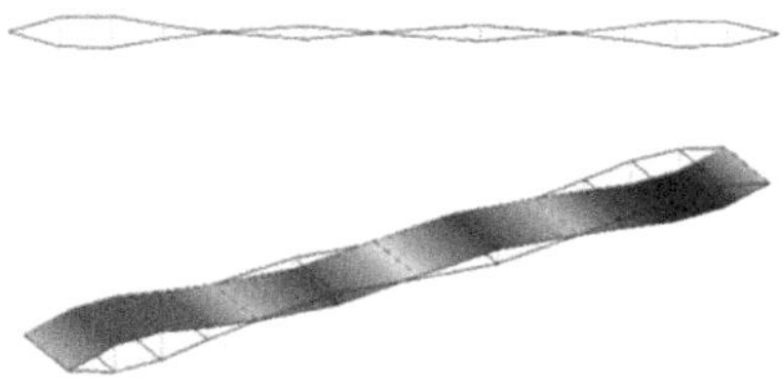

Figure 4. 2-order measured mode.

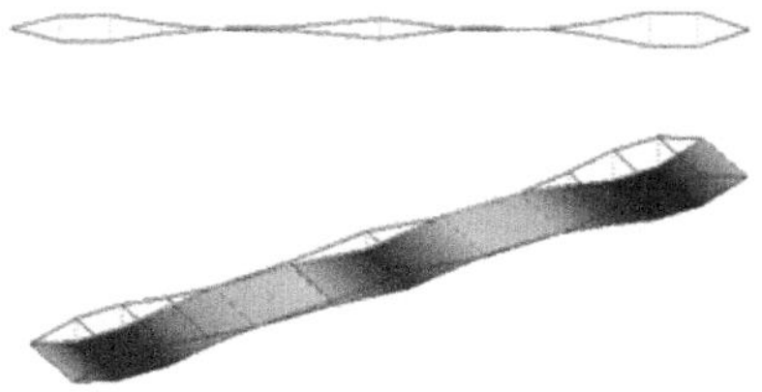

Figure 5. 3-order measured mode.

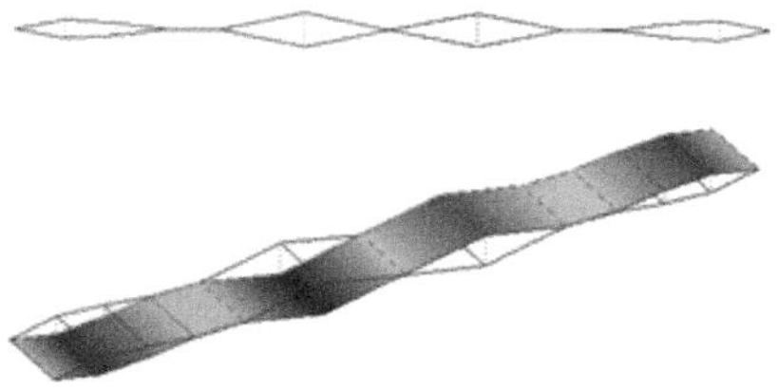

Figure 6. 4-order measured mode.

Detection of Impact Coefficient. A dynamic strain gauge was arranged in the middle of the bridge span, and a loading vehicle passed through the bridge at a speed of 10km/h – 60km/h, and the impact coefficient of the bridge at various driving speeds was measured, as shown in Table 6. The measured impact coefficients of the bridge at various speeds are less than 0.05, calculated according to the specification, which shows that the bridge is in good condition.

4 MAINTENANCE AND REINFORCEMENT MEASURES

The field load test shows that the actual bearing capacity of the bridge can still meet the normal use requirements under design load, but the bridge has already presented some distresses, such as mid-span deflection, transverse cracks in the mid-span bottom plate, and cracks in top web plate and prestressed anchorage plate, so the bridge shall be repaired and reinforced. According to past experience in strengthening long-span continuous girder bridges, the reinforcement work was carried out by applying external prestress and bonding steel plate (Xie 2007; Wang 2006). After demonstration and research, the final reinforcement measures include applying longitudinal external prestress in the box girder, bonding transverse steel plates after sealing cracks in the roof, bonding transverse steel plates after sealing cracks in truss plate, and bonding oblique shear steel plates after sealing cracks in the web.

Table 6. The list of the actual impact factor.

Condition	Type of condition	Speed (km/h)	Impact coefficient
Condition 1	1 loading vehicle Road test	10	0.01
Condition 2		20	0.01
Condition 3		30	0.02
Condition 4		40	0.02
Condition 5		50	0.02
Condition 6		60	0.01

5 CONCLUSION

Through continuous operation safety detection in three years, it is found that: (i) the number of cracks in the bridge increases obviously, but the length and width of cracks at the same position develop slowly; (ii) the deflection in the middle of the main span is greatly changed by seasonal temperature, but the whole bridge has a downward deflection trend; (iii) the material condition of the bridge can meet the specification requirements. The field load test shows that the actual bearing capacity of the bridge before reinforcement can still meet the normal use requirements under design load. Therefore, the operational safety detection method can accurately reflect the actual technical status of bridge operation and is effective and feasible. After the reinforcement work of the bridge was completed in 2016, the operation safety detection was carried out again in 2017. It was found that the bridge was in good technical condition, and the distressing development was effectively controlled.

REFERENCES

Lian Junjiang. *Research on Crack Analysis and Prevention Measures of Wide Box Girder of Continuous Rigid Frame Bridge* [D]. 2015, Chongqing Jiaotong University.

Lu Jie. *Research on Load Test Evaluation Technology of High Pier and Long-span Prestressed Concrete Continuous Rigid Frame Bridge* [D]. 2006, Chang'an University.

Ministry of Transport of the People's Republic of China. *General Specifications for Design of Highway Bridges and Culverts.* (JTG D60-2004) [Z]. 2004-10-01.

Ministry of Transport of the People's Republic of China. *Specification for Inspection and Evaluation of Load-bearing Capacity of Highway Bridges.* (JTG/T J21-2011) [Z]. 2011-11-01.

Ministry of Transport of the People's Republic of China. *Standards for Technical Condition Evaluation of Highway Bridges.* (JTG/T H21-2011) [Z]. 2011-09-01.

Niu Yanwei, Shi Xuefei. and Ruan Xin. Analysis of Long-term Deflection of Long Span Concrete Beam Bridge. *Engineering Mechanics.* 2008, 25(Supplement I): 117–119.
Shi Jia. *Analysis and Research on Cracking Causes of Long Span Prestressed Concrete Continuous Rigid Frame Bridge* [D]. 2013, Southwest Jiaotong University.
Shi Yuping. *Research on Bearing Capacity Evaluation Method and Reinforcement Technology of Concrete Bridges based on Load Test* [D]. 2003, Jilin University.
Wang Fawu, and Shi Xuefei. Research on Long-term Deflection Control of Long-span Concrete Girder Bridge. *Highway*. 2006, 08: 72–76.
Xie Jun, Wang Guoliang and Zheng Xiaohua. Research Status of Long-term Deflection of Long-span Prestressed Concrete Box Girder Bridges [J]. *Journal of Highway and Transportation Research and Development.*, 2007, 25(1): 47–50.

Constructional Engineering and Ecological Environment – Chih-Huang Weng (Ed)
© 2024 The Author(s), ISBN 978-1-032-53198-4

Research on daylighting design of university classrooms based on dynamic annual daylight simulation

Yang Wang* & Lu Bai
South China University of Technology, Guangzhou China
Architectural Design & Research Institute of Scut Co., Ltd, Guangzhou, China

ABSTRACT: Daylighting is an important factor that affects the space quality of university classrooms and has an important impact on the development of teaching activities, students' visual health, and learning efficiency. Under the condition of dynamic annual daylight simulation, the study will discuss the daylighting of the classroom by controlling the design parameters and obtaining the applicable ratio of glazing to floor area so as to optimize the daylighting of the classroom in the teaching building of Jiangxi Arts & Ceramics Technology Institute, and also provide a reference for the design of university classrooms.

1 INTRODUCTION

It has been proved that the successful use of daylight in school buildings is an important element in order to provide visual comfort to users, reducing the use of artificial lighting (Centre for Renewable Energy Sources 2002). Research shows that both students and teachers significantly benefit from the successful use of daylight, as it contributes to health promotion, well-being, and student productivity to the maximum (Mirrahimi *et al.* 1982). Classrooms are where teachers and students have the most frequent activities on campus. The lack of daylight in the classroom is not conducive to the student's concentration, which easily leads to learning fatigue and low efficiency, and will also cause an increase in lighting energy consumption. Therefore, it is necessary to pay attention to the full and effective use of natural light in classroom daylighting design.

Now China takes the daylight factor as the standard of assessment. But it is defined under CIE standard as the overcast sky with limitations in evaluating the daylight of room space in a real daylight climate (Bian & Ma 2017). The daylighting in the building is a dynamic process (Bian & Ma 2018), which is affected by continuously changing daylight climate and design parameters, so dynamic daylighting analysis is more suitable for the actual situation and is the trend of daylighting analysis (Reinhart *et al.* 2006). The dynamic daylight evaluation is mainly done by loading the typical climate data for the area where the research object is located throughout the year, constructing the Perez sky model, and simulating and calculating the building's annual (8760 h) daylight simulation, glare, and other daylight problems, so that the index can more truly reflect the natural daylight situation of the building throughout the year (Kazanasmaz *et al.* 2016).

This paper adopts the Ladybug + Honeybee daylight simulation plug-in to perform dynamic daylight simulation and then conducts correlation analysis and multiple linear regression analysis for the ratio of glazing to floor area and each dynamic daylight evaluation index based on the simulation results to determine the optimal ratio of glazing to floor

*Corresponding Authors: 1030385696@ qq.com and 970102583@qq.com

DOI: 10.1201/9781003410843-6

area. On this basis, finally, this study optimizes the daylighting of the classroom in the teaching building of Jiangxi Arts & Ceramics Technology Institute.

2 METHODOLOGY

2.1 *Research model design*

There are many types of college classrooms, such as ordinary classrooms, ladder classrooms, professional classrooms, etc. Considering that the function, internal space, and lighting requirements of special classrooms involve too many elements, in this simulation, small and medium-sized ordinary classrooms in the Architect Design Data Collection 3rd (China Architecture Publishing & Media Co., Ltd. 2017) are selected as a classical model (Figure 1). The optical parameters of each surface in the room are shown in Table 1 for later simulation.

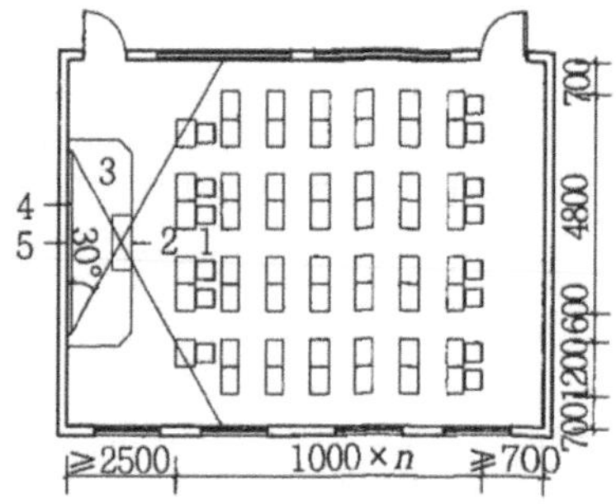

Figure 1. Plan of classical classroom.

Table 1. Optical parameters of classroom surfaces.

Name	Glass	Wall	Ceiling	Floor
Optical Parameters	Transmissivity 0.80	Reflectivity 0.60	Reflectivity 0.75	Reflectivity 0.30

2.2 *Simulation software and dynamic daylighting evaluation metrics*

At present, there is a variety of software with different functions for simulating indoor daylight in buildings. Common daylight analysis software includes Lightscape, Desktop Radiance, Daysim, DIVA, Lumen Micro, Ecotect, and Dialux (Acosta *et al.* 2011). To ensure the accuracy of the daylight simulation and simplify the experiment process, this paper chooses Rhino+Grasshopper as the modeling software, assisted by Ladybug +Honeybee equipped the Radiance+Dayism calculation engine, the built-in building physical environment simulation software in Grasshopper, as the tool for annual dynamic daylight simulation.

With the development of research on architectural daylight, the traditional static daylight evaluation indicators have been unable to meet the design requirements of architectural daylight, and dynamic daylight evaluation has gradually become an important indicator of the quality of natural daylight in buildings (Xue & Liu 2022). Daylight Autonomy (DA) is defined as the percentage of the year when a minimum illuminance threshold is met by daylight alone, which uses work plane illuminance as the representation of whether there is sufficient daylight in a space for people to work by daylight alone (Reinhart *et al.* 2006). Its required minimum illuminance levels for different spaces can be directly taken from reference documents such as "China National Standard of Building Daylight Design" (GB 50033-2013). But it can only stipulate the minimum illumination level, which cannot reflect the uniformity and glare of indoor daylighting. To consider the possible appearance of glare, maximum Daylight Autonomy (DA_{max}) is reported to indicate the percentage of occupied hours when direct sunlight or exceedingly high daylight conditions are present. Assuming that the threshold of potentially glary conditions depends on the space type, DA_{max} was defined as a sliding level equal to ten times the design illuminance of space (Reinhart *et al.* 2006). E.g., for a classroom with a design illuminance of 450lx DA_{max} corresponds to 4500lx. It can measure the probability of direct light or other potential glare and can also point out the location and frequency of large brightness contrast in space. Special Daylight Autonomy (sDA) refers to the ratio of the area in the room that meets a certain DA to the total area of the room, which uses DA as the basis to reflect the advantages and disadvantages of building daylighting capacity (Pu 2021). Useful Daylight Illumination (UDI) is a dynamic daylighting evaluation standard proposed by Nabil and John Mardajevic based on the reference plane illuminance information (Nabil & Mardaljevic 2005). It represents the proportion of effective illuminance data of daylight on the working face within a certain range in a year. It can be divided into three sections: <100lx, 100~2000lx, >2000lx, corresponding to low illuminance, available illuminance range, and high illumination (easy to cause uncomfortable glare and excessive thermal radiation) (Xiang *et al.* 2017).

2.3 *Simulation parameters and conditions*

According to the daily use, the simulation time of the university classroom is set as 8:00-18:00. Based on the provisions of "China National Standard of Building Daylight Design," the minimum illuminance level of DA should be set to 450lx, and other parameter settings for performance simulation are shown in Table 2.

Table 2. Performance simulation parameter settings.

Sky	Climate	Orientation	Analysis plane/m	Shading	Daylighting form	Grid /m
Perez sky	CHN_Jiagxi.Jingdezhen.585270_CSWD.epw	South	0.750	None	One-side daylighting	0.3*0.3

The daylighting of the classroom mainly depends on its size, such as the depth of the space, the storey height, the height of the side window, etc. Therefore, selecting the appropriate size can ensure the daylighting of the classroom. Under the condition of annual dynamic daylighting simulation, this research will study the influence of four design parameters of the classroom, including depth, height, side window height, and windowsill height, on daylighting. The experiment involves 16 working conditions, as shown in Table 3.

Table 3. The setting of working conditions.

Group1 storey height	Condition	1_A	1_B	1_C	1_D
	Storey height/m	3.90	4.20	4.50	4.80
	Depth/m	8.40	8.40	8.40	8.40
	Window area/m^2	16.38	18.72	21.06	24.18
	The ratio of glazing to the floor area	0.203	0.232	0.261	0.300

(*continued*)

Table 3. Continued

Group2 depth	Condition	2_A	2_B	2_C	2_D
	Storey height /m	4.20	4.20	4.20	4.20
	Depth /m	8.40	8.70	9.00	9.30
	Window area /m^2	18.72	18.72	18.72	18.72
	The ratio of glazing to the floor area	0.232	0.224	0.217	0.210
Group3 side window height	**Condition**	**3_A**	**3_B**	**3_C**	**3_D**
	Side window height /m	2.40	2.10	1.80	1.50
	Window area /m^2	18.72	16.38	14.04	11.70
	The ratio of glazing to the floor area	0.217	0.190	0.163	0.135
Group4 windowsill height	**Condition**	**4_A**	**4_B**	**4_C**	**4_D**
Windowsill height /m	0.60	0.75	0.90	1.05	
Window area /m^2	21.06	19.89	18.72	17.55	
The ratio of glazing to the floor area	0.244	0.230	0.217	0.203	

3 DATA ANALYSIS

3.1 *Results*

According to the simulation results, from the perspective of the spatial distribution characteristics of daylighting, the results of each indicator are different (Table 4). For example, in Conditions 2_D and 3_C, on the basis of the simulation results of DA, it can be concluded that the farther away from the side window in the classroom, the worse the daylighting is. However, the trend of $UDI_{100\sim2000lx}$ is different between the two. Only 20%~30% of the time can meet the effective lighting conditions near the side window, and it will rise to 80%~90% with the increase of distance from the side window. According to the trend of $DA_{max, 4500lx}$ and UDI_{2000lx}, it can be seen that this phenomenon occurs because the illuminance value of the classroom near the side window is too high without shading. Therefore, with the increase of the distance from the side window, the percentage of effective daylighting time will first show an increasing trend and then decline at a distance from the side window. This phenomenon is consistent with the use of people's daily space.

The simulation results of all working conditions are shown in Table 5. In the height group, $UDI_{100\sim2000lx}$ changes little with the increase of storey height. Considering DA and sDA, condition 1_D is the best, but from the perspective of DA_{max}, UDI_{100lx} and UDI_{2000lx} 1_D is the worst option. Therefore, working condition 1_B and 1_C is relatively good. In the depth group, it can be found that with the increase of depth, the difference between $DA_{max, 4500lx}$, UDI_{100lx}, $UDI_{100\sim2000lx}$ and UDI_{2000lx} is small. Therefore, from the perspective of DA and sDA, 2_A performs well. In the side window height group, each index has a certain range of change. In general, condition 3_B performs better. In the windowsill height group, comprehensively considering the simulation data, working condition 4_D is a better solution.

3.2 *Regression analysis*

In the above research, we simulated the impact of storey height, depth, side window height, and windowsill height on the natural light environment of the classroom. Next, this paper will extract the information on the window-to-ground area ratio in the simulation data and use the regression analysis method to study the relationship between the window-to-ground area ratio and the natural light environment evaluation standards of university classrooms. The results are shown in Table 6.

Table 4. Simulation diagram of example working conditions.

2_D				
DA_{450lx}	$DA_{max.,4500lx}$	UDI_{100lx}	$UDI_{100\sim2000lx}$	UDI_{2000lx}
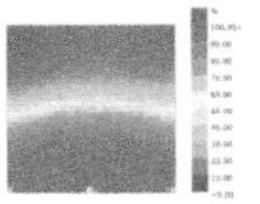	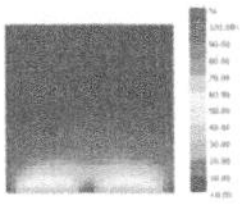	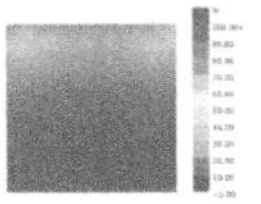	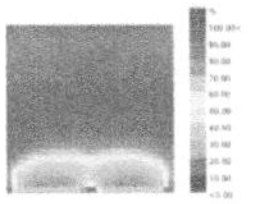	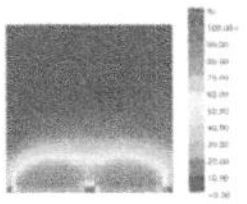

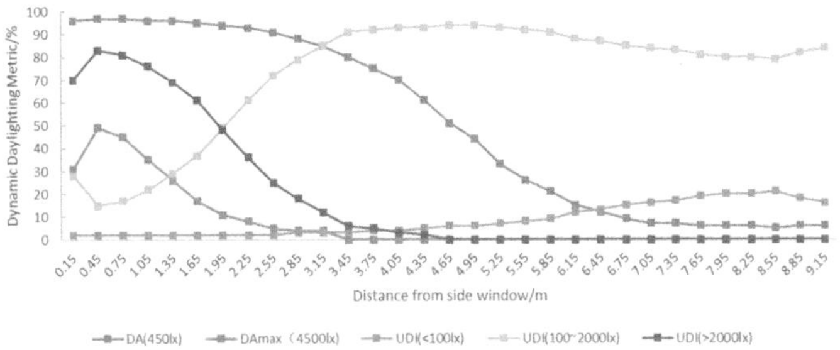

3_C				
DA_{450lx}	$DA_{max.,4500lx}$	UDI_{100lx}	$UDI_{100\sim2000lx}$	UDI_{2000lx}
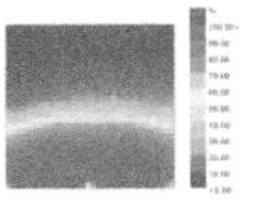	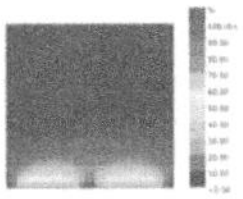	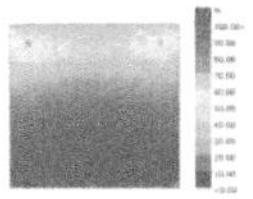	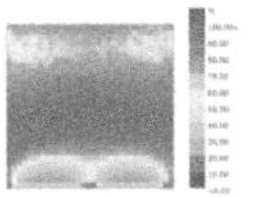	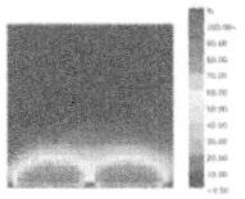

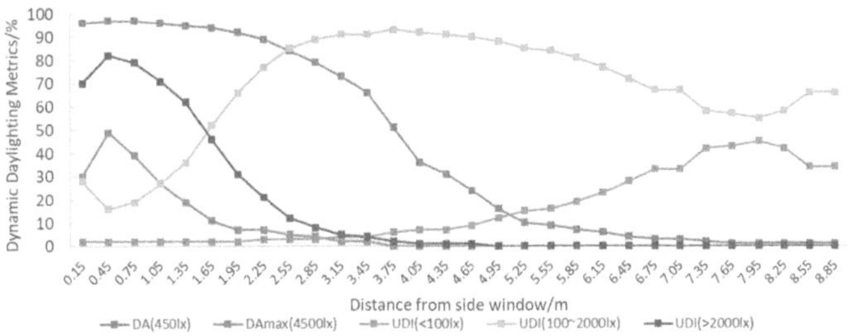

Table 5. Results of working conditions.

Group 1-storey height	Group 2-depth
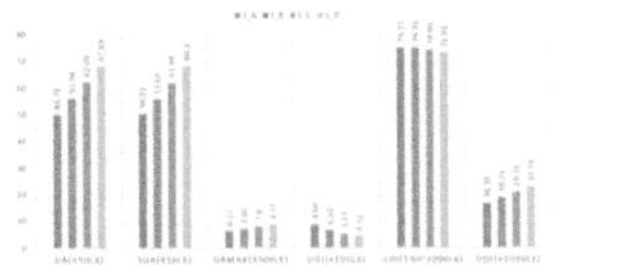	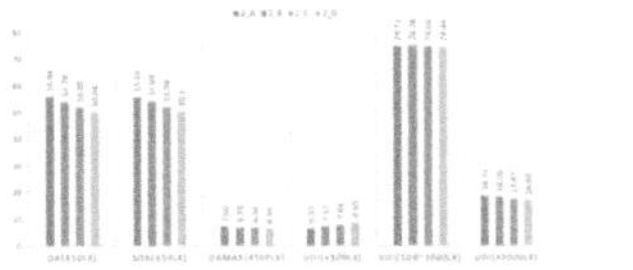
Group 3-side window height	Group 4-windowsill height
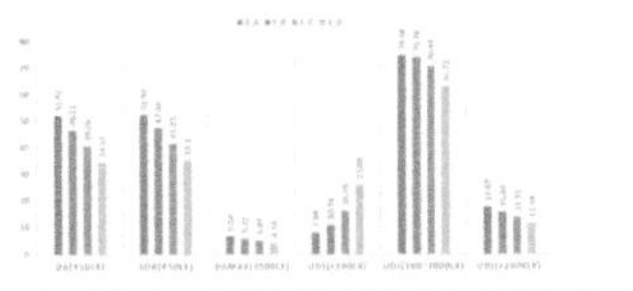	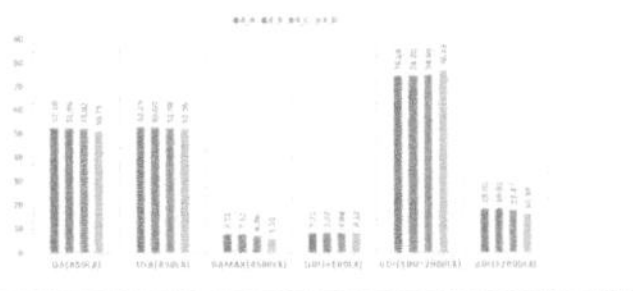

Table 6. Regression analysis of glazing to floor ratio and dynamic daylighting metrics.

DA450lx	sDA450lx
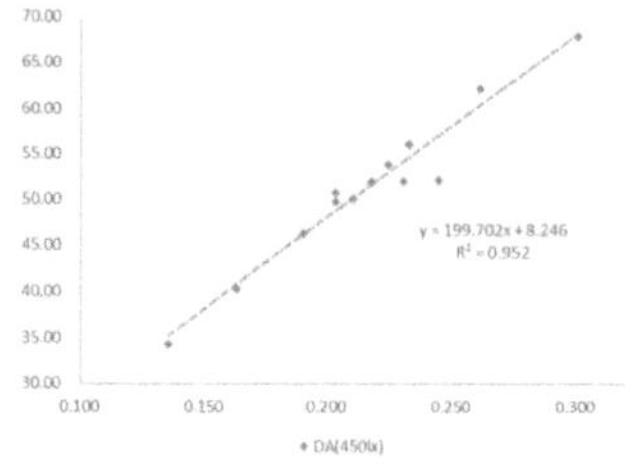	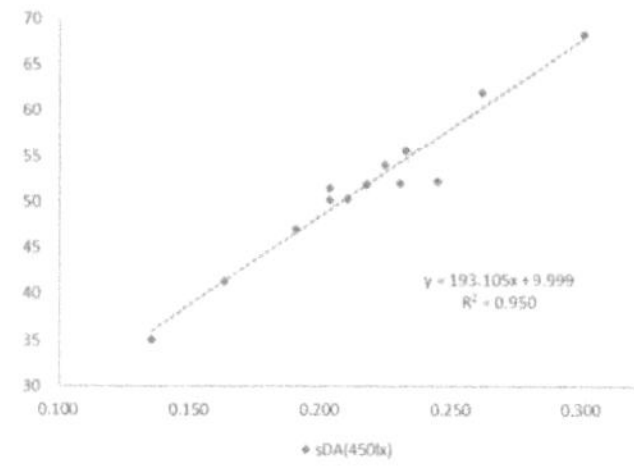
DAmax, 4500lx	UDI
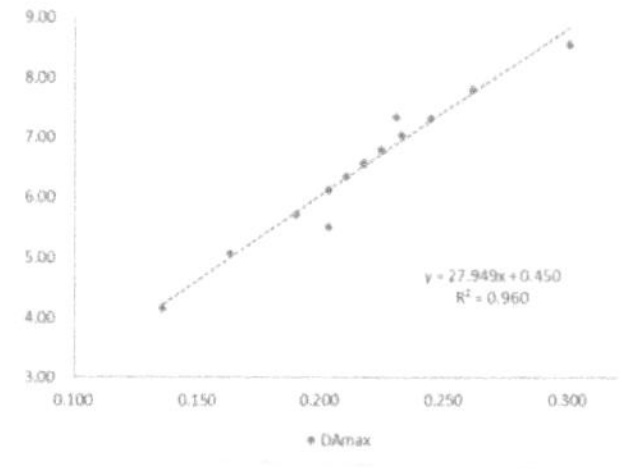	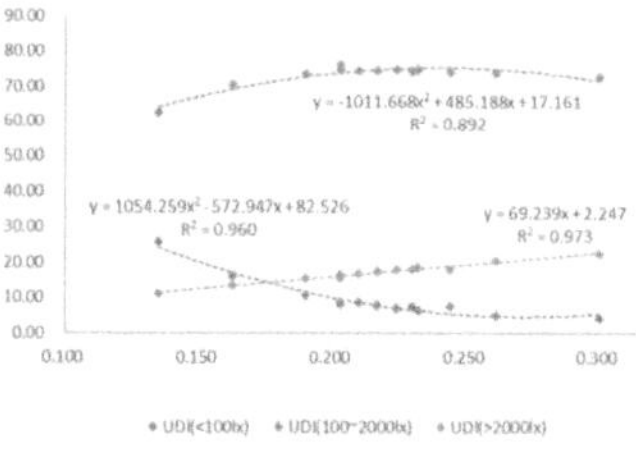

It can be seen that the glazing-to-floor area ratio is linearly related to DA_{450lx}, sDA_{450lx}, $DA_{max,\ 4500lx}$, and the fitting index is high. DA index lower than 50% can be considered insufficient lighting (Bian *et al.* 2017). According to the regression equation, it can be inferred that in this simulation, when the window-to-ground area ratio is greater than 0.209, DA_{450lx} is greater than 50%. When the window-to-ground ratio is greater than 0.207, sDA_{450lx} is greater than 50%. In the three intervals of UDI, the window-to-ground ratio has a good fit. Among them, the window-to-ground ratio has a quadratic relationship with UDI_{100lx}, $UDI_{100\sim2000lx}$ regression analysis, and a linear relationship with UDI_{2000lx} regression analysis. According to the fitting formula obtained from the simulation results of all working conditions in this experiment, when the window-to-ground ratio is 0.238, $UDI_{100\sim200lx}$ will reach the maximum value; When the window-to-ground ratio is 0.272, UDI_{100lx} is the minimum value.

From the perspective of minimizing the adverse daylighting area, the glazing-to-floor ratio and the sum of UDI_{100lx} and UDI_{2000lx} are used for regression analysis, of which results are shown in Figure 2. According to the regression equation obtained, it can be inferred that

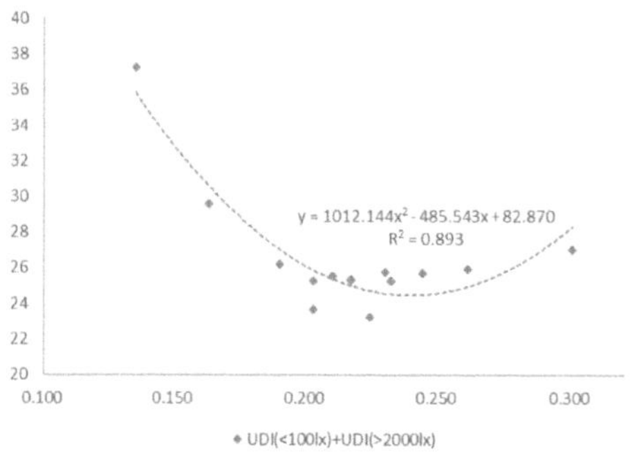

Figure 2. Regression analysis between glazing to ground ratio and UDI_{100lx}+UDI_{2000lx}.

when the ratio is 0.240, UDI_{100lx}+UDI_{2000lx} reaches the lowest value. Therefore, when the window-to-ground ratio is about 0.240, the daylighting of the classroom in this area is excellent, and the unfavorable area is small.

4 CASE OPTIMIZATION

According to the above, the experiment takes the classroom unit of Jiangxi Arts & Ceramics Technology Institute (Figures 3, 4) as an example to simply optimize the daylighting of the classroom. According to the size data, the glazing-to-floor ratio of the classroom is 0.266. The above results show that when the glazing-to-floor ratio is about 0.240, the daylighting of the classroom is excellent. Therefore, after adjusting the design parameters of the space, the two are simulated separately, and the simulation results are shown in Table 7.

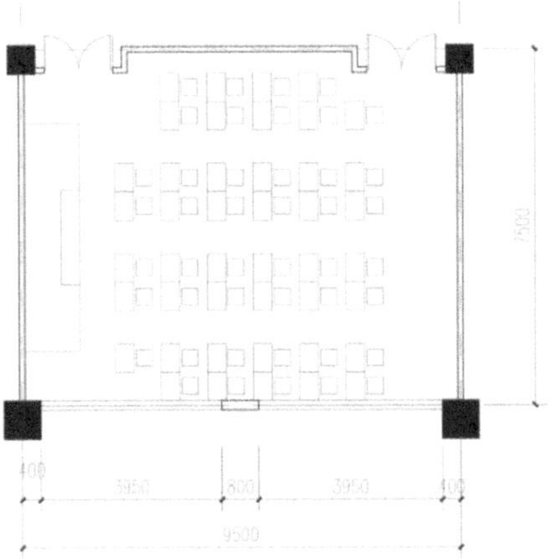

Figure 3. Plan diagram.

Figure 4. Section diagram.

According to the simulation results, after changing the glazing to floor ratio, both DA450lx and sDA450lx are greater than 50%, but UDI100~2000lx has a certain growth. The areas with high illuminance reflected by DAmax, 4500lx, and UDI2000lx that are easy to cause glare all shrink. On the whole, through the adjustment, the effective daylighting of the classroom has been improved, the unfavorable lighting areas have been compressed, and the indoor daylighting has been improved.

Table 7. Case optimization simulation.

Before	Parameter	Width/m	Depth/m	Windowsill height/m	Side window height/m	Glazing-to-floor ratio	
		9.50	7.50	0.90	2.40	0.266	
	Metrics	DA_{450lx}	sDA	$DA_{max.,4500lx}$	UDI_{100lx}	$UDI_{100\sim2000lx}$	UDI_{2000lx}
		63.83	64.13	8.03	4.90	73.65	21.41
	Figure	DA_{450lx}	$DA_{max.,4500lx}$	UDI_{100lx}	$UDI_{100\sim2000lx}$	UDI_{2000lx}	
			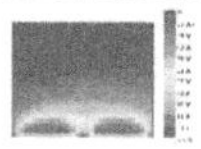		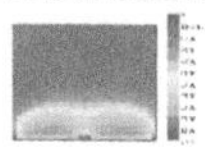	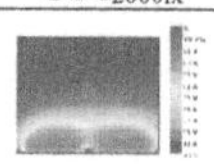	
After	Parameter	Width/m	Depth/m	Windowsill height/m	Side window height/m	Glazing-to-floor ratio	
		9.50	7.50	1.00	2.15	0.238	
	Metrics	DA_{450lx}	sDA	$DA_{max.,4500lx}$	UDI_{100lx}	$UDI_{100\sim2000lx}$	UDI_{2000lx}
		59.85	60.13	6.77	5.86	75.45	18.92
	Figure	DA_{450lx}	$DA_{max.,4500lx}$	UDI_{100lx}	$UDI_{100\sim2000lx}$	UDI_{2000lx}	
		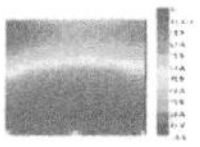	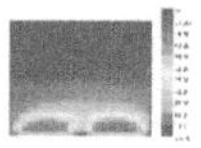		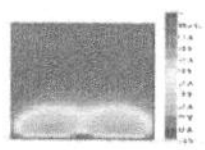	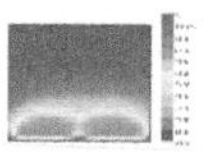	

5 CONCLUSION

The experiment takes the typical classroom space model as the basic research object to use the method of dynamic daylighting simulation and dynamic daylight evaluation indicators DA, sDA, DA_{max} and UDI to discuss the daylighting of university classrooms. Meanwhile, the regression analysis between the glazing-to-floor area ratio and the dynamic daylighting indicators shows a strong mathematical relationship between them. Therefore, from the perspective of reducing unfavorable lighting areas, setting the glazing-to-floor ratio at about 0.240 in this area can compress the unfavorable lighting area to the greatest extent and improve the overall lighting quality. Finally, based on the conclusions above, the classroom of Jiangxi Arts & Ceramics Technology Institute was adjusted and optimized to reduce the area of unfavorable lighting areas and maximize the use of natural light, thereby reducing unnecessary lighting, reducing building energy consumption, and achieving green development.

REFERENCES

Acosta I.; Navarro J. & Sendra J.J. Towards an Analysis of Daylighting Simulation Software. *Energies*. 2011,4,1010–1024

Bian Y. & Ma Y. Analysis of Daylight Metrics of Side-lit Room in Canton, South China: A Comparison Between Daylight Autonomy and Daylight Factor. *Energy and Buildings*, 2017, 138: 347–354. DOI: 10.1016/j.enbuild.2016.12.059

Bian Yu & Ma Yuan. Consumption Dynamic Daylighting Simulation and Lighting Energy Analysis Considering Visual Comfort. *Journal of Zhejiang University (Engineering Science)*, 2018, 52(9): 1638–1643. dx.doi.org/10.3785/j.issn.1008-973X.2018.09.002

Bian Yu, Yuan Lei & Leng Tian-xiang. An Analysis of Dynamic Daylight Performance Metrics & the Daylight Availability of Side-lit Windows. *Journal of Harbin Institute of Technology: Nature Science Edition*, 2017, 49(10): 172–176.

Centre for Renewable Energy Sources (CRES) and Lazaris E. *2002 Bioclimatic Design in Greece, Energy Efficiency and Implementation Directions (Athens)* p 3

China Architecture Publishing & Media Co., Ltd., The Architectural Society of China. Architect Design Data Collection 3rd[M]. China Architecture Publishing & Media Co., Ltd., 2017, 85

GB 50033-2013, *China National Standard of Building Daylight Design* [S].

Kazanasmaz T.; Grobe L. O.; Bauer C.; Krehel M. & Wittkopf S. Three Approaches to Optimize Optical Properties and Size of a South-facing Window for Spatial Daylight Autonomy. *Build. Environ.* 2016,102,243–256.

Mirrahimi S., Lukman N., Ibrahim N. & Surat M. 1982 *Effect of Daylighting on Student Health and Performance, Faculty of Engineering and Built Environment (Malaysia: Department of Architecture National University of Malaysia).* p 130

Nabil A. & Mardaljevic J. Useful Daylight Illuminance: A New Paradigm to Access Daylight in Buildings. *Lighting Res Tech.* 2005.37(1):41–59.

Pu Hongyu. *Optimal Design of the Natural Light Environment in Primary and Secondary School Classrooms in Guangzhou Based on Dynamic Daylighting Simulation.* South China University of Technology, 2021. DOI:10.27151/d.cnki.ghnlu.2021.002166.

Reinhart C., Mardaljevic J. & Rogers Z. Dynamic Daylight Performance Metrics for Sustainable Building Design. *Leukos.* 2006, 3(1): 1–20.

Xiang Luyao Ni Weichao & Wu Enrong, The Parameter Design Study of Typical Reading Space Based on Dynamic Annual Daylight Simulation. *Zhaoming Gongcheng Xuebao*, 2017,28(03):24–29+35.

Xue Y. & Liu, W. A Study on Parametric Design Method for Optimization of Daylight in Commercial Building's Atrium in Cold Regions. *Sustainability.* 2022,14,7667. https://doi.org/10.3390/su14137667

Constructional Engineering and Ecological Environment – Chih-Huang Weng (Ed)
© 2024 The Author(s), ISBN 978-1-032-53198-4

A contemporary wooden architecture design research based on traditional braided timber arch construction method: Taking platform space design of Wuyishan high-speed railway station as an example

Qiang Xu* & Xin Wu
School of Architecture, Fuzhou University, Fujian, China

ABSTRACT: As a special large-span structure in ancient China, the woven timber arch was reflected in the painting "Riverside Scene during Qingming Festival" in the Song Dynasty. But limited materials and the traditional processing technology restrict the development of traditional Chinese timber structures. With the development of new materials, new technology and the demand for modern architectural space span, modern timber architecture inherited the advantages of traditional timber buildings. It began to explore timber large-span public buildings design and construction. Starting from the traditional timber arch structure, this study takes the platform space of Wuyishan high-speed railway station as an example to explore the possibility of the traditional timber arch construction method in the platform design of different types of modern high-speed railway stations.

1 INTRODUCTION

1.1 *Woven timber arch bridge*

As a special type of bridge in ancient China, a woven timber arch consists of flat and straight timber interwoven vertically and horizontally. The members form a restrictive relationship of mutual support and locking, thus realizing the large-span arch structure (Yan 2022) (see Figure 1). Hongqiao in the Riverside Scene during the Qingming Festival by Zhang Zeduan of the Northern Song Dynasty depicts the vivid life of the citizens in Bianjing, the capital city of the Northern Song Dynasty. There are also records of this bridge in Meng Yuanlao's "Tokyo Menghualu", which are the artistic portrayal and powerful proof of the real existence of the Woven timber arch bridge (Min 2017).

In 1979, Mr. Mao Yisheng put forward the Hongqiao bridge many times in his report on the Technical History of Ancient Bridges in China. Experts inspected the wooden arch Bridges in the mountainous areas of Zhejiang and Fujian after the symposium. They confirmed that the Hongqiao technology prevailing in the Central Plains during the Northern Song Dynasty was not lost. At this point, the wooden arch bridge has been paid attention to again. The existing hundreds of wooden arch Bridges are mainly concentrated in remote mountainous areas at the junction of Fujian and Zhejiang provinces, ranging from Lishui City in the north, Fuzhou in the south, Wuyishan in the west, and coastal areas in the east. In 2009, Min-Zhe wood-woven arch bridge construction technique was listed as a "traditional Chinese wood-arch bridge construction technique" in UNESCO's "Intangible Cultural Heritage in urgent need of Protection" (Liu 2006).

*Corresponding Author: wuxin@fzu.edu.cn

DOI: 10.1201/9781003410843-7

Figure 1. Hongqiao in the Riverside Scene during Qingming Festival by Zhang Zeduan of the Northern Song Dynasty (photo source: Riverside Scene during Qingming Festival, Collection of the Palace Museum, Beijing).

Fang Yong (1995) discussed the spreading process of Hongqiao, the characteristics of the wooden arch structure and its significance in the history of architecture through field investigation of Hongqiao in Fujian and Zhejiang mountains (Yan 2022). And based on that, Liu Jie (2004) officially named the Bianshui Hongqiao and the existing wooden Arch Bridge in Fujian and Zhejiang areas "Woven Timber Arch Bridge" and "woven timber Arch Beam Bridge" (Fang 1995). Yang Shijin (2002) analyzed the structural characteristics and stress characteristics of laminated beams and arches, extended this structural form to steel structures, and explored the possibility of serving the past for the present (YANG 2002). Zhang Cheng (2007) made a comparative study on the structural forms of the Hongqiao bridge in Bianshui and the wooden arcade bridge in Fujian and Zhejiang provinces and discussed the similarities and differences between the Hongqiao bridge and wooden arcade bridge (Zhang 2007). Qin Chiquan (2008) proposed that the main structures of the Hongqiao and Min-Zhe woven timber arch bridge belong to the mutual bearing structure system based on the structure geometric-generation mechanism (wong 2008). On this basis, Min Tianyi (2017) looked at the "woven timber" structure from the construction method of Bianshui Hongqiao in Shanghe Scene at Qingming Festival and gave a detailed description of the construction system of the woven timber arch bridge in graphic mode. Compared with Bianshui Hongqiao, corridor Bridges in Fujian and Zhejiang provinces are covered with corridor houses, which are transformed in terms of function and significance and combine the functions of different types of buildings such as Bridges, pavilions and temples so as to achieve the interaction between the inevitability of traffic and the non-inevitability of villagers' daily life (Min 2017). Liu Junping (2017), by studying the mechanical calculation method of the woven timber arch, found that the woven timber arch structure was different from the structural design dominated by Western mechanics. The woven timber arch did not belong to the ideal two-dimensional plane or three-dimensional space truss structure in structural mechanics, and there was no corresponding structural, mechanical calculation model (LIU 2017). On this basis, Xiong Junfeng (2019) used modern construction technology to build Qingyangzhou woven timber arch bridge, with a total length of 14.4 m, a clear span of 13.6m and a width of 3.8m. The structural stability was verified by MIDAS /Civil model calculation. It explores the combination of modern timber structure construction technology and traditional woven timber arch structure, creating the possibility for the application of traditional woven timber arch structure in modern Bridges (Xiong 2019).

At present, the research on the woven timber arch technology mainly focuses on the inheritance of the technology, and the practice mainly applies it to landscape Bridges. There are few studies and examples of the application of woven timber arches to large public buildings. Woven timber arch technology is a kind of log structure, and the building volume is affected by the size of the logwood. In addition, China's current fire standards restrict the construction scale and technology application of wood structures (Cui 2022) Therefore, this study starts from the high-speed railway platform part and sees the big from the small. Taking the platform space of Wuyishan high-speed Railway station as an example, this paper discusses the possibility of the construction method of traditional brought-timber arch structure in the platform design of different types of modern high-speed railway stations.

1.2 *Platform space in different types of high-speed railway stations*

With the rapid development of transportation infrastructure around the country, many cities have built different types of high-speed railway stations. As an accessory building of the station house and waiting hall, the high-speed railway platform provides shelter for passengers to stay for a short time. In general, the scale of the platform is much smaller than that of the station house and waiting hall. The form of the platform is also mainly affected by the type of station. The selection of station types should start from the whole, adopt corresponding design strategies according to different objective conditions, and constantly adjust the components of railway passenger stations in the design process so as to achieve the goal of matching the internal functional relationship with the external objective conditions (Zhang 2022). Current high-speed railway station waiting rooms in China are mainly divided into three types: upper-going-line type, down-going-line type and side-going-line type. Side-going-line type can be subdivided into line-side flat type, line-side down type and line-side upper type according to the elevation relationship between the station room and platform (Peng 2012).

1.2.1 *Upper-going-line type*

Upper-going-line type means that the main body of the station is built on the track line, and the platform belongs to the subordinate part of the station. This kind of high-speed railway station does not need an overpass, underground passage or other redundant horizontal traffic. Directly through the corresponding ticket gate can take the escalator down to the corresponding platform (see Figure 2), reducing the traffic cost of passengers in the process of taking a ride. The passenger capacity of high-speed railway stations can also be improved by increasing the number of ticket gates. Therefore, this form is mostly used in large-scale high-speed railway stations, such as Hangzhou East Railway Station and Chengdu Railway Station

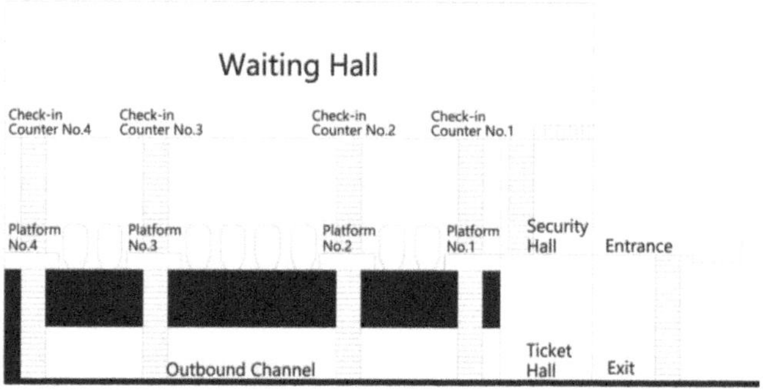

Figure 2. The layout of the Upper-going-line type station (picture source: author).

1.2.2 *Down-going-line type*

The Down-going-line type is very similar to the upper-going-line type. With the rapid development of more and more small and medium-sized cities in China, when new high-speed railway stations are built in some areas, the planning of high-speed railway lines may precede the construction of high-speed railway stations. Moreover, high-speed railway lines are affected by natural topography, and some high-speed railway stations are mainly built under the track lines. The streamlining of this type of high-speed railway station is as simple and clear as the upper-going-line type station. After checking in, passengers can go up to the platform through the corresponding escalator. Such as Yuyao North Station, and Huaihua South Station.

1.2.3 *Side-going-line type*

The side-going-line type station house means that the main body of the station house is on the side of the platform, and the square in front of the station and the station house is at the same level. Passengers check in and enter the platform through the overpass or underpass. Because of its simple and clear flow line, and the station room and platform are independent of each other, most of the small and medium-sized high-speed railway stations in China choose this type. Side-going-line type station room according to the high and low relationship between the platform and the station room, and divided into line-side down type, line-side flat type, and line-side upper type.

Line-side down type

The Line-side down type refers to the station in the track line under the side, platform higher than the main body of the station. Using the bottom side down type, passengers can go up to the platform through the escalator after the ticket check (see Figure 3). For example, Anhui Lu'an Station.

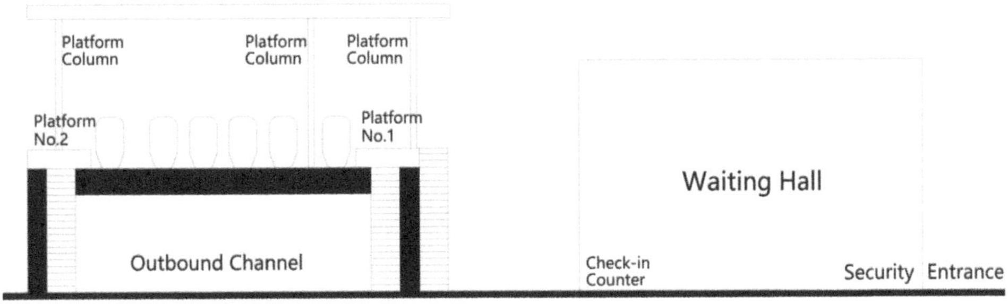

Figure 3. The layout of the line-side down type station (picture source: author).

When the scale is expanded, and a two-storey waiting hall is set up, in order to avoid unnecessary reentry streamlines caused by passengers from the second floor returning to the first floor, an overpass should be set up, and a streamline should be adopted (Zhang 2021).

Line-side flat type

The Line-side flat type means that the main body of the station room is at the same height as the platform. After checking in, passengers can choose to directly enter the platform on the same side horizontally or arrive at the side platform through the underground passage and escalator (see Figure 4). For example, Wuyishan North Station.

Line-side upper type

The Line-side upper type refers to the main body of the station being higher than the platform. Passengers, after checking in, can go down to the same side platform through the escalator or the overpass bridge and the escalator to the side platform (see Figure 5).

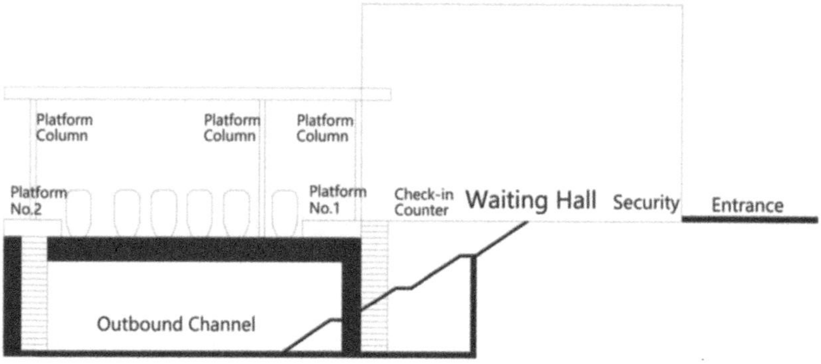

Figure 4. The layout of line-side flat type station (picture source: author).

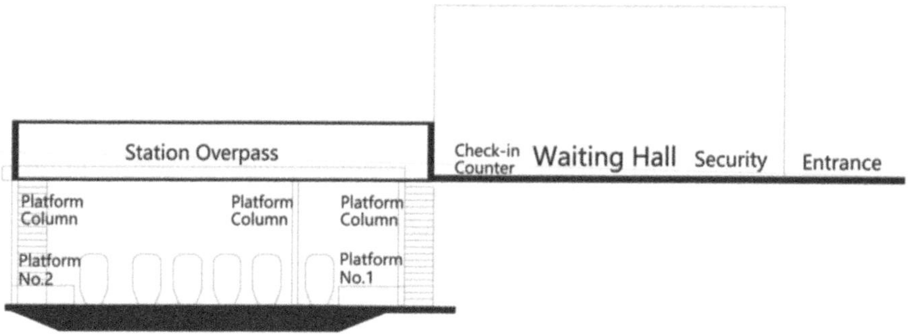

Figure 5. The layout of line-side upper type station (picture source: author).

2 APPLICATION EXPLORATION OF WOVEN TIMBER ARCH IN HIGH-SPEED RAILWAY PLATFORM

The traditional platform usually uses the column platform canopy. This kind of platform structure is simple with low cost. However, the platform pillar occupies part of the platform space area, which obstructs the passenger streamline and line of sight. With the increasing travel demand of Chinese railway passengers, higher requirements are put forward for the distributing capacity and carrying capacity of platforms. In order to make more efficient use of the platform space area, the column-free station canopy came into being. The first column-free station canopy in China was applied in the renovation of Beijing Station in 2004. In the past ten years, more and more column-free stations have been used in the construction or renovation of high-speed railway stations.

The main structural forms of the canopy without a pillar in China are: tube truss structure, cable-stayed structure, stringed beam structure, section steel structure, truss structure and suspension cable structure, among which tube truss structure is in the majority (Zheng 2010). The platform pillars of column-free platforms usually fall between the tracks or span multiple platforms. The platform space is characterized by a large span and high clearance. The special column-free space of the woven timber arch meets the space requirements of modern high-speed railway platforms. Li Bohan (2022) calculated the platform width of the current high-speed railway platform in China and concluded that the span of the column-free platform canopy across a platform generally reached 20.38m, which could be fully realized under the existing steel structure technology (li 2022). For example, Chongqing West Railway Station adopts a fair-faced concrete platform canopy with a maximum span of

26.85m (see Figure 6) (Li 2021). The span of traditional woven timber arch Bridges in Fujian and Zhejiang regions can usually reach 20m~35m (Li 2020). The development of new materials and technologies also creates more possibilities for the application of woven timber arch technology in modern high-speed railway platforms.

In the down-going-line and upper-going-line type stations, the stations and platforms are integrated as a whole in terms of shape and structure. The scale of the platform is affected by the number of track lines. When the two types of high-speed railway stations are combined with the woven timber arch technology, the shape and structure of the woven timber arch technology can be reflected to a large extent. The upper-going-line type is highlighted in the station part, while the down-going-line type is highlighted in the platform part. However, as the main body of the upper-going-line type station is elevated above the rail line, the distribution of a large number of people puts forward higher requirements on the bearing capacity of the structure, and the construction cost is also the highest.

In contrast, the Side-going-line type station has a higher degree of freedom when combined with the woven timber arch technology because the main body of the station is relatively independent of the platform part, which can also largely reflect the complete shape and structure of the woven timber arch as the platform part. The overpass section can also be designed by combining the corridor space on the upper part of the woven timber arch bridge to reflect the rationality of its function, which is a suitable type for the combination of woven timber arch technology and a high-speed railway platform. In addition, the platform space of high-speed railway stations is the first impression of passengers after arriving in a city and the last memory of passengers after leaving the city (Li 2009). Although the main body of the station and the platform are separated in structure, they should be an organic whole. The design should consider the shape of the platform and the overall style of the station building to maintain unity. In this study, Wuyishan, a city with the westernmost distribution of existing wooden arch Bridges in the Fujian and Zhejiang regions, is selected. Wuyishan high-speed Railway Station is taken as the study case. The use of woven timber arch technology in the platform part of the station can also reflect the regional architectural characteristics of the Fujian and Zhejiang regions.

Figure 6. Chongqing West Railway Station fair-faced concrete pillar-free canopy (Photo source: Application of fair-faced concrete in platform canopy of passenger station _ Li Jiaqi 2021 (Li 2021).

3 CASE STUDY OF WUYISHAN HIGH-SPEED RAILWAY STATION

Wuyishan high-speed Railway Station (Wuyishan North Station), located in Nanping City, Fujian Province, belongs to the middle station of the Hefu high-speed railway. It is designed as a medium-sized station with a total construction scale of 5999 m^2. The main body of the building is 2 floors, with a reinforced concrete structure, and the roof of the waiting hall is shaped steel structure. Wuyishan high-speed Railway station adopts line-side flat type, and the platform scale is 2 sets 4 lines, one basic platform and one side platform, two arrival and departure lines and two intermediate lines. The canopy adopts a steel structure with a platform column and cast-in-place concrete composite roof, covering a total area of 8,100 m^2. The platform scale of the selected case is two stations and four tracks, which is similar to the span calculated by Li Bohan (2022). It has the conditions for the construction of a canopy without platform pillars. It is within the span that can be reached by the traditional construction method of timber woven arch bridge.

The main body of Wuyishan North Railway Station extracts unique classical elements from Wuyi's characteristics. Wood grain metal louvers are used on the wall body and curtain wall details, reflecting strong local characteristics (see Figure 7). From a distance, Wuyishan North Station is like a simple house (Wuyishanbei Railway Station 2019). However, due to the influence of the line side layout, the station house is partially separated from the side platform. The platform part of the canopy adopts the steel structure with the platform column and the cast-in-place concrete composite roof (see Figure 8), so the overall style is quite different from that of the station house.

This research applied the traditional timber woven arch technology to the selected platform canopy part of the case, through the platform side of the retaining wall as the wooden arch starting point, through three bars and a crossbar connecting the first structural system, on this basis, by five bars and four bar of the second structure system and complete timber woven arch structure (see Figure 9).

Figure 7. Elevation of Wuyishan North Railway Station (photo source: Network (azure 2017).

Figure 8. Platform canopy of Wuyishan North Railway Station (photo source: Internet (khcc 2016).

4 CONCLUSION

The traditional timber woven arch technology has been inherited for hundreds of years in Fujian and Zhejiang. Due to its remote geographical location, it was buried deep in the mountains. The application of traditional timber woven arch technology to the platform canopy of Wuyishan high-speed Railway station can not only improve the utilization rate of platform space but also reflect the artisan tradition of timber woven arch bridges in the Fujian and Zhejiang regions.

Figure 9. The process of building the model of braided wood arch structure (Picture source: author).

To sum up, the side-going-line type is more suitable for combining with the timber woven arch technology in terms of scale and span compared with other station types. However, the appropriate form of platform canopy should be selected scientifically and reasonably according to the actual situation of the main body of the station hall and the scale of the track line. The overall form of the platform canopy should be unified with the main body of the station house when the space needs are met, and the structure is reasonable.

REFERENCES

Binbin Li. *A Study on Station Space Design without Platform Pillar and Canopy in Railway Passenger Station.* SCUT(South China University of Technology): 2009).

Fang-fang Peng. Choice of Train Station Type. *The World of Building Materials*, 33(5): 4. (2012).

Jie Liu. *The Origin of Jiangnan Timber Structure Problems in the Process of Research and Development.* Doctoral Dissertation, Tongji University: 2006).

Jiaqi Li. Application of Fair-Faced Concrete in Platform Canopy of Railway Passenger Station. *Railway Construction Technology*, (12): 5. (2021).

Junfeng Xiong. *New Technology in Design and Construction of Qingyangzhou Woven Timber Arch Bridge of International Horticultural Exhibition 2019*, Beijing, China. Building Structure, 49(17): 5. (2019).

Jian Zheng. *Railway Passenger Station Detail Design*: People's Communications Publishing House; 2010.

Jia-lu Zhang. Station Type Selection and Control Factors of Medium and Small Passenger Stations. *Architecture Technology*, 52(1): 5.(2021).

Kai Cui. Establish the Concept of High-quality Development – Promote the Development of Modern Timber Architecture, *Architecture* (16): 5. (2022).

khcc. The Platform of Wuyi Shan North Railway Station 2016 [Available from: https://you.ctrip.com/traffic/wuyishan22/h22853032-dianping94158738.html.

Liu Junping; Yang Yan; Chen Baochun; College of Civil Engineering FU. Timber arch Bridge Structural Type and the Value of China Woven Timber Arch Bridge Technology. *Fujian Architecture & Construction*, 9: 6. (2017).

Li B-h. Design Methods of Interior Spaces of Modern Large Highspeed Railway Station. *Architecture & Culture*, (09): 4. (2022).

Mr.azure. *Wuyishan North Railway Station* 2017 [Available from: http://www.dianping.com/photos/881044777.

Tianyi Min. From Arch Bridge over the Bian River to Lounge Bridge in the region of Zhejiang and Fujian Provinces–The Illustration of Tectonic Mode in Wooden Arched Bridge. *Architecture & Culture*, 12: 3. (2017).

Wong QsI. A Preliminary Study on the Structure Types of Hongqiao. *China Water Transport*, 06: 2. (2008).

Wuyishanbei Railway Station 2019 [Available from: https://baike.baidu.com/item/%E6%AD%A6%E5%A4%B7%E5%B1%B1%E5%8C%97%E7%AB%99/9270260?fr=aladdin.

Yan Liu. The Challenge of Span: Constructional Thoughts of Historical Timber Structure in China and Europe. *Journal of Architecture*, 4: 8. (2022).

Yong Fang. *Hongqiao Test Architectural Journal*, 11: 6. (1995).

Yang Shijin (Tongji University S, China). Research on Laminated Beam Arch Bridge. *Highway*, 12: 6. (2002).

Yue Zhang. *Study on Station Type Selection and Scheme Design of New Large-scale Railway Passenger Station Interior Architecture of China*, (4): 2. (2022).

Yan Liu. Thresholds and Secrets of Bridge-Building Craftsmanship: an Anthropological Study on the Building Technology of Woven Arch Bridges in Southeast China. *Architectural Journal*, (06): 6. (2020).

Zhang Cheng Guan Ruiming (College of Arch..Univ..of Huaqiao Q, Fujian,362021). Comparative Research between Rainbow Bridge and Timber-arched Lounge Bridge. *Fujian Architecture & Construction*, 04: 4. (2007).

Constructional Engineering and Ecological Environment – Chih-Huang Weng (Ed)
© 2024 The Author(s), ISBN 978-1-032-53198-4

Segmentation and modeling method of building structure point cloud based on BIM model

Guilin Zhang, Yang Han*, Lijie Meng, Dong Wei & Yanhui Yang
State Grid Hebei Electric Power Construction Co., Ltd., Shijiazhuang, Hebei Province, China
Hebei Electric Power Engineering Supervision Co., Ltd., Shijiazhuang, China

ABSTRACT: Updating point cloud data into a building information model (BIM) is a new, efficient and fine management method for monitoring construction and operation status. The accurate segmentation, modelling method and implementation process of the unit point cloud based on the BIM model are proposed. First, we use a clustering algorithm to cluster point clouds based on the geometric information of building elements in the BIM model and give them different semantic information; Then, the coarse extraction of various planes is realized by the region growth algorithm, and the fine segmentation of adjacent planes is realized by combining the distance algorithm of boundary points; Finally, ABM indoor is used to model the segmented data based on the three-dimensional triangle algorithm and associate it with the component semantic information in BIM, so as to achieve a high degree of coordination between the two types of data. Taking a substation project under construction as an example, the proposed method is used to scan, segment and model the substation structure under construction. The engineering application results show that this method can automatically segment and model the plane according to the BIM model element information. The segmentation accuracy is higher than that of the traditional method, which has strong practical value.

1 INTRODUCTION

BIM technology plays an important role in the whole life cycle of the construction project (Liu 2019). The difficulty of transforming the point cloud data obtained by laser scanners into a BIM model is that the point cloud data is chaotic and not semantic (Awwad et al. 2010). How to match the point cloud data with the BIM model is the focus of scholars at home and abroad. Liu (He *et al.* 2021) matched the building mesh model with the point cloud data based on the characteristic points, realizing the real-time monitoring and management of the building construction progress. Awwad (Meyer et al. 2022) used the traditional RANSAC algorithm to segment the point cloud plane. However, the segmentation efficiency and the unstructured plane segmentation are poor, causing excessive or insufficient segmentation. He (Yang et al. 2022) used a deep learning algorithm for the input of the 3D point cloud and got its BIM model parameters. However, the algorithm requires the structural characteristics of the target building. Before the bounding box of each unit is established, the excess is cumbersome and complex, and the final generated BIM model has low accuracy. It can be seen above that in the transformation of the two kinds of data, and the existing methods do not implement fine segmentation of the point clouds according to the geometric characteristics of the BIM model. The transformation process of the model is

*Corresponding Author: 13672067512@163.com

DOI: 10.1201/9781003410843-8

complicated, and the results are rough, which cannot be better combined with BIM. Based on the needs of construction progress monitoring and the research of point cloud automatic segmentation and modelling. For the above problems, a set of whole-process methods from point cloud data to the BIM model is proposed. A preliminary semantic segmentation of the point cloud data is first performed by a clustering algorithm. Then introduce the concept of common pastry and the point cloud essence Detailed segmentation. Finally, the automatic modelling of the segmented data was implemented using the ABM-indoor.

2 POINT CLOUD SEGMENTATION AND MODELING

2.1 *Point cloud clustering*

This paper clusters the geometric information in the point cloud data only. The reflection intensity, colour and other information in the point cloud data have no significance in this study, so it is not used. After analyzing the geometric characteristics of different point cloud elements, plane-type elements are established based on the difference in constraints (Romero & Arranz 2021; Wang *et al.* 2021). Based on the general structural characteristics of industrial buildings, the elements of the plane type are divided into three types according to the geometric dimensions: floor and ceiling, walls, and columns.

According to the geometric characteristics of the floor and ceiling, the height information in the point cloud data is used as the clustering standard. First, the elevation value of all the coordinate points in the point cloud data is calculated. By setting an upper and lower boundary, finding the points in this range are marked as a cluster. This process is constantly iterated in the point-cloud data to identify other repeated elevation values, and new clusters are built by setting up new upper and lower boundaries.

Due to the disorder and disorder of point clouds, although point clouds with similar high information are clustered, it is still difficult to recognize their semantic information. The 3D surfaces were projected from these clustered point cloud data using the Delaunay Triangulation algorithm. The geometry and size of the surface were further filtered for the categories with the ceiling and floor semantic information. For set shapes, the point cloud data with complex shapes are removed, while for geometric dimensions, the area and perimeter of each polygon are calculated by calculating the proportional circle area (APC), where the APC is:

$$APC_i = \frac{Ap_i}{Rt_i^2}, \quad Rt_i = \frac{R_{2-i}}{R_{1-i}} \tag{1}$$

Where, Ap_i represents the area of the polygon i, Rt_i represents the ratio of the circumference to the area, and R_{2-i} and R_{1-i} indicate the radius of the circle proportional to the circumference and the area of the polygon i, respectively. The algorithm will be filtered based on the degree of compactness of the polygon and the calculated area. After many experiments, the size of the APC must be greater than or equal to 1 square meter before the point cloud can correctly express the semantics of the floor and the ceiling.

After clustering the floor and ceiling, the clustering algorithm continues to identify the walls. Walls are considered vertical planes, the same as in section 2.1.1. We use the Delaunay Triangulation algorithm to generate polygons and establish the preliminary vertical plane through the edge direction and length of the polygon. Based on this, different vertical planes are repeatedly adjusted to find the clusters most suitable for the vertical distribution of the point cloud data. Among them, the red plane ψ is the final plane that is best suitable for a specific point group, and the best plane of the adapted point group is found by repeatedly adjusting the π, σ, ρ. To adjust the position of the plane for each iteration, the algorithm extracts the feature vector orientation vectors X_π, X_σ, and X_ρ, and the orientation angles

$\theta_X^{X_\pi}$, $\theta_X^{X_\sigma}$, and $\theta_X^{X_\rho}$ of the adjusted vertical plane. Finally, the point clouds were clustered by vertical clustering.

Although the cylinder also belongs to the vertical plane, its direction is diverse, with the last section method being tedious and unreasonable. Therefore, the clustering method of the column surface can help to cluster the vertical plane of the column position by detecting the regular gaps on the ground and the ceiling. Since the columns are connected to the floor and ceiling, each gap in the floor should have an equal gap in the parallel ceiling in the upper part. When these gaps are identified, similar to the clustering method of the vertical plane of the upper segment, the algorithm generates a bounded plane to cluster the point cloud data in the plane.

2.2 *Refined segmentation of flat point clouds*

After the clustering is completed, the concept of co-pastry and its constraints can be introduced based on the regional growth algorithm to realize the point cloud fine segmentation. First, the area growth algorithm is used to divide the point cloud data of the building plane roughly; then extract the boundary points from the maximum angle between the point in the cut plane and the adjacent point of the rough-cut plane to select the common boundary point where the two planes intersect. Then, the best segmentation result of the plane is obtained.

Since the constraints of clustering are constrained only to geometrical features, there are no clear bounds between point clouds of different semantic features, and the segmentation of different types of point cloud data is required. Regional growth algorithms are common methods for coarse segmentation of cluster-completed point cloud data. The normal vector and the curvature are the basic features of the surface (Su et al. 2021). The proposed algorithm first calculates the normal vector of each point and then obtains the corresponding curvature and selects the point with the minimum curvature as the initial seed point. The adjacent points of the seed point need to meet the specific angle value between the normal vector of each point and the normal vector of the current seed point, and the seed point and its neighbourhood point are grouped into a single plane. When their curvature values are smaller than the set curvature value, these points will not be treated as neighbours of that seed point but are classified as new seed points. In this paper, a surface composed of the target point and its adjacent points is used to calculate the normal vector and curvature of a point. The normal vector of a point is calculated using PCA or using the eigenvector and eigenvalues of the covariance matrix established from neighbouring points. Let the target point be p_i, and the corresponding covariance matrix C is:

$$C = \frac{1}{n}\sum_{i=1}^{n}(p_i - p_s)(p_i - p_s)^T, \quad p_s = \frac{1}{n}\sum_{i=1}^{n} p_i \tag{2}$$

Where, n is the number of adjacent points of p_i, and p_s represents the centre of the adjacent points. The normal vector of the point p_i is:

$$C \cdot \vec{v_j} = \lambda_j \cdot \vec{v_j}, \quad j \in \{0, 1, 2\} \tag{3}$$

Where, λ_j represents the j-th eigenvalue of the covariance matrix, and $\vec{v_j}$ represents the j-th eigenvector. The normal vector is the eigenvector corresponding to the minimum eigenvalues.

Since the eigenvalue $\lambda_0 < \lambda_1 < \lambda_2$ represents the degree of change of a point along the three main directions, the curvature σ of the point p_i can be approximated as:

$$\sigma = \frac{\lambda_0}{\lambda_0 + \lambda_1 + \lambda_2} \tag{4}$$

Where λ_0, λ_1 and λ_2 are the eigenvalues of the covariance matrix, and σ is the curvature of the point. After obtaining the normal vector and curvature of all points, the translation angle value between the normal vector of the current seed point and its adjacent points is fixed, which is considered co-pastry when they are within the threshold. The operation steps of this process are: (1) Calculate the partial curvature, and add the minimum curvature point to the seed point set; (2) Calculate the normal vector of the current seed points and the adjacent points. If the adjacent point of the current seed point not only meets the set translation angle value but also the curvature value is small, and the set curvature value is small, the adjacent value is included in the seed point sequence; (3) Delete the current seed point. Adjacent points were treated as new seed points that continue to enable regional growth; (4) Iterate this process continuously until the seed point sequence is empty.

After obtaining the boundary points, the distance threshold of the non-rough extraction plane point to the plane passing through the rough segmentation plane boundary points is then used to extract the common surface where the two planes intersect. Among them, the non-coarse extracted plane points are those obtained from the original point clouds, removed from section 2.2.1 and filtered using the regional growth algorithm. Points in the plane after non-coarse extraction are represented as P(x， y， z), and the boundary points in the coarse extraction plane are expressed as Q(l， m， n). The distance from the plane to the boundary point d(P, Q)is:

$$d(P, Q) = \sqrt{(x-l)^2 + (y-m)^2 + (z-n)^2} \tag{5}$$

By setting different distance thresholds for the boundary points of the noncoarse extracted plane, different segmentation results can be obtained, and it is manifested as the difference in the segmentation distance length between the two adjacent model units after the segmentation.

2.3 *Generation of 3D models*

How to establish regular planes and find their interrelationships from chaotic and discontinuous point cloud data (Rausch & Haas 2021) is the difficulty of 3D model generation. With different geometric characteristics of the plane, the 3D plane model generation method is also different. The plane is divided into horizontal plane modelling, vertical plane modelling and non-plane element modelling.

The plane of the 3D model is surrounded by regular line segments, while the plane edges composed of point clouds are irregular and messy. Therefore, the first step in modelling is to fine-tune the edges of the polygons, called parallel internal polygons. The purpose of this step is to eliminate the bumps of the polygonal boundaries. ABM-indoor Each node for setting a polygon in the ABM-indoor is formed by the connection of the two edges, $n_i, \ldots, n_j$. First, from each node of the existing polygon, an arc with a radius of 10 cm is established, and the intersection of the two connecting edge bisectors on the existing polygon is defined as the homnate node $(n_i', , n_j')$ of the internal auxiliary polygon (IAP). Subsequently, the algorithm removed existing nodes more than 20 cm apart from the homologous nodes. Finally, a new polygon was created in the ABM-indoor, forming a 3-dimensional surface with regular edges and fewer nodes.

3 ENGINEERING EXAMPLE

Zanxi 220 kV substation is an above-ground indoor substation. In order to facilitate the construction progress of the substation, the 3D laser scanning technology is adopted, and the point cloud is updated to the building information model (BIM) to realize the rapid identification of the project progress.

The point cloud data obtained from laser scanning was taken as input and the BIM 3D model as output. The results of labelling the different semantic components and combining them with the refined segmentation point cloud data based on the common pastry extraction are shown in Figure 1. The point cloud data statistics obtained from each direction segmentation are detailed in Table 1.

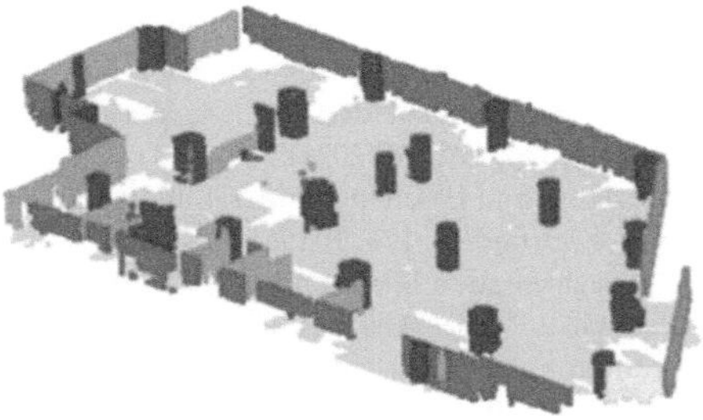

Figure 1. Segmentation results of point cloud data.

Table 1. Summary of point cloud segmentation results of various building components.

Building elements	Point to the number of clouds	Building elements	Point to the number of clouds
The earth's surface	395,673	Wall (direction y)	15,487
Ceiling	693,285	Wall (direction j)	78,140
Walls (direction x)	95,735	Non-planar elements	8,539

The imported model underground facility is 4230, and the above-ground facility is 6790. A comprehensive accounting of the total civil progress of the new 220 kV substation is 65%, and the actual progress of the project is 69%, with an error of 4%. The proposed algorithm greatly reduces the scanning time while ensuring the quality of the point cloud and realizes the dynamic monitoring of the building milestone progress identification.

4 CONCLUSION

With the construction progress monitoring task of a substation project in Hebei province, this paper puts forward the process method from point cloud data to the BIM model. Based on the classification algorithm, the building objects with different semantic features are extracted for the point cloud data based on BIM model features, proposing a refined segmentation method based on common pastry extraction. A 3D plane model is automatically established based on the point cloud data in the ABM-indoor environment. By conducting the actual scanning of a substation building under construction, the obtained point cloud data is accurately divided, and the BIM 3D model is generated. Finally, it shows that this process not only greatly improves the segmentation accuracy and modelling efficiency of the point cloud but also ensures the success rate of point cloud identification and data quality. It provides algorithm and process support for the transformation method of point cloud data to the BIM model.

REFERENCES

Awwad T M et al. An Improved Segmentation Approach for Planar Surfaces From Unstructured 3D Point Clouds. *Photogrammetric Record.* 25(129), 5–23 (2010).

He Wenjing, Yang Jian, Xiong Wuyue. Automatic BIM Model Generation from 3D Point Cloud Based on Depth Learning. *Journal of Civil Engineering and Management.* 38(03), 133–139 (2021).

Liu Shasha. *Technical Method of Building Construction Progress Monitoring Based on Point Cloud Data and BIM Integration.* Southwest Jiaotong University. (2019).

Meyer T, Brunn A and Stilla U. Change Detection for Indoor Construction Progress Monitoring Based on BIM, Point Clouds and Uncertainties. *Automation in Construction.* 141, 104442 (2022).

Romero J R and Arranz J J. Automatic Segmentation and Classification of BIM Elements From Point Clouds. *Automation in Construction.* 124, 103576 (2021).

Rausch C and Haas C. Automated Shape and Pose Updating of Building Information Model Elements From 3D Point Clouds. *Automation in Construction.* 124, 103561 (2021).

Su Zhonghua et al. Building Plane Segmentation Based on Point Clouds. *Remote Sensing.* 1, 95–115 (2021).

Wang Mengjun, Qiu Qi and Wang Qian. Visual Analysis of the Research Frontier of Construction Project BIM and Point Cloud Integration – based on the Web of Science Core Database. *Journal of Civil Engineering and Management.* 6, 9–16 (2021).

Yang Wenbin, et al. Geometric Relation Based Point Clouds Classification and Segmentation. *Concurrency and Computation: Practice and Experience.* 34(11) (2022).

Constructional Engineering and Ecological Environment – Chih-Huang Weng (Ed)
© 2024 The Author(s), ISBN 978-1-032-53198-4

Load test and reinforcement measures for a through-type reinforced concrete arch bridge

Xijian Liang*
Henan Provincial Communications Planning & Design Institute Co., Ltd, Zhengzhou, China

Qiuyan Sun
Henan Technical College of Construction, Zhengzhou, China

ABSTRACT: Static load test is carried out on the bridge to test the deflection and strain, and a dynamic load test is performed to test natural frequency, damping ratio, and vibration mode. For those bridges that do not meet the normal use requirements of the design load, it is recommended to adopt the methods of arch springing reinforcement, reconstruction of transverse and longitudinal connection, and replacement of suspenders.

1 INTRODUCTION

The arch bridge is the bridge that takes the arch as the main load-bearing member. The through-type arch bridge is a reinforced concrete structure. Its arch rib has a reinforced concrete box section, and its beam has a reinforced concrete rectangular section. The arch bridge has a net span of 170 m, a rise of 35 m, an arch rib height of 3.5 m, and a width of 1.5 m. The bridge has great flexibility. The upper ends of the left and right suspenders are anchored to the top of the arch rib, the lower ends are anchored to the bottom of the beam, and the bridge deck is placed on the beam. The mode of structural force transmission is load → bridge deck → beam → arch rib → arch springing → foundation. The design load of the bridge is as follows: automobile over 20, trailer-120, and crowd 3.5 kN/m^2.

2 CONTENTS OF LOAD TEST

2.1 *Contents of static load test*

A static load test is a field test that can test the mechanical effect of control sections under an equivalent static load. The process of the load test is as follows: (i) decision of tested sections; (ii) installation of test equipment; (iii) loading and data collection; (iv) supply of testing result. According to Specification for Inspection and Evaluation of Load-bearing Capacity of Highway Bridges (JTG/T J21-2011), the static load test contents of the bridge are as follows: (i) the maximum positive bending moment effect of the crown section, which is tested under symmetrical and eccentric loading conditions; (ii) the maximum negative bending moment effect of the springing section, which is tested under eccentric loading conditions; and (iii) the maximum positive bending moment effect of the section at 1/4 L of the arch rib, which is tested under symmetrical and eccentric loading conditions. The test section is shown in Figure 1 and Table 1. The loading efficiency coefficient is in the range of 0.96–1.03, which meets the requirements of [0.95, 1.05] specified in the specification (JTG/T J21-2011).

*Corresponding Author: 275310524@qq.com

DOI: 10.1201/9781003410843-9

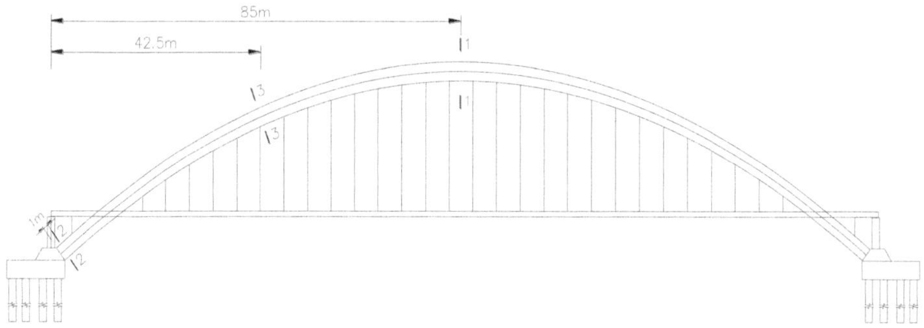

Figure 1. Schematic diagram of tested section.

Table 1. Table of tested sections and items.

Section number	Location	Tested Item	Condition
1-1	Crown	Deflection, strain, suspender c able force	Condition 1 (symmetrical load) and Condition 2 (eccentric load)
2-2	Springing	Strain, suspender cable force	Condition 5 (eccentric load)
3-3	1/4 L of arch rib	Deflection, strain, suspender cable force	Condition 3 (symmetrical load) and Condition 4 (eccentric load)

Strain sensors are arranged for Sections 1-1, 2-2, and 3-3 to test the strain of the arch rib during the static load test. Prisms are provided for sections 1-1 and 3-3, and the total station is used to test the deflection during a static load test, and the cable force of suspenders at the main force bearing points in the process of the load test is tested by cable force dynamic tester (Chen 2022; China Communications Press 2015; Luo 2015). Nine three-axle engineering vehicles are used for the load test. The total weight of each vehicle is less than 35 t, and the axle load of each vehicle is less than 14 t (Lu 2013).

2.2 *Contents of dynamic load test*

Dynamic load test is a field test that can test vibration characteristics under dynamic load and environmental load. The dynamic load test is carried out to test the vibration frequency, damping ratio, and vibration mode of the bridge. Vertical acceleration sensors are transversely arranged on both sides of the bridge deck and at the top of the arch rib (Qiu 2012; Wu 2011; Xu 2012).

3 RESULTS OF LOAD TEST

3.1 *Results of static load test*

From the deflection test results and analysis, as shown in Table 2, most of the measured deflection values are close to the theoretical ones, and the measured deflection values of some measuring points are greater than the theoretical deflection values. The deflection calibration coefficient is between 0.93 and 1.09, which does not meet the requirement specified in the specification that the calibration coefficient shall be less than 1. The test results show that the stiffness of the bridge does not meet the normal use requirements of the design load.

Table 2. List of deflection test results under various conditions.

Condition	Cross section	Measuring point	Measured elastic deflection (mm)	Theoretical deflection value (mm)	Calibration coefficient
Condition 1	Section 1-1	1/2 L of left arch rib	−21.2	−20.5	1.03
		1/2 L of right arch rib	−22.4	−20.5	1.09
Condition 2	Section 1-1	1/2 L of left arch rib	−29.3	−28.3	1.04
		1/2 L of right arch rib	−19.9	−19.7	1.01
Condition 3	Section 3-3	1/2 L of left arch rib	−42.2	−40.6	1.04
		1/2 L of right arch rib	−41.6	−40.6	1.03
Condition 4	Section 3-3	1/2 L of left arch rib	−42.9	−45.9	0.93
		1/2 L of right arch rib	−35.2	−36.1	0.97

From the residual deflection test results and analysis, as shown in Table 3, the measured relative residual deflection is between 5.9% and 12.3%, which meets the requirement specified in the specification that the relative residual deflection shall be no more than 20%, indicating that the bridge can basically return to the initial state and be in elastic working state after being stressed.

Table 3. List of test results of relative residual deflection under various conditions.

Condition	Cross section	Measuring point	Residual deflection (mm)	Measured deflection (mm)	Relative residual deflection (%)
Condition 1	Section 1-1	1/2 L of left arch rib	−2.5	−23.7	10.6
		1/2 L of right arch rib	−3.1	−25.6	12.3
Condition 2	Section 1-1	1/2 L of left arch rib	−2.5	−31.9	7.9
		1/2 L of right arch rib	−1.4	−21.3	6.6
Condition 3	Section 3-3	1/2 L of left arch rib	−3.4	−45.6	7.5
		1/2 L of right arch rib	−3.8	−45.4	8.3
Condition 4	Section 3-3	1/2 L of left arch rib	−3.2	−46.1	6.9
		1/2 L of right arch rib	−2.2	−37.4	5.9

From the strain rest results and analysis, as shown in Table 4, most of the measured strain values are close to the theoretical deflection values, and the measured strain values of some measuring points are greater than the theoretical ones. The strain calibration coefficient is between 0.82 and 1.11, which does not meet the requirement specified in the specification that the calibration coefficient shall be less than 1. The test results show that the strength of the bridge does not meet the normal use requirements of the design load.

From the residual strain test results and analysis, as shown in Table 5, the measured relative residual strain ranges from 3.5% to 13.2%, which meets the requirement specified in the specification that the relative residual strain shall be no more than 20%, indicating that the bridge can basically return to the initial state and be in elastic working state after being stressed.

From the cable force test results and analysis, Table 6 shows that the measured value of suspender cable force is between 946kN and 1695kN, which is less than the standard value of 2200 kN.

Table 4. List of strain test results under various conditions.

Condition	Cross section	Location of the measuring point	Measured elastic strain (με)	Theoretical strain (με)	Calibration coefficient
Condition 1	Section 1-1	The bottom surface of the left arch rib	85.0	80.6	1.05
		The top surface of the left arch rib	−123.0	−120.5	1.02
		The top surface of the right arch rib	−118.3	−120.5	0.98
Condition 2	Section 1-1	The bottom surface of the left arch rib	100.2	90.6	1.11
		The top surface of the left arch rib	−150.2	−150.5	1.00
		The top surface of the right arch rib	87.6	90.6	0.97
Condition 3	Section 3-3	The bottom surface of the left arch rib	125.4	135.1	0.93
		The top surface of the left arch rib	−137.9	−159.6	0.86
		The top surface of the right arch rib	−130.8	−159.6	0.82
Condition 4	Section 3-3	The bottom surface of the left arch rib	140.7	152.8	0.92
		The top surface of the left arch rib	−174.5	−190.7	0.92
		The top surface of the right arch rib	−119.6	−128.6	0.93
Condition 5	Section 2-2	The top surface of the left springing	102.2	100.8	1.01

Table 5. List of relative residual strain test results under various conditions.

Condition	Cross section	Location of the measuring point	Residual strain (με)	Measured strain (με)	Relative residual strain (%)
Condition 1	Section 1-1	The bottom surface of the left arch rib	10.3	95.3	10.8
		The top surface of the left arch rib	−7.9	−130.9	6.0
		The top surface of the right arch rib	−7.4	−125.7	5.9
Condition 2	Section 1-1	The bottom surface of the left arch rib	4.9	105.1	4.7
		The top surface of the left arch rib	−9.8	−160.0	6.1
		The top surface of the right arch rib	3.2	90.8	3.5
Condition 3	Section 3-3	The bottom surface of the left arch rib	5.6	131.0	4.3
		The top surface of the left arch rib	−10.5	−148.4	7.1
		The top surface of the right arch rib	−8.1	−138.9	5.8
Condition 4	Section 3-3	The bottom surface of the left arch rib	21.3	162.0	13.2
		The top surface of the left arch rib	−21.9	−196.4	11.1
		The top surface of the right arch rib	−5.1	−124.7	4.1
Condition 5	Section 2-2	The top surface of the left springing	7.7	109.9	7.0

Table 6. List of cable force test results under various conditions.

Condition	Suspender number	Suspender cable force of left arch (kN)	Suspender cable force of right arch (kN)
Condition 1	12#	1312	1159
	13#	1055	1041
	14#	1209	1127
	15#	1133	1296
	16#	1199	1177
	17#	1165	1152
	18#	1019	1028
Condition 2	12#	1512	1052
	13#	1065	946
	14#	1286	1037
	15#	1211	1298
	16#	1272	1041
	17#	1276	1040
	18#	1076	1006
Condition 3	4#	1326	1344
	5#	1287	1187
	6#	1291	1279
	7#	1278	1248
	8#	1353	1287
	9#	1165	1119
Condition 4	4#	1434	1203
	5#	1279	1123
	6#	1443	1179
	7#	1327	1131
	8#	1386	1122
	9#	1270	1105
Condition 5	2#	1589	1486
	3#	1695	1380
	4#	1585	1347
	5#	1561	1232
	6#	1441	1219
	7#	1362	1187
	8#	1090	995

3.2 *Results of dynamic load test*

From the natural frequency test results and analysis, the measured first-order vertical bending vibration frequency of the bridge identified from the amplitude spectrum of the measuring points is 0.80 Hz, which is 0.96 times the theoretically calculated value of 0.83 Hz. The test results show that the vertical dynamic stiffness of the bridge is weak.

From the Damping Ratio Test Results and Analysis, the damping ratio of structural vibration signals can be calculated by the time domain method and frequency domain method. The free attenuation waveform method and envelope fitting method are time domain analysis methods, which must be able to measure the free attenuation vibration of structures at a certain natural frequency (usually the lowest natural frequency). The frequency domain method is used to calculate the damping ratio with the frequency spectrum of the signal. Half-power bandwidth method is the most common one to calculate the damping ratio by using the frequency spectrum, which calculates the damping ratio according to the half-power bandwidth of a spectral peak in the spectrum. The damping ratio of this bridge is calculated by the half-power bandwidth method. The test results show that the damping ratio of this bridge lies between 1.5% and 1.7%, which is within the normal range of the damping ratio applicable to this kind of bridge.

From the vibration mode test results and analysis, the calculation model of the bridge is established by a finite element program. According to the data collected by the signal acquisition and processing analyzer of the dynamic measurement system, it can be seen that the measured results of the vibration modes of the bridge are basically consistent with the theoretically calculated results.

4 REINFORCEMENT MEASURES

According to test results, the bearing capacity of the bridge cannot meet the normal use requirements under design load, and it needs to be reinforced. The following reinforcement measures are taken on the basis of the structural stress characteristics of the bridge and the common reinforcement methods (Xu 2021; Zhang 2021; Zhao 2011; Zhou 2008).

Reinforcement of arch springing. The methods of enlarging the section and bonding the steel plate for reinforcement are adopted, which means that the steel box is wrapped outside, and the micro-expansion concrete is filled inside. The main construction processes are as follows: (i) dismantle the bridge deck system to make the arch rib in a bare state; (ii) treat the cracks (if any) at the arch springing; (iii) chisel the concrete surface within the reinforcement range of the arch springing into a rough surface with concave and convex, not less than 6 mm, and clean it; (iv) install steel bars at the arch springing; (v) install steel box; (vi) pour and fill micro-expansion concrete.

Transverse connection and bridge deck reinforcement. The transverse beams and decks of the original deck system are generally cracked, so it is considered to replace the whole deck system. The steel-concrete composite structure is adopted for the new deck system. The transverse steel beam has a box section, and longitudinal and vertical stiffeners and transverse diaphragms are provided to ensure the stability of the transverse beam. The longitudinal beam has an I-shaped section. Shear connecting keys adopt shear nails. Cast-in-place steel fiber C50 concrete is used for the bridge deck, and asphalt concrete is poured on the bridge deck. The main construction processes are as follows: (i) dismantle the bridge deck system to make the arch rib in a bare state; (ii) reinforce the arch springing; (iii) install suspenders; (iv) lift the transverse steel beams and longitudinal steel beams of the bridge deck, which are prefabricated in advance in the factory; (v) pour concrete of bridge deck (steel fiber C50 concrete); (vi) provide waterproof layer, pavement, sidewalk brick, railing, etc.

In terms of the replacement of suspenders, the suspenders are replaced because the anchor heads of the suspenders of the whole bridge have water accumulation, corrosion, and other distress. If the dismantled suspenders have no distress after strict inspection, they can be considered to be used again.

5 CONCLUSION

The deflection from the static load test can be used to evaluate whether the static stiffness of the bridge meets operating requirements. The strain values from static load can be applied to evaluate whether the strength of the bridge meets operating requirements. The natural frequency from the dynamic load test can be adopted to assess the dynamic stiffness of the whole bridge. The damping ratio and vibration mode obtained from the dynamic load test can be a reference for the overall condition of the bridge.

When the bearing capacity of a through-type reinforced concrete arch bridge cannot meet the requirements of normal use, reinforcement measures shall be taken. For this kind of reinforced concrete arch bridge without tie members, it is recommended to strengthen the deck by adding transverse and longitudinal connections. At the same time, the condition of the arch springing is very important to the arch bridge, so the arch springing shall be reinforced at the same time when the bridge is strengthened. Defective suspenders and anchor heads shall be replaced without delay.

REFERENCES

Chen Min. Strengthening Design of Composite Rigid Frame Arch Bridge. *Transport Construction & Management*, 2022(4):94–95.

JTG/T J21-01-2015 *Specification for Highway Bridge Load Test [S]*. China Communications Press, 2015.

Lu Fei & Peng Cheng. Analysis, Comparison, and Application of Common Methods for Bridge Reinforcement. *Highway*. 2013(10):121–123.

Luo Xiaoying. Static and Dynamic Load Test of a Through Tied Arch Bridge. *Shanxi Science & Technology of Communications*, 2015(3):65–69.

Qiu Ge. The Load Test Analysis of a Tied Arch Bridge. *China Exploration & Design*. 2012(08):84–89.

Wu Jianqi, Zheng Xiao & Zhang Tingting. Research on Dynamic Load Test of Highway Bridge Engineering. *Railway Engineering[J]*. 2011(3):26–28.

Xu Dayu, Zheng Riliang & Cao Qian. Load Test and Evaluation of Through Concrete Tied Arch Bridge. *Journal of Hunan City University (Natural Science)*, 2012(1):1–4.

Xu Zhixing. Reinforcement and Load Test of a Reinforced Concrete Truss Arch Bridge. *Fujian Transportation Technology*, 2021(4): 74–77.

Zhang Zhiwei. Key Construction Technology and Load Test of Long span Concrete Arch Bridge. *Northwest Hydropower*, 2021(3):79–82.

Zhao Rui & Zhu Yuexin. Talking About the Construction Technology of Highway Bridge Reinforcement. *China New Technologies and Products*. 2011(5):81.

Zhou Haijun, Wu Yongchang, Tan Yeping & Zhuang Yan. Summary of Bridge Load Test Research. *Journal of China and Foreign Highway*, 2008(4):164–166.

Constructional Engineering and Ecological Environment – Chih-Huang Weng (Ed)
© 2024 The Author(s), ISBN 978-1-032-53198-4

Numerical simulation of hydration and microstructure evolution of steam-cured cement paste

Chao Zou* & Xin Bai
School of Civil Engineering, University of South China, Hengyang, China

Jiahui Zhu
School of Civil Engineering, Central South University, Changsha, China

Saifei Zheng
School of Civil Engineering, University of South China, Hengyang, China

ABSTRACT: To intuitively show the formation and development of cement paste during the steam-curing process, a hydration kinetic model of cement paste in the unsteady steam-curing process was established based on the relationship between the degree of hydration and water-to-cement ratio, temperature and time. The time-varying model of the porosity of cement pastes considering the temperature effect was proposed. The test results show that the established model can effectively predict the degree of hydration and porosity of cement pastes with different water-to-cement ratios under different steam-curing temperatures and times. The hydration rate of cement paste steam-cured under 80°C is 10 times faster than that under 20°C.

1 INTRODUCTION

Steam-cured concrete precast components have the advantages of high early strength, fast demolding time, and high production efficiency. Therefore, they are adopted in large quantities in railway infrastructure construction projects in China. However, the performance of steam-cured concrete at the later stage is significantly different compared with that of normal temperature-cured concrete. The 28-day compressive strength is 10% ~ 15% lower than the latter (He 2012: Peng 2007; Wu *et al.* 2010). The hydration reactions and temperature field changes of concrete during steam-curing are more complex and drastic, and the formation and evolution of microstructure are more complicated (Feng *et al.* 2014; Gallucci *et al.* 2013; Li *et al.* 2019; Ma *et al.* 2017). In fact, the author has studied the changes of free water in cement paste during the steam-curing process using the low-field nuclear magnetic resonance technique and revealed the hydration characteristics of three stages in the cement paste by the content of chemically bound water (Zou *et al.* 2021). However, the hydration process and microstructure formation of cement paste are difficult to be observed visually, especially under the non-steady-state conditions of steam-curing.

In this study, a kinetic hydration model of cement paste in the unsteady steam-curing process was established based on the relationship between the degree of hydration and water-to-cement ratio, temperature and time, and a time-varying model of the porosity of cement paste considering the temperature effect was proposed. Besides, the reliability of the established models was verified by MIP tests.

*Corresponding Author: zouchao2021@usc.edu.cn

 DOI: 10.1201/9781003410843-10

2 MODEL CONSTRUCTION

2.1 *Modeling the hydration kinetics of cement during steam-curing*

Based on the hydration temperature rise test under adiabatic conditions, Breugel (Van Breugel 1991) proposed an approximate analytical expression for the exothermic hydration heat of cement, as shown in Equation. 1:

$$Q(t) = Q_{max} \cdot [1 - \exp(-\gamma t)] \tag{1}$$

Where Q_{max} is the final heat release after the cement hydration is completed; t is the hydration time, and γ is the shape parameter of this function.

Since the amount of hydrated cement in time t is proportional to the amount of heat Q(t) it releases, it defines the degree of hydration $\alpha(t)$ as:

$$\alpha(t) = \frac{Q(t)}{Q_{max}} = 1 - \exp(-\gamma t) \tag{2}$$

By isothermal calorimetric tests, the Q_{max} and γ of cement paste with a water-cement ratio of 0.3 at 20°C can be fitted to 268.4 J/mol and 0.0387, respectively. While the rate of cement hydration reaction accelerates with the increase of temperature and obeys the well-known Arrhenius Equation. Bazant (1986) has proposed a maturity Equation based on the Arrhenius Equation for the calculation of the equivalent time t_e at different temperatures, corresponding to the reference temperature T_{ref}:

$$t_e = \int_0^t \exp\left[\frac{Ea}{R}\left(\frac{1}{T_{ref}} - \frac{1}{T}\right)\right] dt \tag{3}$$

Therefore, the theoretical relationship of hydration degree with hydration time and curing temperature can be expressed by the following Equation:

$$\alpha(t, T) = 1 - \exp\left\{-\gamma\omega_0 \int_0^t \exp\left[\frac{Ea}{R}\left(\frac{1}{T_{ref}} - \frac{1}{T}\right)\right] dt\right\} \tag{4}$$

Where t denotes the curing time; ω_0 represents the initial water-to-cement ratio of the cement paste; γ is the shape parameter of the Equation; E_a denotes the activation energy of the cement; R is the universal gas constant; and T_{ref} and T denote the reference temperature (293 K) and the actual curing temperature, respectively.

On the other hand, the variation in the curing temperature also has a significant effect on the morphology and structure of the cement paste. Bentur *et al.* measured the capillary porosity of C_3S after hydration at different conditioning temperatures and found that the higher the temperature, the higher the capillary porosity of the paste at the same degree of hydration (Van Breugel 1991). The ratio of the hydration products to the volume of the cement particles is noted as ν, which gradually increases with increasing temperature. The increase in ν indicates an increase in the density of the protective layer on the surface of the cement particles, which increases the resistance to the diffusion of the hydration products and thus leads to a decrease in the rate of the hydration reaction. Breugel derived the relationship between $\nu - T$ from experiments with the mathematical expression (Van Breugel 1991):

$$\nu(T) = 2.22 \cdot \exp\left(-2.8 \cdot 10^{-5} \cdot T^2\right) \tag{5}$$

where the unit of T is °C.

Therefore, in order to more accurately describe the hydration development of the cement paste during the evaporation process, the effect of temperature on the hydration rate of the cement paste must also be considered from a morphological and microstructural point of view when modeling the hydration of cement during steam-curing, i.e.:

$$\alpha_{ste}(t,T) = \Omega(T) \cdot \alpha(t,T) \tag{6}$$

Where $\alpha_{ste}(t,T)$ is the hydration degree of the cement paste during the steam-curing process; $\Omega(T)$ is the influence factor of temperature on the microstructure of the cement paste, which can be calculated according to Eq. 7:

$$\Omega(T) = \left[\frac{\nu(T)}{\nu(T_{ref})}\right]^{\beta} \tag{7}$$

Where β is the shape parameter.

2.2 *Modeling the microstructural evolution of cement during steam-curing*

In 1947, Powers (Powers & Brownyard 1947) introduced the concept of gel-to-air ratio. They derived a calculated expression for the porosity of cement paste related to the degree of hydration and water-to-cement ratio based on theoretical calculations, i.e., the volume fraction of water and air in the capillaries of cement paste:

$$\varphi = f_w + f_a = \frac{63\omega_0 - 23.15\alpha}{20 + 63\omega_0} \tag{8}$$

Where φ is the porosity of the hydrated cement paste; f_w and f_a are the volume fraction of water and air, respectively; w_0 is the initial water-to-cement ratio; α is the degree of hydration of the cement.

However, this Equation only considers the change in pore structure from the perspective of volume change due to chemical reactions. It does not consider the effect of ambient temperature on the pore structure during cement hydration.

As mentioned before, during the heating stage of the steam-curing process, the temperature change increases the partial pressure of gas in the cement paste and expands the pore volume, while the expansion is not recoverable when cooling down, which results in the coarsening of the pores of the steam-cured cement paste. Therefore, the pore structure development of cement paste during the steam-curing process needs to consider the temperature effect. Thus, based on the study of Powers, this study improves the Equation for the porosity evolution of cement paste with the hydration degree under steam-curing conditions and proposes the evolution Equation for the porosity of steam-cured cement paste as follows:

$$\varphi(t,T) = \frac{1}{\Omega(T)} \cdot \frac{63\omega_0 - 23.15\alpha_{ste}(t,T)}{20 + 63\omega_0} \tag{9}$$

3 RAW MATERIALS AND EXPERIMENTAL METHODS

The cement used was Portland cement of grade 42.5, produced by the China Building Material Institute. Deionized water was used for mixing.

Specimen casting and maintenance: Cement and deionized water were mixed at a water-cement ratio of 0.3 and 0.5 in a laboratory environment at 20°C. Then they were pre-cured at 20°C for 3 hours (the preheating stage) and placed in different water bath heating pans for steam-curing, respectively. The water baths were heated at a uniform rate to the target temperature (the heating stage) within 2 hours, with a target temperature of 60°C. The target temperature was maintained for 8 hours (the constant temperature stage). Finally, the temperature was controlled to room temperature of 20°C within 2 hours (the cooling stage).

Mercury intrusion porosimetry (MIP): The test was conducted by using Auto Pore IV 9500 mercury-in-pressure instrument manufactured by Micromeritics. Before the test, the specimens with deserved age were taken out and immersed in an isopropyl alcohol solution for 7 days to terminate hydration and then dried in a vacuum desiccator at 60°C for 48 hours. Finally, several specimens of green bean size (about 1.5 g) were selected for the test.

4 RESULTS AND DISCUSSION

Figure 1 gives the development of the hydration degree with time for a cement paste with a water-to-cement ratio of 0.3 during the steam-curing process at different temperatures based on the model simulation. It can be seen that the higher the steam-curing temperature is, the higher the hydration degree of the cement paste at the same age. The hydration degree of cement paste grew slowly in the stationary stage; and began to accelerate in the heating stage with the rapid growth of hydration degree; after 4 hours of the constant temperature stage, the hydration rate tended to slow down again. When steam curing at 40°C, the hydration rate of cement is about 2.4 times that at 20°C; at 60°C, the hydration rate is about 5 times that at 20°C; and at 80°C, the hydration rate can reach 10 times of that at 20°C. Figure 2 shows the evolution of the cement paste porosity with water-to-cement ratio of 0.3 during the steam-curing process at different temperatures obtained from the simulation. It can be seen that the porosity in the cement paste gradually decreases with the growth of time. The porosity decreases slowly in the stationary stage and the heating stage; in the constant temperature stage, the porosity decreases rapidly; until the cooling stage, the rate of porosity decrease slows down again. It can be seen that the water-to-cement ratio has a significant effect on the development of the hydration degree of the cement paste during the steam-curing process. The larger the water-to-cement ratio is, the faster the cement hydration rate is and the higher the hydration degree at the end of steam curing. At the early stage of hydration, all the capillary pores in the cement paste are filled with free water. With the consumption of free water, more hydration products are formed, and the porosity is gradually reduced. The porosity of the cement paste decreases at a slow rate during the preheating stage and the heating stage; at the constant temperature stage, the porosity decreases rapidly; and at the cooling stage, the porosity decreases at a slow rate again. By comparing the simulated results with the measured results of MIP, it is found that the model of porosity evolution of steam-cured cement paste through hydration time, considering the effect of temperature, can be more accurately predicted.

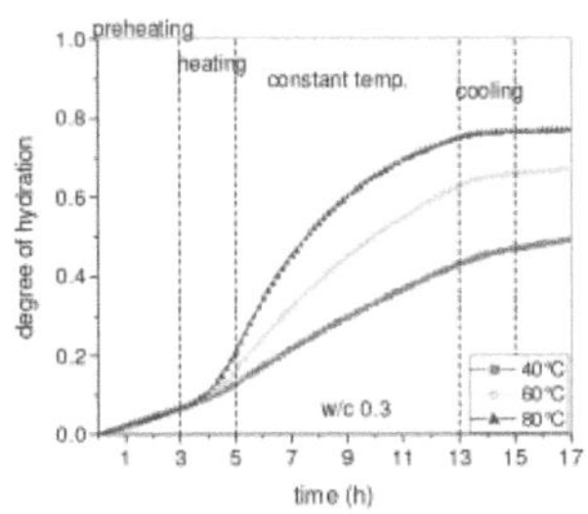

Figure 1. Development of hydration degree of cement paste during the steam-curing process.

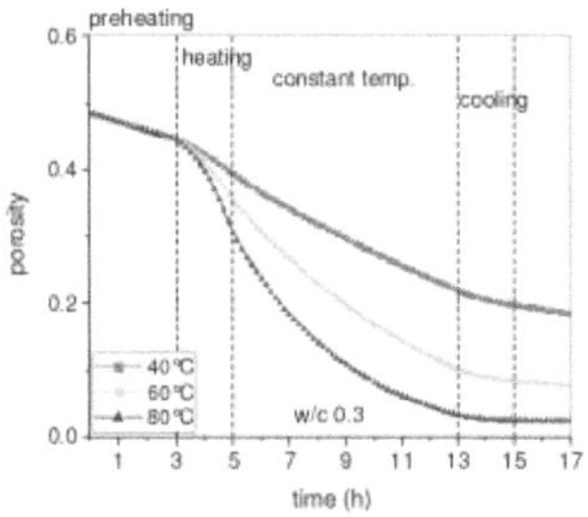

Figure 2. Evolution of porosity of the cement paste during the steam-curing process.

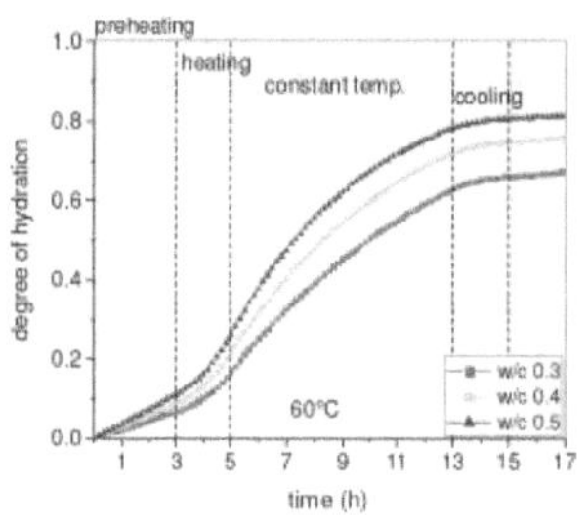

Figure 3. Development of hydration degree of cement pastes during steam-curing of 60°C.

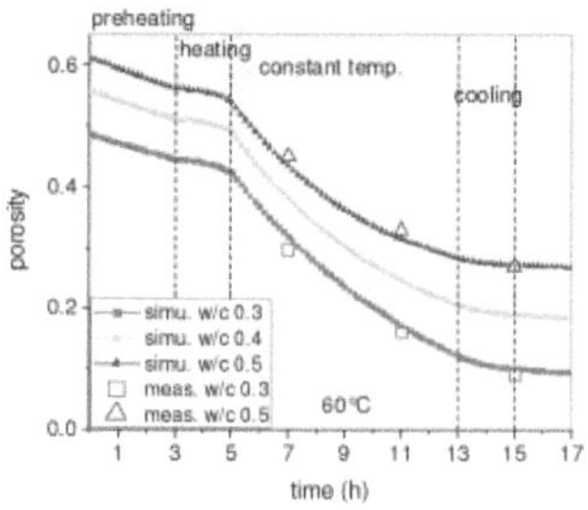

Figure 4. Evolution of capillary porosity of cement pastes during steam-curing of 60°C.

5 CONCLUSIONS

The main conclusions in this paper were summarized as follows:

(1) The hydration reactions under steam-curing conditions are basically the same as those at room temperature. But at 60°C, the hydration rate is about 5 times that at 20°C, and at 80°C, the hydration rate can reach 10 times that at 20°C.
(2) The heating stage is the main period causing the structural damage of the steam-curing cement paste.
(3) The evolution model of hydration degree and porosity of cement paste during the non-stationary steam-curing process established can effectively predict the hydration degree and porosity of cement paste with different water-to-cement ratios under different steam-curing temperatures and curing times.

REFERENCES

Bazant Z.P. *Creep and Shrinkage of Concrete, Mathematical Modeling*: Fourth Rilem International Symposium, Evanstone, Illinois, USA. (1986).

Feng P., Miao C. and Bullard J. A Model of Phase Stability, Microstructure and Properties During Leaching of Portland Cement Binders. *Cement & Concrete Composites*. 49(12), 9–19 (2014).

Gallucci E., Zhang X. and Scrivener K. Effect of Temperature on the Microstructure of Calcium Silicate Hydrate (C-S-H). *Cement and Concrete Research*. 53, 185–195 (2013).

He Z. *Heat Damage Effects of Steam Curing on Concrete and Corresponding Improvement Measures*. Central South University. (2012).

Li L., Long G., Liu F. et al. Deformation Behavior of Concrete During Steam Curing. *Materials Report*. 33 (08), 1322–1327 (2019).

Ma K., He J., Long G. et al.: Steam-curing Temperature Effect and its Influence on Heat Damage of Cement-based Material. *Materials Report*. 31(23), 171–176 (2017).

Peng B.: *Influence of Steam-curing System on the Performance of High Strength Concrete*. Wuhan University of Technology. (2007).

Powers T.C. and Brownyard T.L. Studies of the Physical Properties of Hardened Portland Cement Paste. *Journal of Thermal Analysis and Calorimetry*. 43(2), 669–680 (1947).

Van Breugel K. *Simulation of Hydration and Formation of Structure in Hardening Cement-based Materials*. Delft University of Technology. (1991).

Wu J., Liu S. and Deng S.: Effect of Forepart Strength of Cured-concrete. *Journal of Ningbo University*. 23 (04), 112–116 (2010).

Zou C., Long G., Zeng X. et al.: Water Evolution and Hydration Kinetics of Cement Paste Under Steam-curing Condition Based on Low-field NMR Method. *Construction and Building Materials*. 271,121583 (2021).

Constructional Engineering and Ecological Environment – Chih-Huang Weng (Ed)
© 2024 The Author(s), ISBN 978-1-032-53198-4

Design and application of floating underwater foundation bed riprap leveling ship for inland river immersed tunnel

Tao Li*
CCCC Second Harbor Engineering Co., Ltd, Shanghai, China
Key Laboratory of Large-span Bridge Construction Technology, Wuhan, Hubei, China
Research and Development Center of Transport Industry of Intelligent Manufacturing Technologies of Transport Infrastructure, Wuhan, China

Chenyang Fan & Jinxu Wang
CCCC Second Harbor Engineering Co., Ltd., Shanghai, China

ABSTRACT: Aiming at the problems that the existing immersed pipe foundation bed equipment is restricted by inland rivers, it is unable to pass, the construction cost is high, and the ship moving efficiency is low. Based on the Yuliangzhou section of the Xiangyang east-west axis road project and based on automatic centralized control and satellite positioning technology, a foundation bed leveling construction equipment using rotatable riprap pipe floating barge + automatic anchoring system + leveling equipment is designed, which realizes the high-efficiency and high-precision construction objectives of free navigation in the inner river basin and foundation bed leveling of the immersed tunnel. The field application shows that the leveling equipment can reduce the displacement time of the leveling ship from the conventional 20 min to 3 min, and improve the construction efficiency by about 5 times. The positioning accuracy of the construction ship can be controlled within 10 cm, and the leveling accuracy can be controlled within ±4cm, which provides technical support for the project.

1 INTRODUCTION

Immersed pipe tunnel is a tunnel form in which prefabricated pipe joints are installed on the seabed foundation mattress layer by floating transportation and successively connected through (Eelco & Patrick 2022; Li *et al.* 2014). The foundation needs to be processed before the installation of the immersed tube, and the control accuracy of the gravel foundation bed laying directly affects the installation accuracy of the immersed tube in the later stage, which is also the guaranteed process for the successful installation of the immersed tube (Zhang 2018). At present, there are mainly three kinds of construction technology and equipment for immersed tube tunnel foundation bed. The floating leveling vessel with fixed positioning pile is adopted, such as the "Scradcway" floating leveling vessel with fixed positioning pile developed by Boscalis Company in the Netherlands (Yang *et al.* 2014). However, after each pebble ridge construction, the leveling vessel needs to pull out the pile and position again, so the continuous laying of the gravel ridge cannot be realized. Self-elevating platforms are used as carriers, such as the "Jinping No. 1" used in the foundation bed leveling construction of the immersed tunnel of the Hong Kong-Zhuhai-Macao Bridge (Su *et al.* 2018; Yang *et al.*

*Corresponding Author: 1771902930@qq.com

DOI: 10.1201/9781003410843-11

2014). However, because the equipment carrier is a self-elevating platform, the overall dimensions of the leveling equipment are large, which cannot meet the navigation conditions of inland waters, and the investment is huge; Riprap leveling frame technology is adopted, such as "Yihang Jinping 2" used in Shenzhen-Zhongshan Channel (He 2020; Ma & Song 2018), but this technology also has the disadvantages of high cost and low ship moving efficiency. With the development of inland river pipe sinking project construction, the above construction methods have some problems when applied to inland rivers, such as limited application scope, low efficiency, or high investment cost. Therefore, new construction equipment for foundation bed leveling is urgently needed.

2 ENGINEERING BACKGROUND

The Yuliangzhou Section of the Xiangyang East-West Axis Road Project has a total tunnel length of 5400 m. The pipe sinking scheme is adopted to cross the Hanjiang River twice, with a water depth of 15 m~26 m. The total length of the pipe sinking section is 1011 m, and a total of 10 pipe joints are set up (Ren & Zhao 2019), which belong to the inland river pipe sinking tunnel. The pipe sinking cushion is laid with a pebble cushion as the foundation. The diagram of the pebble cushion is shown in Figure 1. In order to avoid local high points of the pipe sinking foundation and ensure that its bottom plate is evenly stressed, a high-precision leveling foundation cushion is set between the bottom plate, and the foundation as the foundation bed, so special ships with high leveling accuracy are required for construction (Hua *et al.* 2021). The river runoff in this area is gentle, and the water level changes little. The annual flow velocity is less than 0.3 m/s, and the flow velocity in flood season is less than 0.8 m/s. The wave, flow, and other conditions are ideal. Therefore, based on the good inland river hydrological conditions and navigation restrictions of the ship lock, the project adopts the floating riprap leveling ship independently developed by CCCC Second Harbor Engineering Co., Ltd. for construction.

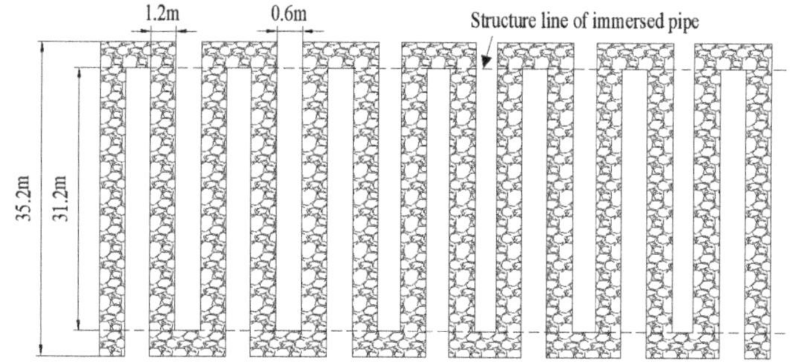

Figure 1. Schematic diagram of pebble cushion.

3 COMPOSITION OF LEVELING EQUIPMENT

The floating underwater riprap leveling ship for subgrade bed is mainly composed of a floating barge, riprap leveling equipment, stone conveying equipment, and construction management system, which is used to lay the gravel cushion for immersed pipes within the water depth of 15 m~26 m, as shown in Figure 2. The floating barge adopts a flat barge. The automatic anchoring system moves the hull to adjust the scope of gravel paving and leveling operations. Riprap leveling equipment and stone conveying equipment are installed on the floating barge.

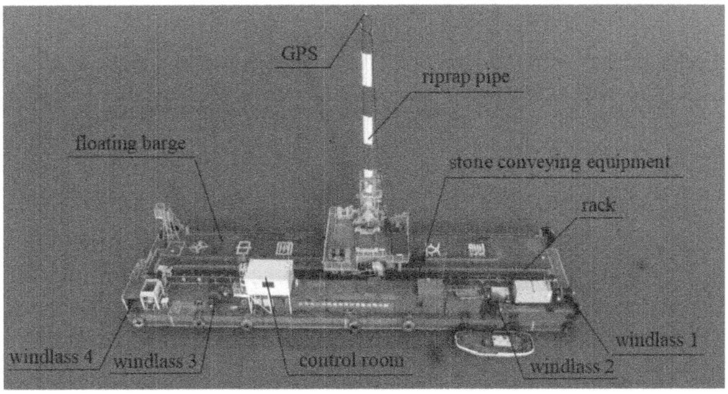

Figure 2. Floating underwater foundation bed riprap leveling ship.

3.1 *Floating barge*

The width of gravel laying for pipe sinking of the supporting project is 31.2 m, and the downstream is Cuijiaying Shiplock. It is required that the width of the boat passing through should not exceed 18 m, and the height of the boat should not exceed 10 m. Therefore, the floating barge adopts the "Jingrun 106" vessel, with a length of 55 m, a width of 15.3 m, and a draft of 2 m.

3.2 *Riprap leveling equipment*

The riprap leveling equipment is composed of riprap pipe, riprap pipe lifting winch, riprap pipe support and clamping system, traveling trolley, etc., as shown in Figure 3, mainly to achieve high-precision riprap leveling of the foundation bed. A riprap pipe with a diameter of about Φ1200 mm round pipe, whose length is adjusted through the joint action of four elevation balance oil cylinders at the bottom to control the accuracy of riprap leveling; The riprap pipe lifting winch is installed on the mobile trolley, with a working load of 50 kN and a nominal speed of 12 m/min. Its main function is to lift the riprap pipe to achieve leveling in different water depths. The riprap pipe support and clamping system are installed at the upper and lower positions of the riprap pipe fixing frame. The riprap pipe can be compressed or loosened through the stroke change of the intermediate oil cylinder. The traveling trolley moves on the side of the ship. It is driven by gear racks to drive the riprap pipe to move longitudinally, complete the movement of the riprap pipe and realize the leveling of different ridge widths.

Figure 3. Riprap leveling equipment.

3.3 *Stone conveying equipment*

The stone conveying equipment adopts the form of a belt conveyor, and the conveyor belt joint is vulcanized. The belt conveyor sends the stone in the hopper to the riprap pipe at a speed of 300 t/h.

3.4 *Construction management system*

The riprap management system is located in the cab, which can control the traveling trolley, belt conveyor, riprap pipe, etc. It can display the voltage, current, running time, height of the riprap pipe, the position of the telescopic oil cylinder, and other parameters and information of the motor of the riprap pipe, as well as the overload, overheating, and disconnection faults of the motor. The system is equipped with an emergency stop button to shut down the system in case of an emergency.

4 CONSTRUCTION TECHNOLOGY

This process flows mainly includes the moving and positioning of leveling ship, berthing and replenishing of the replenishment ship, lowering and checking the elevation of riprap pipe, laying of pebble cushion, leveling quality inspection, and recycling riprap pipe.

The first step is moving and positioning of leveling ship. After each ridge of crushed stone is constructed by the leveling ship, the ship needs to move 2m laterally by operating 4 hydraulic winches on the leveling ship. The leveling ship moving to position adopts the real-time feedback convergence control method based on the position information. The control system outputs the winch action signal according to the current position coordinate information and the target position coordinate information. It drives the ship to move along the given track through the retraction and release of the four winches, thus realizing the automatic ship-moving function. The closed-loop control system is shown in Figure 4.

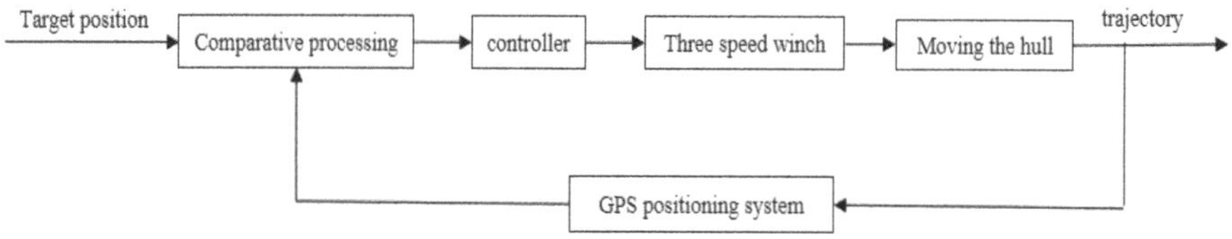

Figure 4. Riprap leveling equipment.

In the process of automatic control, the ship position information is detected by the two GPS sensors at the bow and stern, and the rope length of the four winches to be retracted is calculated. The ship moving control strategy adopts rough positioning and fine positioning. The length of the control rope for rough positioning is greater than 1m, and a 1.5 m/min high-speed control winch is used for rough positioning. The length of the control rope during fine positioning is less than 0.4 m, and the winch is controlled to rotate at a low speed of 0.5 m/min during fine positioning; the length of the control rope for intermediate positioning is 0.4 m~1 m, and the average speed of winch at high speed and low speed is adopted for intermediate positioning. Through field verification, the automatic ship moving system can reduce the ship moving time from 20 minutes to 3 minutes, and the positioning accuracy can be controlled within 10 cm (Hua *et al.* 2021).

The second step is replenishment by barge berthing. After replenishing materials at the wharf, the replenishment ship will lean against the non-leveling side of the leveling ship, and then moor it on the mooring bollards of the non-leveling side of the leveling ship with ropes.

Finally, the cushion pebbles will be transported to the riprap pipe through the belt conveyor of the leveling ship.

The Third step is to lower the riprap tube and check the elevation. The cage rotates the riprap pipe from the horizontal state to the vertical state and then releases the leveling winch cable through the lifting winch to lower the riprap pipe. During lowering, the GPS 3 on the riprap pipe monitors the lower surface elevation of the leveling scraper. Assuming the length of the riprap pipe is L_1, the length of the elevation adjustment cylinder at the bottom of the riprap pipe is L_2, and the elevation of the target subgrade bed is H_1, the elevation control of GPS 3 when the riprap pipe is lowered is:

$$H = L_1 + L_2 + H_1 \tag{1}$$

The fourth step is laying pebble bedding. The leveling trolley moves at the side of the ship, driving the riprap pipe to move longitudinally and completing the riprap pipe displacement during the riprap leveling operation. To ensure the leveling accuracy, a pebble leveling layer, an 80 cm thick leveling layer, and a 40 cm thin leveling layer are used for laying the pipe sinking foundation bed.

The fifth step is to check the leveling quality. After the single pebble ridge is filled, the remaining pebbles in the riprap pipe shall be emptied. Recheck the bottom elevation of leveling scraper again to ensure that it is consistent with the top elevation of the pebble ridge. Moving the leveling trolley to make the riprap pipe move according to the original leveling track. The scraper shall level the paved cushion again to ensure that the top elevation of the pebble ridge is controlled within ± 4cm.

The sixth step is to recover the riprap pipe. After the construction of the leveling ship is completed, the riprap pipe shall be lifted and recovered to the inboard side of the leveling ship.

5 RESPONSE ANALYSIS TO WAVES

The wave response of the floating leveling ship is analyzed to select a better window period during construction and ensure leveling accuracy. The floating leveling ship is modeled and analyzed in Sesam. The wave frequency is 0.4 rad/s to 2 rad/s, with an interval of 0.1 s, and there are 17 ships in total. The wind wave direction is from 0° to 45° and 120° to 180°, with an interval of 15°. There are 7 wind waves in total.

Under the action of the wave spectrum of unit wave height, the short-term prediction of the motion amplitude of ship heave, pitch, and roll is shown in Figures 5 to 7.

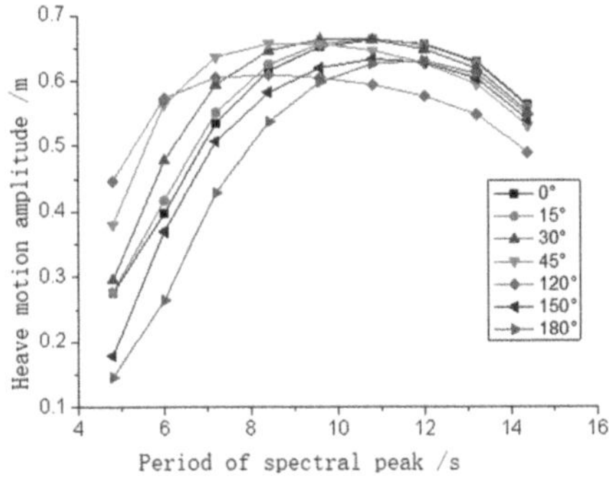

Figure 5. Ship heave response at unit wave height.

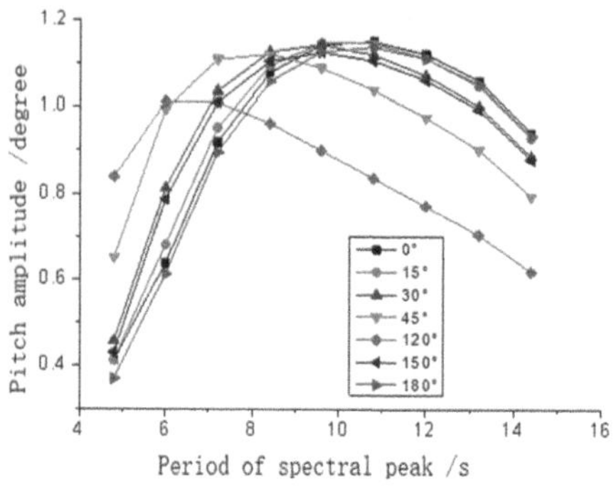

Figure 6. Ship pitch response at unit wave height.

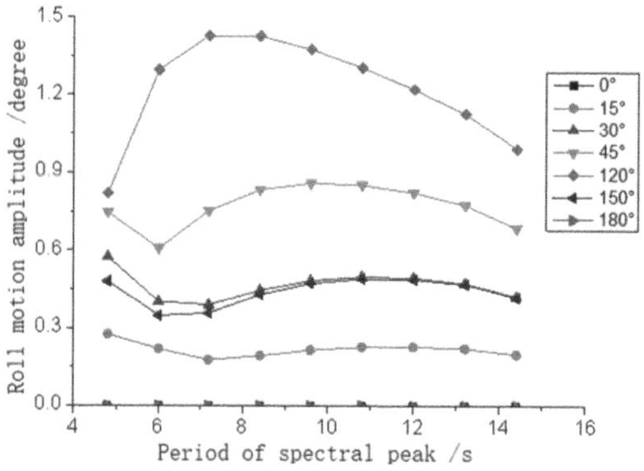

Figure 7. Ship roll response at unit wave height.

The effective wave height of general waves in the construction water area is 0.03m, and the period is about 8s. It can be seen from the above figure that when the period is about 8s, the amplitude of heave motion of the ship at different wave direction angles at unit wave height is 0.646 m, the maximum amplitude of pitch motion is 1.126 degrees, and the maximum amplitude of roll motion is 0.445 degrees. In this water area, the maximum amplitude of the heave motion of the ship at different wave direction angles is 0.02m, pitch motion amplitude is 0.034 degrees, and roll motion amplitude is 0.013 degrees. Therefore, the waves in the waters have little impact on the construction of floating leveling vessels.

6 FIELD APPLICATION

The floating bed riprap leveling equipment has been successfully applied to the pipe sinking bed leveling of the Yuliangzhou Section of the Xiangyang East-West Axis Road Project, as shown in Figure 8. During the construction, the total station+survey line and multi-beam scanning survey results shall be used to guide the subgrade bed leveling construction. Taking the leveling operation of the E2 pipe joint applied by the floating leveling barge in the project as an example, there are 65 rows of pebble beds to be leveled for the E2 pipe joint.

Figure 8. Application of floating bed leveling vessel.

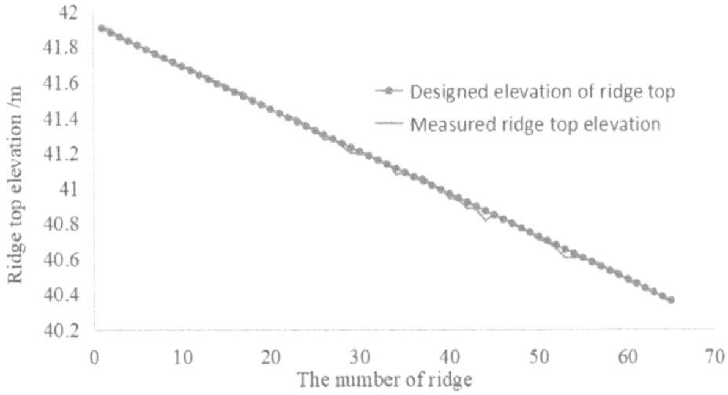

Figure 9. Subgrade bed elevation measured by total station.

6.1 *Measurement of total station and measuring line*

The point on the central axis of the pebble ridge is selected at a distance of 3 meters by total station and measuring rope for measurement. The measurement results are shown in Figure 9. The average deviation range of single ridge elevation is -3.6 cm~+2.9 cm. Within the design allowable deviation (±4cm), the qualified rate of original data of measuring points on the measuring line is 97.09%.

6.2 *Measurement of multi-beam*

The distance between data points generated by multi-beam measurement is 0.1 m. A scatter map is generated by data processing, and color contrast is set. A total of 575683 data are collected.

The statistical results are shown in Table 1. 89% of the data meet the design requirements. The allowable deviation between the top edge lines on both sides of the cushion and the design position plane is 0 cm~20 cm, the longitudinal deviation of the pebble ridge is 0cm~ +5 cm, and the longitudinal width deviation of the pebble ridge is 0 cm~10 cm, meeting the design requirements.

Table 1. Statistical analysis of multi-beam data.

Total data	Data within ± 4cm	Data below -4cm	Data larger than 4cm	the sum
575683	512356	63327	0	575683
100%	89.00%	11.00%	0%	100%

The qualification rate of the foundation bed measured by the two measurement methods is 89%, so the development of the floating leveling ship is successful. Generally, the leveling accuracy of the foundation bed can be controlled within ±4 cm, which can meet the construction requirements of the foundation bed leveling of inland river-immersed tunnels.

7 CONCLUSION

The floating underwater foundation bed riprap leveling ship for immersed tunnel realizes the high-precision leveling construction of the foundation bed using the floating leveling construction technology of rotatable riprap pipe. The use effect is good. It solves the problem that the traditional leveling equipment is limited by the inland river ship lock, the construction cost is high, and the ship moving efficiency is low. It has the following characteristics:

First, it has a one-button automatic anchoring and positioning function, which realizes the fast, efficient, and high-precision moving and positioning of the barge and can reduce the moving time of the leveling ship from 20 minutes of conventional manual work to 3 minutes.

Second, the leveling equipment adopts a rotatable riprap pipe structure, which avoids the disadvantage that large leveling equipment is difficult to adapt to the restricted conditions of inland navigation and realizes the construction of barges in inland rivers.

Third, the automatic feeding belt conveyor and automatic elevation compensation technology are adopted to realize the automatic continuous supply of stone and high-precision laying and leveling of the foundation bed.

At the same time, in the process of subgrade bed leveling, strict control of stone screening proportion, reduction of debris, and other adverse situations, as well as strengthening the monitoring of the back silting of the subgrade bed, will help to improve the leveling accuracy.

REFERENCES

He Jing.: Jack Up Gravel Paving and Leveling Ship "Yihang Jinping 2". *Ship Design Communication*. (01), 107(2020).

Hua Xiaotao, Li Tao, Wang Jinxu. Design and Application of Automatic Ship Moving Control System for Riprap Leveling Ship on Foundation Bed of Immersed Tunnel. *China Harbour Construction*. 41(01), 62–66 (2021).

Li Jialin, Lin Fangang, Liu Yong-jun, Li Youzhi. Development and Application of Ultra-large Immersed Pipe Ballast Water System. *China Harbour Engineering*. 40(1), 64–68+73(2014).

Ma Zonghao and Song Jiangwei. Application and Development of Pre-paving Method for Foundation Bed Leveling in the Immersed Tunnel. *China Harbour Construction*. 38(02), 16–19(2018).

Ren Yaopu and Zhao Zhiwu. Study on the Selection of Dry Dock Site for Xiangyang Hanjiang Immersed Tunnel. *Engineering Construction and Design*. 12, 121–122(2019).

Su Faqiang, Li Jialin and Wang Mingxiang. Construction and Application of Offshore Deepwater High-precision Gravel Bed Paving Leveling Ship. *Highway*. 63(08), 43–46(2018).

van Putten Eelco and van Os Patrick. The A24 Blankenburg Connection: An Innovative Design Concept for an Immersed Tunnel Project in a Busy Port. *Tunnelling and Underground Space Technology*. (06), 124 (2022).

Yang Xiuli, Shao Manhua and Xu Jie. Type Selection of Overall Scheme for Riprap Leveling of Gravel Cushion Paving Ship (Platform) for Hong Kong Zhuhai Macao Bridge Sinking Pipe. *Construction Technology*. 43(11), 17–19+35(2014).

Zhang Jianjun. High Precision Control Technology and Application of Crushed Stone Foundation Bed of Offshore Deep-water Immersed Tunnel. *China Harbour Construction*. 38(12), 52–54+72(2018).

Constructional Engineering and Ecological Environment – Chih-Huang Weng (Ed)
© 2024 The Author(s), ISBN 978-1-032-53198-4

Spatial accessibility analysis of the prefab building suppliers using Gaussian based 2-step floating catchment area method: A case study of Fuzhou city

Jinghan Pan, Qianru Bi & Xin Wu*
Fuzhou University, Fuzhou, China

ABSTRACT: In the green building research area, more and more studies have been carried out on prefabricated buildings, as the life cycle assessment (LCA) has found that the carbon footprint and primary energy footprint of prefabricated buildings during their entire life cycle are much lower than the traditional concrete constructions. Few current prefab building studies have focused on the supply of components, which would increase carbon emissions and the cost of delivery. Therefore, measuring the spatial accessibility of prefab suppliers to the areas in demand would be a key step in planning the prefab suppliers for carbon emission reduction, an integral part of conducting urban service facility planning and promoting.

A case study has been carried out in Fuzhou city to address the issue of supply chain and the spatial accessibility of prefab building-related suppliers. A quantitative analysis is carried out with the Gaussian two-step moving search method, and the output is visualized with the ArcGIS software package.

Three significant results emerged. Firstly, the core high-value areas are located in the central regions, including Gulou District, Taijiang District, and Cangshan District. Secondly, there are significant differences in the spatial configuration of the supply of prefab building materials in each region of the city. Finally, from the perspective of industrial development planning, the supply and demand imbalance areas are identified. Correspondently, three planning strategies are proposed to provide a reference for urban operation managers, which would enhance the development of the building industry heading towards the goal of "double carbon."

1 INTRODUCTION

Nowadays, roughly 50% of the greenhouse gas (GHG) emissions in the atmosphere are attributable to building construction and operations (Bonamente & Cotana 2015). As the World Resources Institute (WRI) carbon emissions statistics by industry showed in 2017 in China that the carbon emissions of the building construction sectors were the second most (23.2%), only after the power sectors (41.6%). Therefore, low carbon studies in the construction sector are of great importance to achieve the national "double carbon" goal. At present, rapid urbanization, implementing and carrying forward the advantages of prefab building construction, which are labor intensification, building materials saving, industrialized production, high construction efficiency, and high construction quality, have important practical and theoretical significance for the current adjustment of resource allocation in the construction field and low-carbon city construction.

The carbon footprint, cost, and energy consumption of the transportation of building materials have a pivotal impact on the development and popularity of prefab buildings in the

*Corresponding Author: wuxin@fzu.edu.cn

DOI: 10.1201/9781003410843-12

four stages of the whole life cycle of the building. However, due to the hilly topography of Fujian Province (Zhang 2007), and the relatively unbalanced development of building materials logistics (Chen 2018), the supply problem between the scattered construction sites and prefabricated components production points has not been solved, which hinders the social demand for the green transformation of the construction industry.

Spatial accessibility is one of the important indicators for evaluating whether the spatial facilities are reasonably configured (Li *et al.* 2016). Distance, time, and other indicators are usually used to judge how easy it is for one location to overcome spatial barriers to reach another location (Ma & Cao 2006), which can visually reflect whether the spatial configuration of a particular city is balanced. Using spatial accessibility to describe the supply and demand of prefab building materials can quantify the spatial allocation of building materials suppliers more accurately and evaluate the feasibility of building materials transportation between regions and cities for prefab construction or optimization. Aiming at the evaluation advantages of spatial accessibility and the practical problems and research vacancies in the reality of urban development of assembly building materials supply, this paper analyzed the accessibility of assembly building materials supply in Fuzhou city by Gaussian two-step moving search method. After considering the current transportation situation in the development of assembly buildings in Fujian province and obtaining the accessibility of the whole study area by ordinary kriging interpolation method, an urban assembly industry layout is provided. The accessibility of the whole study area was obtained by ordinary kriging interpolation, which provides reference and reference for the layout of the urban assembly industry.

2 REVIEW

As research in China continues to leap forward in related fields, such as the transportation of assembled buildings, many domestic scholars have conducted valuable research on the transportation of assembled building components. Chen Wei *et al.* (2017) took assembled building projects as the object of their research, providing an in-depth analysis of the heterogeneous and non-synchronous characteristics of the three processes of on-site assembly, component production, and transport logistics. Luo Jie *et al.* (2016) further refined the research area, focusing on the characteristics and difficulties of this process in building construction, and analyzed some key points of safety management in the construction of assembled buildings. In terms of information flow, Chang Chunguang (Chang & Wu 2015) combined BIM with RFID by establishing an information-sharing platform and discussed the application of an integrated BIM-RFID system in the transportation process. Li Tianhua (2011) summarized the challenges of poor information transfer during the transportation of components, providing new ideas to solve the critical problem of lacking a technical platform to link the various stages in the project life cycle. On the issue of risk avoidance in the transportation process, Liu Jingai (2016) analyzed the main risk factors of building components in transportation and stacking stages from the perspective of building component manufacturers and then proposed to solve the deepening design problem by improving design standardization and modularization. At the present stage, the research on the transportation process of prefab buildings is extensive, but the research on the service scope and resource allocation accessibility of prefab companies at the city level is less involved.

3 DATA SOURCES AND PROCESSING

3.1 *Data sources*

(1) Fuzhou city administrative division map. The data is derived from Baidu Map vectorization and OpenStreetMap (OpenStreetMap Homepage 2022) website.

(2) Population census data of each township (town, street) in Fuzhou City. The data was obtained from the seventh national census data released by the Fuzhou City Bureau of Statistics (Fuzhou Municipal Bureau of Statistics Homepage 2022) on November 15, 2021.
(3) Data of suppliers of prefab building materials in Fuzhou City. The cut-off date for the data information is October 4, 2022. In this paper, 1065 valid suppliers of assembly building materials were extracted and screened as research objects through the website of Aiqicha (Aiqicha Homepage 2022).
(4) After comprehensive consideration of road speed limit and road construction in the study area, logistics transportation time cost, cargo damage, energy consumption, and other factors (Yang 2014), the transportation mode of prefab building materials and the distance threshold were clarified. The service radius (distance threshold) of this study was divided into 8 levels (Table 1).

Table 1. Service radius levels. [Source: Self-drawn].

Level 1	Level 2	Level 3	Level 4	Level 5	Level 6	Level 7	Level 8
5 km	8 km	10 km	12 km	14 km	16 km	18 km	20 km

3.2 *ArcGIS data pre-processing*

(1) Determination and import of supply and demand points: ArcGIS software is used for digitizing the collected coordinate information in space to form the supply point layer and demand point layer, respectively (Figures 1 and 2).
(2) Road topology network: construct a network dataset based on the road network map of Fuzhou City in ArcGIS.
(3) Network analysis method: use the network analysis module in ArcGIS, construct the OD matrix, and calculate the road network distance and reachability from the demand point to the supply point and supply point to the demand point, respectively.

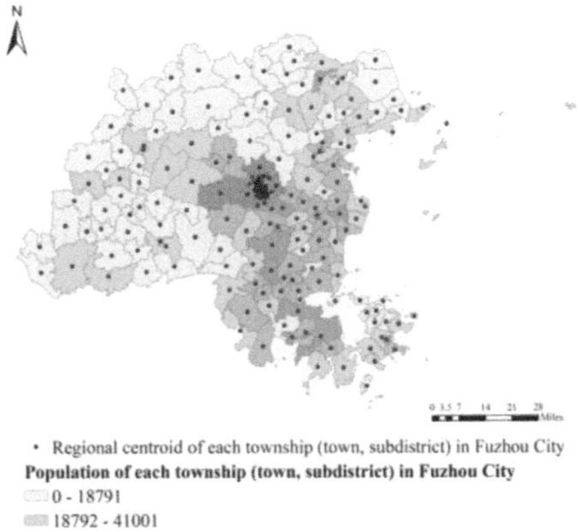

Figure 1. Population indexing class and location of mass center distribution [Source: Self-drawn].

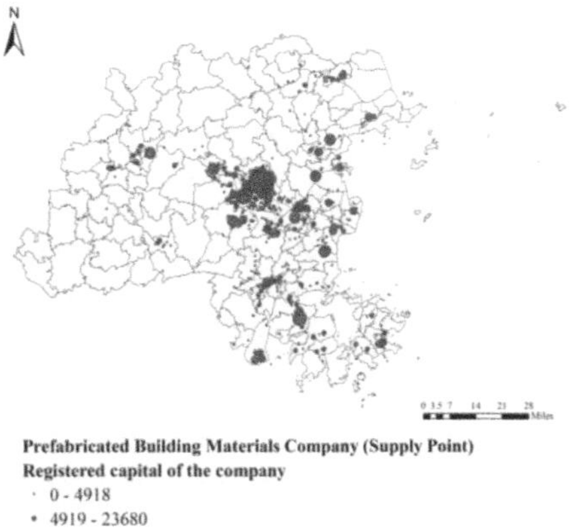

Figure 2. Distribution location and registered capital level of supply points of prefab building materials [Source: Self-drawn].

4 RESEARCH METHODOLOGY

4.1 *Spatial analysis*

Kernel density analysis is a nonparametric method for estimating probability density functions, which is heavily applied in the analysis of the spatial distribution characteristics of geographic elements, responding to the density characteristics of point elements (Xu & Gao 2016; Yu & Ai 2015). In this study, this method is used to develop an analysis of the sparse density of the distribution of assembly building materials supply points within Fuzhou City.

4.2 *Standard deviation ellipse analysis*

The standard deviation ellipse analysis tool is one of the classic methods for analyzing the directional characteristics of spatial distribution, which can obtain the spatial characteristics of series data, including central trend, discrete and directional trend, through the analysis of aggregated geographical element points (Li *et al.* 2017). In this study, this method is used to analyze the changes in the scope and directional distributivity of the spatial distribution agglomeration area of the assembly building materials supply points in Fuzhou City.

4.3 *Analysis of Gaussian-based 2-step floating catchment area method*

1) In the first step, the search domain of supply point j is determined by taking the supply point j of the building material company obtained by screening within Fuzhou City as the center. The respective corresponding distance threshold d_0 (Table 1) as the radius, searching the set of demand points k falling within the search domain and calculating the supply-demand ratio R_j of supply point j.

$$R_j = \frac{S_j}{\sum\limits_{k \in \{d_{kj} \leq d_0\}} P_k} \tag{1}$$

Where, R_j is the supply-demand ratio of supply point j of the building material company, i.e., the service capacity of this supply point; S_j is the total supply quantity of point j, which we measure by the registered capital of the building material company in the experiment; d_{kj} denotes the road network distance from demand point k to supply point j; P_k is the demand quantity of the search domain point k, which is measured by the corresponding population size.

2) In the second step, given the spatial action threshold d_0 for each township center of mass demand point i, a new spatial action domain is also formed, and the supply ratio R_l of each building material company point l falling in it is assigned a weight using Gaussian equation. Then the weighted supply ratio R_l is summed up and calculated to obtain the accessibility of prefab building materials Ai for each township center of mass i. The size of A_i reflects a certain extent, the accessibility of building materials in the experimental area. The magnitude of A_i reflects the per capita possession of prefab building materials in the experimental area to a certain extent.

$$A_i = \sum\nolimits_{l \in \{d_{ii} \leq d_0\}} G(d_{il}, d_0) R_l \tag{2}$$

In Equation (2), R_l represents the supply ratio of building materials company l in the spatial scope of the demand point; $G(d_{il}, d_0)$is a Gaussian equation considering the spatial impedance problem, and the calculation method is shown in Equation (3). Other indicators are described in Equation (2).

$$G(d_{kj}, d_0) = \begin{cases} \dfrac{e^{-\left(\frac{1}{2}\right)\times\left(\frac{d_{kj}}{d_0}\right)} - e^{-\left(\frac{1}{2}\right)}}{1 - e^{-\left(\frac{1}{2}\right)}}, & \text{if } d_{kj} \leq d_0 \\ 0, & \text{if } d_{kj} > d_0 \end{cases} \tag{3}$$

3) Finally, we use the ordinary kriging interpolation method to interpolate the calculated accessibility values. The histogram and QQ plot of the accessibility values were observed in the "Explore Data" of ArcGIS to see whether they conformed to the normal distribution, and then the appropriate numerical transformations were selected for ordinary kriging interpolation, and finally, the accessibility results of the whole Fuzhou city domain were obtained

4.4 *Supply and demand evaluation grading*

On the basis of the accessibility results, the results are measured in terms of supply and demand through Equation (4) to facilitate further classification of the supply and demand evaluation levels across Fuzhou.

$$Q_i = A_i \times \frac{\max R_j}{\max(A_i)} \tag{4}$$

In Equation (4), Q_i represents the equilibrium degree of supply and demand in space, which represents the matching relationship between the supply and demand between the supply point of building materials and the demand point of townships. max (R_j) is the maximum value of the supply-demand ratio of supply point j of building materials companies, and max(A_i) is the maximum value of the accessibility of prefab building materials in each township plenum i. According to the value of Q_i, the supply and demand situation in various parts of Fuzhou City was classified into five categories of sufficient supply, balanced supply and demand, insufficient supply, lack of supply, and no supply by natural interruption point analysis (Jenks) (Table 2). Among them, $Q_i = 0$ means that the area is beyond the scope of transportation of prefab building materials and there is no supply; $0<Q_i\leqq 1$ means the supply is less than the demand, and $1<Q_i$ means the supply is greater than the demand (there is no such situation in the data).

Table 2. Classification of supply and demand levels in Fuzhou City. [Source: Self-drawn].

Reachability level	Q_i	Supply and demand
1	$0.5831\leqq Q_i\leqq 1.0000$	Sufficient supply
2	$0.2219\leqq Q_i<0.5830$	Balanced supply and demand
3	$0.0474\leqq Q_i<0.0473$	Insufficient supply
4	$0.0001\leqq Q_i<0.5000$	Lack of supply
5	$Q_i = 0.0000$	No supply

5 EXPERIMENT RESULTS

5.1 *Distribution of supply points of prefab building materials*

The kernel density analysis method was used to analyze the distribution of assembly building materials supply points, and the natural breakpoint method was used to divide the kernel density results into five classes: high-value area, medium-high value area, medium-value area, medium-low value area, and low-value area (Figure 3). From Figure 3, the 1065 supply points were screened to produce a large cluster and numerous small clusters in the study area, and the small clusters were distributed around the large clusters in an east-to-west pattern. The large high-value clusters are located in Gulou and Cang Shan districts. Gulou District, located in the main city, is the political, economic, and cultural center of the city, and the development of the assembly exhibition industry is more complete, attracting many companies to locate there. CangShan District, near the center of the main city, absorbs many building material companies that were originally scheduled for Gulou District but did not lay out in it due to high costs or lack of quality land sources. As a result, the distribution of supply points is equally extremely high. With the median zone as the boundary, the geographic areas below the median zone are mainly presented in the southwest and northwest. In summary, the spatial clustering characteristics of the supply points of prefab building materials in Fuzhou are obvious, but the unbalanced geographical distribution is equally significant.

From the standard deviation ellipse analysis, the distribution characteristics of the clustering range and direction of prefab building materials supply points are obtained. The ellipse with 29776 m as the long semi-axis and 21211 m as the short semi-axis at 119.352572°E, 25.98939°N can encompass most of the indoor supply points in Fuzhou (Figure 3). The clustering area of prefab building materials supply points is roughly distributed northwest-southeast, covering Gulou District, Taijiang District, and Cangshan District, and some areas in Jinan District, Mawei District, Changle District, Minhou County, and Fuqing City. The distribution results are generally consistent with the spatial pattern of Fuzhou's economic development.

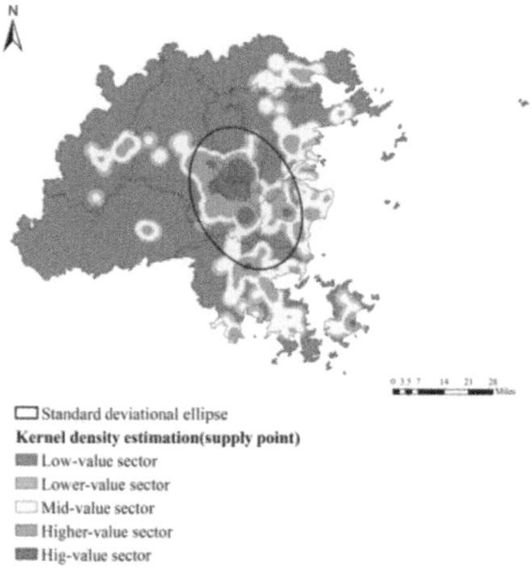

Figure 3. Directional distribution and Spatial Analysis of supply points of prefab building materials [Source: Self-drawn].

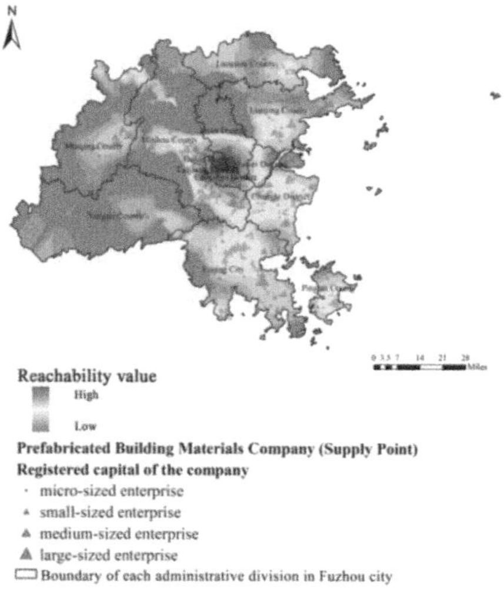

Figure 4. Spatial accessibility overlay of different levels of supply points and prefab building materials [Source: Self-drawn].

5.2 *Analysis of spatial accessibility of prefab building materials*

Analysis of the Accessibility of Suppliers of Different Grades of Assembly Building Materials. Processing the information of assembly building materials suppliers in Fuzhou

City, it is found that the development scale of each company varies greatly, with as much as 350,000,000 yuan of registered capital and as little as 100,000 yuan, with a wide gap. To make the regional accessibility analysis supported by accurate and perfect data on the current situation, this study uses the registered capital of suppliers of assembly building materials as the division standard and divides the suppliers into four grades large, medium, small, and micro, for accessibility analysis, and obtains the superposition diagram of spatial accessibility of supply points and assembly building materials of different grades in each district (Figure 4). From the figure, it can be seen that the spatial accessibility of large suppliers in Fuzhou is slightly better than that of medium suppliers and significantly higher than that of small and micro suppliers. Although the three levels of suppliers in the downtown area of Fuzhou and the eastern coastal area are dense, the accessibility of suppliers in the downtown area of Fuzhou is significantly higher than that in the eastern coastal area. The main reason is that the city center area is more mature in urban planning and construction, with a high level of economic development, and the land available for construction is basically saturated. The level of supply and demand is more stable. In contrast, the eastern coastal area has responded to the national call of "marching to the sea" in recent years, and has made great efforts to promote infrastructure construction, with a large demand for construction. The accessibility of suppliers of all grades is slightly inferior to that of the city Center area.

Spatial Distribution of Accessibility of Suppliers of Prefab Building Materials. The accessibility results obtained by the Gaussian two-step moving search method showed that the accessibility of prefab building materials suppliers in the study area showed the overall spatial characteristics of high in the east and low in the west (Figure 5). Most of the high accessibility areas are located in the eastern coastal area, forming six high-value areas. The scale of the high-value area in the central part of Fuzhou city is the largest; Changle city and Fuqing city are the second largest; the scale of northeast Luoyuan County is smaller in size. The low-value area is mainly distributed in the western part of the study area. The accessibility value of prefab building materials suppliers is mainly determined by 2 factors: the registered capital of building materials companies and the population size in the corresponding search area. On the whole, the spatial accessibility of prefab building materials and the population overlay of each township (Figure 5) shows that the high accessibility areas are centered on the townships with high population density and spread outward in an irregular ring, and the accessibility gradually decreases; the low accessibility areas are mainly located in Yongtai County, Minqing County, western Minhou County, and northern Jin'an District, and the population density in these areas is generally low. It can be seen that the registered capital of the supply point has a higher degree of influence on accessibility than its corresponding population size, i.e., the registered capital (company size) is the dominant factor in terms of the degree of influence.

5.3 *Analysis of spatial accessibility of prefab building materials*

To further analyze the supply-demand relationship between the supply of prefab building materials and the demand of residents in each township (town and street) of Fuzhou City, the supply-demand ranking analysis (Figure 6) and data statistics (Table 3) for each area were conducted based on the analysis of spatial accessibility based on the spatial equilibrium degree Q_i.

From the perspective of the whole study area, the area of the study area with balanced supply and demand and sufficient supply accounts for 21.8% of the total area, which is relatively low. Some of the areas are in a situation of scarcity of supply. Along with the existence of a small number of areas with or without supply, the two situations total 61.6%. The results show that the overall supply and demand of prefab buildings in Fuzhou are not very optimistic.

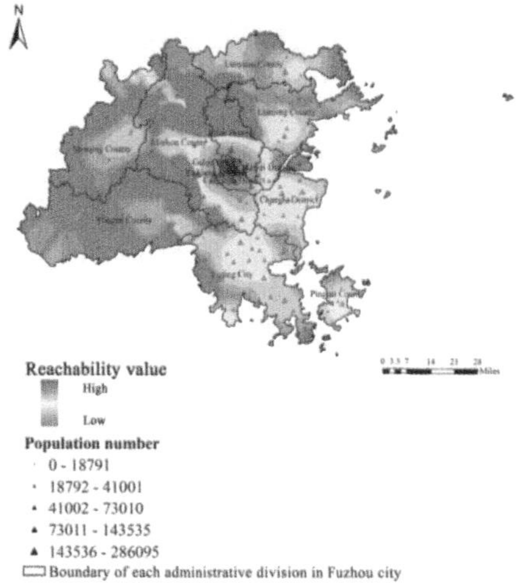

Figure 5. Spatial accessibility of prefab building materials and population overlay of each township [Source: Self-drawn].

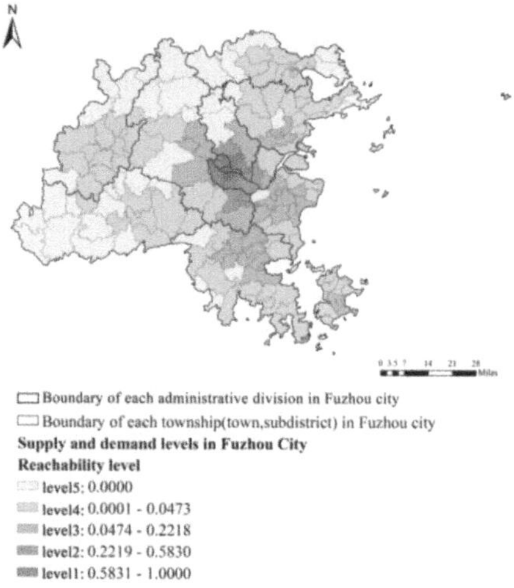

Figure 6. Fuzhou City, each township (town, street) assembly supply point supply and demand grading [Source: Self-drawn].

Table 3. Percentage of supply and demand levels. [Source: Self-drawn].

Zone	Level1	Level2	Level3	Level4	Level5	Total
Cangshan District	78.6%	21.4%	0.0%	0.0%	0.0%	100.0%
Fuqing City	0.0%	0.0%	38.5%	50.0%	11.5%	100.0%
Gulou District	100.0%	0.0%	0.0%	0.0%	0.0%	100.0%
Jin'an District	55.6%	11.1%	11.1%	0.0%	22.2%	100.0%
Lianjing County	0.0%	0.0%	13.6%	54.5%	31.8%	100.0%
Luoyuan County	0.0%	0.0%	0.0%	58.3%	41.7%	100.0%
Mawei District	0.0%	0.0%	50.0%	50.0%	0.0%	100.0%
Minhou County	0.0%	12.5%	25.0%	31.3%	31.3%	100.0%
Minqing County	0.0%	0.0%	0.0%	68.8%	31.3%	100.0%
Pingtan County	0.0%	0.0%	20.0%	53.3%	26.7%	100.0%
Taijiang District	100.0%	0.0%	0.0%	0.0%	0.0%	100.0%
Yongtai County	0.0%	0.0%	0.0%	38.1%	61.9%	100.0%
Changle District	0.0%	0.0%	50.0%	44.4%	5.6%	100.0%
Total	18.7%	3.1%	16.6%	38.3%	23.3%	100.0%

From a township perspective, it was observed that the area of the well-supplied area in Gulou District, Cangshan District, and Taijiang District was greater than 70% of their respective total areas, and nearly half (55.6%) of the well-supplied area in Jinan District, where the regional assembly industry is better developed. Fuqing City, Lianjiang County, Luoyuan County, Mawei District, Minqing County, Pingtan County, Yongtai County, and Changle District, the supply and demand level of eight places are in the level three (and below), the regional supply and demand imbalance is more significant. Among them, Minqing County and Yongtai County supply scarcity or no supply areas accounted for 100% of the county, and almost no supply points of assembly building materials in the region should be combined with the actual industrial development layout requirements of Fuzhou City as soon as possible, able to put forward the corresponding adaptive strategy.

6 CONCLUSIONS

After experiments, it is concluded that the supply accessibility of prefab building materials in Fuzhou is low, and the resistance to the development of the prefab industry is high. The supply scarcity areas are concentrated in the western and eastern parcels away from the central city. At the same time, the imbalance between supply and demand within the region is significant. As such, it seems important to first improve the service capacity and spatial layout of the supply. Secondly, macro-regulation and micro-adjustment are both needed in regional industrial planning to promote the smooth development of the assembly industry in each region.

REFERENCES

Aiqicha Homepage, https://aiqicha.baidu.com, last accessed 2022/10/25.

Chang Chunguang & Wu Feifei. Construction Process Management of Assembled Buildings Based on BIM and RFID Technology. *Journal of Shenyang University of Architecture (Social Science Edition)*,17 (02),170–174(2015).

Chen Huangxin.: *Research on the Constraints of the Development of Prefab Building and its Countermeasures in Fujian Province* [D]. Fujian Engineering College, 2018.

Chen W., Qin H.L. & Tong M.D. Resource Scheduling for Assembly Construction Projects Under Multidimensional Workspace. *Journal of Civil Engineering*,50(03),115–122 (2017).

Fuzhou Municipal Bureau of Statistics Homepage. http://tjj.fuzhou.gov.cn, last accessed 2022/10/15.

Li Mengtong, Yang Lingbin & Wei Ye.: Model Research of Gauss Two-step Mobile Search Method: A Case Study of Green Space Accessibility in Shanghai. *Progress in Geography*. 35(08),990–996(2016).

Li Tianhua. *Research on the Application of BIM and RFID in the Life Cycle Management of Assembled Buildings* [D]. Dalian University of Technology, 2011.

Li Deren, Yu Hanruo, Li Xi.: Spatial and Temporal Analysis of Urban Development in Countries Along the "Belt and Road" Based on Luminous Remote Sensing Images. *Geomatics and Information Science of Wuhan University*. 42(06), 711–720(2017).

Liu Jingai.: Research on Quality Risk Management of Prefab Building Parts (Components) Production–Jinan as an Example. *Construction Economics*. 37(11), 114–117 (2016).

Luo Jie, Song Fabo, Shen Lizhi & Yan Wei.: Study on Some Key Points of Safety Management in the Construction Of Prefab Buildings. *Construction Safety* 31(8), 19–25 (2016).

Ma Linbing, Cao Xiaoshu.: Evaluation Method of Urban Public Green Space Landscape Accessibility Based on GIS. *Journal of Sun Yat-sen University (Natural Science Edition)* (06),111–115(2006).

OpenStreetMap Homepage. https://www.openstreetmap.org, last accessed 2022/10/25.

Xu Zening, Gao Xiaolu.: Boundary Recognition Method of Urban Built-up Area Based on Electronic Map Point Of Interest. *Acta Geographica Sinica* 71(06), 928–939(2016).

Yu Wenhao, Ai Tinghua.: Visualization and Analysis of POI Points in Cyberspace Supported by Kernel Density Estimation Method. *Journal of Surveying and Mapping*. 44(01), 82–90(2015).

Yang Shilong.: *Research on Economic Service Radius of Overland Freight Transport Based on Carbon Emission Cost* [D]. Beijing Jiaotong University. (2014).

Zhang Jiyang.: Present Situation, Causes and Prospect of Road Traffic Development in Fujian Province. *Development Research*. (04), 35–36(2007).

Constructional Engineering and Ecological Environment – Chih-Huang Weng (Ed)
© 2024 The Author(s), ISBN 978-1-032-53198-4

Research status and trend analysis of smart building based on citeSpace

Bing Dong*
The Second Hospital of Jilin University, Changchun, China

ABSTRACT: Smart building is a significant direction of future urban development, which is related to our life and well-being. However, there is still a lack of bibliometric analysis of its research status and development trend. This study retrieved 12,321 articles related to smart buildings in the Scopus database in the past ten years (2012-2022). We used CiteSpace for visualizing and analyzing international cooperation, research hotspots, and research frontiers.

The results show that academic cooperation in the field of smart building can be divided into three clusters, which are largely influenced by geography. Internet of Things, Energy Management, and smart power grid are the research hotspots recently. Ambient Intelligence, Deep Learning, Solar Building, Blockchain, and other technologies are the research frontiers at present and may be the next research hotspots. This study provides a reference for the further development and application of smart buildings.

1 INTRODUCTION

In the context of a new round of industrial revolution, about 58% of the world's population is expected to live in urban areas by 2025. Information technology and artificial intelligence are disrupting various industries, and smart buildings are getting closer and closer to us. The Intelligent Building Institution defines Smart Building as an "Intelligent building", which integrates various systems to effectively manage resources in a coordinated mode to maximize: technical performance; investment and operating cost savings, and flexibility" (Basic et al. 2019). As an important development direction of the construction industry, it is necessary to timely analyze the academic achievements in the smart building field to determine the research status and development trend of this field.

CiteSpace is mainly used for visual analysis of research documents based on mapping knowledge domains (Chen 2006). The software, based on the analysis of a quantity of literature, collects quantifiable information such as literature theme, keywords, published time, and citation status of the research filed. And analyzing the information by using statistical methods. Then present knowledge structure, research development trends, research hotspots, etc., by using a network diagram. Yan Huiquan combined CiteSpace and TRIZ to analyze the field of cancer, clustering factors such as countries, authors, and institutes and obtained the hotspots, trends, and distribution of research in the field of cancer treatment (Yan et al. 2020). Wang used CiteSpace for visual analysis and predicted the key research directions and future development trends in the field of big data (Wang & Lu 2020). Jiang used CiteSpace to track the development process and trend of cleaner production and compared the differences in research hotspots between CNKI and WOS databases (Jiang et al. 2022). Guo et al.,

*Corresponding Author: 61500649@qq.com

 DOI: 10.1201/9781003410843-13

according to the research on Epilepsy with Suicide, revealed the hotpots and frontier of the research fields and figured out some main reasons which caused Epilepsy with Suicide by using CiteSpace (Guo et al. 2022). Li et al. explored the knowledge structure and development trend in the field of information security by using the built-in frequency statistics and centrality calculation methods of CiteSpace (Li 2018). Overall, CiteSpace can analyze the overall research trend of a certain field. This study applies CiteSpace in the field of smart buildings to present the international cooperation, research progress, and development trend in the field of smart buildings in a visual way.

2 MATERIALS AND METHODS

Scopus is the world's largest peer-reviewed indexed abstracts database (Singh et al. 2021). Compared with other databases, Scopus has a more comprehensive collection of engineering literature. Therefore, Scopus was selected as the literature retrieval platform in this study. The Data processing methods in this study are shown in Figure 1, including database search, data filter, data format conversion, parameter setting, and data importing.

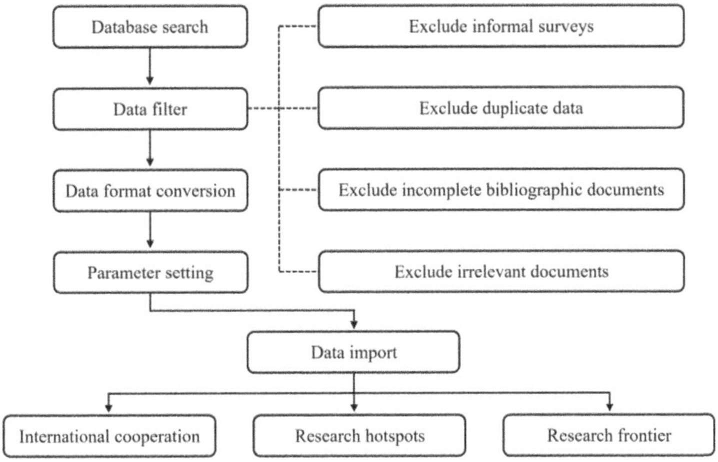

Figure 1. Data processing flow.

In the process of literature search, using "smart building" as the keyword, academic journals, reviews, conference papers, conference reviews, and editorials were selected as the database for literature search. It was published from 2012 to 2022 (retrieved October 27 2022). A total of 12,321 articles were retrieved.

The retrieved literature information is exported in RefWorks format, including author, title, year, keywords, literature type, source publication, etc. After removing informal surveys, duplicate data, incomplete catalogs, and irrelevant articles, 12288 pieces of valid information were finally obtained. Since CiteSpace is developed based on the data format of Web of Science, format conversion is required (Chen 2017). Save the RefWorks data as a CSV file. Open them in Excel and save them as tab-delimited files named "download_*.txt."

The Citespace version used in this study is 6.1.R3, and the parameters are set in the following way: Set the time partition to 2012-2022, year per slice to 1, node types to Country and keyword, and g-index to k=10.

3 DISCUSSION

3.1 *International cooperation status*

Smart Building locates in the cross-filed of civil engineering and information engineering, and related research is used to be finished by several scholars. With globalization's deepening, scientific research has gradually changed from closed innovation to open cooperation. Scholars live in different countries/regions, leading to differences in their academic environment, scientific research resources, service system, and informatization level. International academic cooperation can not only provide complementary resources for both parties and accelerate the breakthrough of technical problems but also facilitate the integration of industrial chains between countries, share global technological resources, and more easily convert academic achievements into products.

To have an in-depth understanding of the international academic cooperation in the field of smart buildings, Citespace was used for visualized analysis, and countries/regions with more than 200 publications were reserved. The log-likelihood ratio (LLR) test algorithm is used for the clustering analysis of literature keywords, and two or three effective clusters are formed. The results show that the average contour value of clustering is $S=0.7499$, $S>0.5$ means the clustering is reasonable, and $S>0.7$ means the clustering is convincing. Figure 2 shows the international cooperation network, the scientific research situation, and the cooperative relationship among countries/regions. In Figure 2, each node represents a country/region. The size of the node is related to the number of publications, and the color and density of connections are related to the closeness of cooperation between different countries/regions.

The statistical function of Citespace shows that a total of 145 countries/regions in the world have participated in the relevant research. Among them, the most significant number of published countries are the United States (1897), China (1779), India (887), Italy (814), and the United Kingdom (760). As seen from the national (regional) cooperation network diagram, academic cooperation in the field of smart buildings is greatly influenced by geography. The UNITED STATES and CANADA, located in the Americas, have rich scientific research achievements and have formed close research cooperation relations. CHINA,

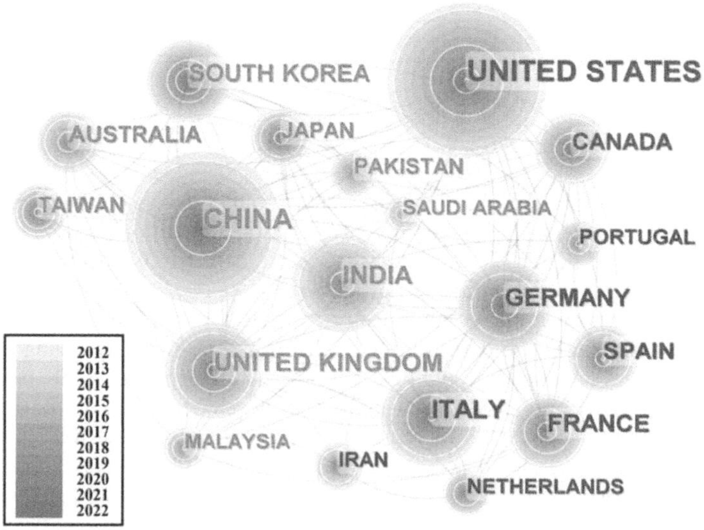

Figure 2. Network diagram of international cooperation in the field of smart buildings.

INDIA, UNITED KINGDOM, SOUTH KOREA, AUSTRALIA, TAIWAN, JAPAN, and other countries are mostly located in the Asia-Pacific region. In the three clusters, the number of publications is the largest, which significantly impacts the international scale. As representatives of European countries, ITALY, GERMANY, FRANCE, SPAIN, PORTUGAL, IRAN, and NETHERLANDS have conducted extensive research with each other, although their publications are fewest.

3.2 *Research hotspots*

Keywords play a very important role in bibliometrics, which highly generalize the thesis topic. The keywords with high frequency and strong centrality can reflect the hot issues in this research field to a certain extent. Co-occurrence analysis takes keywords as nodes and studies the co-occurrence times of two different keywords in a literature group to get this research field's hot content and development trend. The co-occurrence analysis function of CiteSpace and the critical path algorithm are used to form the co-occurrence map composed of nodes and lines, as shown in Figure 3. This figure can intuitively reflect the research hotspots in the field of smart architecture.

In the figure, each node's size represents the keyword's frequency, the node's color represents the keyword's time and the connection between nodes' association strength (Chen 2020). After sorting, the following categories of main high-frequency words are obtained:

Internet of Things: Smart Home, Internet of Thing, Smart HM, Automation;

Building Energy Management: Energy Management, Energy Storage, Energy Utilization, Renewable Energy Resources, Energy Management System, Energy Efficiency;

Smart Power Grid: Smart Power Grid, Smart Grid, Demand Side Management, Demand Response, Electric Power Transmission Network;

Other Key Words related to building: Scheduling, Optimization, Building, Electric Vehicle, Heating, Cost, Intelligent Building.

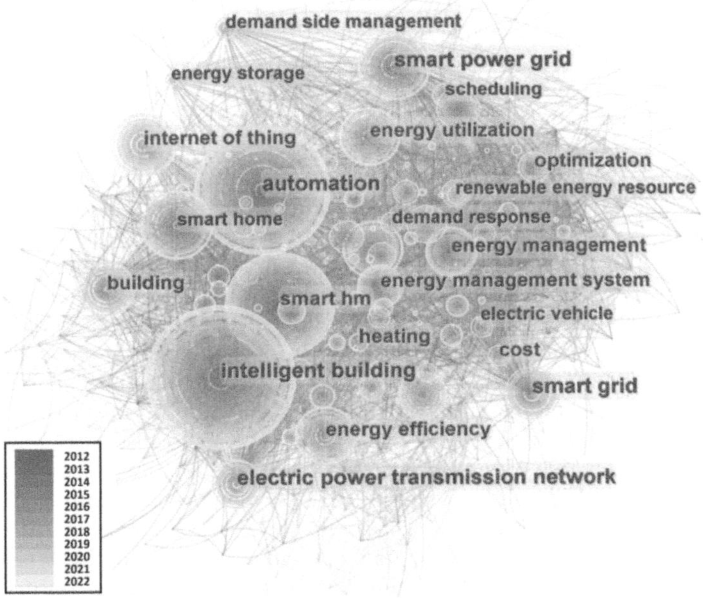

Figure 3. Hotspots diagram of the smart building field.

As can be seen from the figure, keywords in the field of Smart building research mainly revolve around three directions: Internet of things technology, represented by "Smart Home," Building Energy Management represented by "energy Management," and smart grid represented by "Demand Side Management."

The Internet of Things (IoT) is based on the Internet, which can connect terminal sensor devices through communication networks at a distance and realize the integration of "management, control, and operation" of each device (Tang et al. 2019). IoT has been applied to many smart home products, such as access control systems, lighting, central air conditioning, smart curtains, etc. These products are mainly composed of terminal products, mobile apps, the Internet of Things cloud, and communication networks. Some products can be directly connected to the network; Other devices are connected to the cloud through gateways or smartphones. The device will send the collected data to the cloud, and the cloud will store, manage and analyze the data to bring users a comfortable and convenient living environment. However, these products are still far from the goal of being intelligent and easy to use. At the same time, they also face severe privacy and security problems and even may bring serious property losses. How to improve the security of user data is the focus of current research in this field. The combination of IoT and construction could lead to the next revolution in construction.

Building energy management minimizes energy consumption and pollution through intelligent technology by tracking the energy supply and consumption of buildings and ensuring users' comfort (Mariano-Hernandez et al. 2021). In recent years, with the increasing global Energy crisis, Renewable energy resources have been more and more accepted in the construction field. Construction has become a "prosumer"; that is, it not only consumes energy, but also produces energy, which poses new challenges to managing building energy. To maintain the system's stable operation and continuously provide safe and stable electric power for the building, the Energy Storage system based on the battery is also incorporated into the building's energy management system due to the uncontrolled energy supply of Renewable Energy Resources. The Energy Storage system can not only reduce the fluctuations caused by Renewable Energy Resource supply, but also relieve the power supply pressure of the power grid and reduce the peak-valley difference of power load through the operation of power storage and discharge during the peak period in the off-load period. In addition to the passive management represented by Energy Storage, Renewable Energy resources have penetrated the distribution system on a large scale due to the need for energy transformation in recent years. Achieving the immediate response of the building side and dynamic supply and demand balance is also an essential responsibility of building energy management and a problem that researchers need to face.

A smart grid is a fully automated power transmission network that can be connected to each grid user and node to ensure the bidirectional flow of information and electricity between all nodes in the transmission and distribution process from power plants to end users (Dileep 2019). Due to the application of energy network and communication network technology, a smart grid has faster transmission speed, a stronger economy, and a safer transmission process than a traditional grid. Since the 21st century, the electricity demand has been increasing daily with the change has been production mode. Suppose the traditional power grid generally adopts ways such as increasing unit output and building generating equipment to solve the shortage of power supply. In that case, this way will further deteriorate the load curve, increase the peak-valley difference, and cause more pressure on the power system. Demand Side Management refers to the method by which the government and power enterprises guide users through price, policy, and other means to change users' original consumption behavior (Siano 2014). As an important part of the smart grid, Demand Side Management can realize the optimal scheduling of load resources and generation-side resources to effectively optimize the demand-side power consumption mode and improve the energy utilization rate. It is a hot topic in the field of smart building research.

3.3 *Research frontier*

Emergent words refer to keywords with a significant increase in frequency in a short period of time. Through the analysis of emergent words, the research hotspots and research frontiers in this field can be judged in a period. The emergence keyword detection algorithm is used to calculate the top 25 keywords with mutation intensity, and the generated keyword emergence diagram in the field of a smart building is shown in Figure 4. In the figure, "begin" represents the beginning time of the emergent word, "end" represents the end time of the emergent word, and "strength" refers to the mutation intensity of the keyword. The higher the value, the stronger the frontier of the emergent word.

By analyzing the emergence degree and time node of the above keywords, the research frontier in the field of smart architecture shows obvious migration and change, which can be roughly divided into two stages:

From 2012 to 2016. This period is the outbreak period of the Internet, mobile Internet, and the rapid growth period of the Internet of Things, and more emergent words emerge in this stage. The research frontier focuses on the application of the Internet and the Internet of Things in the field of architecture. As an important part of Smart architecture, Smart Home appeared as an emergent word many times in this stage, which proved that Smart Home was the frontier research direction in this stage.

From 2019 to now. In the early stage of the outbreak of artificial intelligence technology, data storage technology, represented by Blockchain, and Smart contract and data processing

Top 25 Keywords with the Strongest Citation Bursts

Keywords	Strength	Begin	End	2012-2022
internet	78.46	2012	2016	
smart hm	60.59	2012	2013	
ubiquitous computing	48.98	2012	2016	
information technology	44.42	2012	2014	
sensor	44.32	2012	2014	
wireless telecommunication system	42.57	2012	2016	
algorithm	38.1	2012	2015	
zigbee	37.47	2012	2016	
smart home system	32.41	2012	2014	
communication	31.47	2012	2013	
computer simulation	29.05	2012	2014	
middleware	26.72	2012	2015	
smart grid	25.72	2012	2013	
mobile device	25.29	2012	2016	
smart house	24.24	2012	2015	
smart home	26.91	2013	2014	
electric power distribution	25.05	2013	2015	
blockchain	36.2	2019	2022	
machine learning	26.12	2019	2022	
hvac	24.67	2019	2022	
smart contract	24.2	2019	2022	
ambient intelligence	124.19	2020	2022	
deep learning	50.51	2020	2022	
solar building	42.58	2020	2022	
privacy by design	34.4	2020	2022	

Figure 4. The keyword emergence diagram in the field of smart building.

technology, represented by Deep Learning and Machine Learning, have become new frontier directions. The research frontier in the field of smart buildings is shifting in the direction of intelligence. It is worth noting that, as an emergent word starting from 2020, Ambient Intelligence has a mutation intensity of 124.19, which is the keyword with the highest degree of emergence. Ambient Intelligence senses the state of the environment and users through sensors and establishes a sensitive, responsive, and adaptive environment system with the help of various artificial intelligence algorithms (Cook et al. 2009). With the progress of artificial Intelligence technology, the research on Ambient Intelligence is maturing, which is an extremely important frontier in the field of intelligent buildings. In addition, Privacy by Design, Solar Building, and HVAC (Heating, Ventilation, and Air Conditioning) also provide new modes and scenarios for the field of a smart building.

4 CONCLUSIONS

In recent years, with the development of information technology and artificial intelligence, research on smart buildings has also achieved great results. Based on the data of 12,321 articles related to smart buildings in the Scopus database in recent ten years (2012-2022), this study drew the international cooperation network map, hotspots diagram, and keyword emergence map in the field of smart buildings. It also summarizes international cooperation, research hotspots, and research frontiers in the field of smart architecture. The current academic cooperation in the field of smart architecture is divided into three clusters, which are divided into the Asia-Pacific, the Americas, and Europe, according to the geography course. The main research hotspots in the field of smart buildings are Internet of Things technology represented by Smart Home, smart grid, represented by "Demand Side Management" and building energy management. Blockchain, Machine Learning, HVAC (Heating, Ventilation, and Air Conditioning), smart contracts, ambient intelligence, deep learning, solar building, and privacy by design are currently the frontier research directions. In the future, these directions may become new research hotspots and drive the continuous development of the field of smart architecture.

REFERENCES

Basic S., Strmo N.V., Sladoljev M.: Smart Cities and Buildings. *Gradevinar*. 71(10), 949–964 (2019).

Chen C. CiteSpace II: Detecting and Visualizing Emerging Trends and Transient Patterns in Scientific Literature. *Journal of the American Society for Information Science and Technology*. 57(3), 359–377 (2006).

Chen C.: Science Mapping: A Systematic Review of the Literature. *Journal of Data and Information Science*. 2 (2), 1–40 (2017).

Chen C. A Glimpse of the First Eight Months of the COVID-19 Literature on Microsoft Academic Graph: Themes, Citation Contexts, and Uncertainties. *Frontiers in Research Metrics and Analytics*. 5, 607286–607286 (2020).

Cook D.J., Augusto J.C. and Jakkula V.R.: Ambient Intelligence: Technologies, Applications, and Opportunities. *Pervasive and Mobile Computing*. 5(4), 277–298 (2009).

Dileep G.: A Survey on Smart Grid Technologies and Applications. *Renewable Energy* 146, 2589–2625 (2019).

Guo Y., Xu Z.Y.R., Cai M.T., Gong W.X. and Shen C.H.: Epilepsy with Suicide: A Bibliometrics Study and Visualization Analysis via CiteSpace. *Frontiers in Neurology*. 12, 823474 (2022).

Jiang Y.G., Li M., Dennis A., Liao X., Ampaw EM.: The Hotspots and Trends in the Literature on Cleaner Production: A Visualized Analysis Based on Citespace. *Sustainability*. 14(15), 9002 (2022).

Li X.T., Li H.: A Visual Analysis of Research on Information Security Risk by Using CiteSpace. *IEEE Access*. 6, 63243–63257 (2018).

Mariano-Hernandez D., Hernandez-Callejo L., Zorita-Lamadrid A., Duque-Perez O. and Garcia FS.: A Review of Strategies for Building Energy Management System: Model Predictive Control, Demand Side

Management, Optimization, and Fault Detect & Diagnosis. *Journal of Building Engineering*. 33, 101692 (2021).
Singh V.K., Singh P., Karmakar M., Leta J., Mayr P.: The Journal Coverage of Web of Science, Scopus and Dimensions: A Comparative Analysis. *Scientometrics*. 126(6), 5113–5142 (2021).
Siano P.: Demand Response and Smart Grids-A Survey. *Renewable & Sustainable Energy Reviews*. 30(2), 461–478 (2014).
Tang S., Shelden D.R., Eastman C.M., Pishdad-Bozorgi P., Gao X.H.: A Review of Building Information Modeling (BIM) and the Internet of Things (IoT) Devices Integration: Present Status and Future. *Automation in Construction*. 101, 127–139 (2019).
Wang W.H. and Lu C.: Visualization Analysis of Big Data Research Based on Citespace. *Soft Computing*. 24 (11), 8173–8186 (2020).
Yan H.Q., Lyu P.H., Wang L., Yu Z.M.: TRIZ Theory and the Method of Cancer Document Selection for Chemical Complexes and Innovation Schemes of Meta-Analysis with Lymphomas as an Example. Journal of Chemistry. 2020, 6294613 (2020).

Constructional Engineering and Ecological Environment – Chih-Huang Weng (Ed)
© 2024 The Author(s), ISBN 978-1-032-53198-4

Key technologies for rapid retrofit of urban bridges with confined spaces

Ximeng Sun
Beijing Municipal Road & Bridge Co., Ltd., Beijing, China

Yanhui Cao*
Beijing Municipal Road & Bridge Co., Ltd., Beijing, China
Beijing University of Technology, Beijing, China

Jinhua Ye & Rui Ma
Beijing Municipal Road & Bridge Co., Ltd., Beijing, China

ABSTRACT: In order to push forward retrofit technology for domestic urban bridges, the application and development of retrofit technology for urban bridges have been investigated and researched. At the same time, the key technical points of a bridge retrofit project in Kunming are described in this paper, with emphasis on the construction method of replacing hollow slabs with I-steel composite beams for large-scale viaducts under space-constrained conditions and the construction method of replacing long-span concrete box girders with steel box girders for the upper-span closed-frame passage. The research results can provide a reference for the design and construction of urban bridge retrofit engineering.

1 INTRODUCTION

China boasts the country with the largest number of bridges in the world, comprising 961,100 highway bridges and 73,802,100 linear meters as of the end of 2021, according to the Statistical Bulletin on the Development of the Transportation Industry in 2021 issued by the Ministry of Transport (Ministry of Transport 2021). Since the reform and opening up, China has witnessed rapid economic and social development, and domestic cities are subject to increasing traffic flow and traffic demand. Also, with the advancement of design concepts and technology, China's bridge design standards have undergone multiple changes. Many bridges built 20 years ago no longer meet the requirements of today's specifications.

Currently, urban traffic congestion in China has been worsening, especially in large and megacities. As a result, many urban infrastructures can no longer cater to the current traffic operation requirements, and an abundance of bridges incurred varying degrees of damage due to long-term overload operation. With serious cases even showing cracking and deflection in the span of the main girders, concrete spalling, and corrosion of the reinforcement. Portions of the bridges are inevitably in need of different degrees of maintenance, reinforcement, or transformation. Some bridges have overturned superstructures due to overload and overrun during operation, so it is necessary to retrofit the bridges. If the bridges with serious damage in this part are demolished and rebuilt thoroughly, it will cost a lot of economic costs and take a long construction period, thereby aggravating the already overwhelmed urban traffic. In the meantime, it is difficult to estimate the resource consumption, greenhouse gas emissions, and waste of time caused by the massive vehicle detours. Therefore, as far as possible to reduce the

*Corresponding Author: 15650751272@163.com

DOI: 10.1201/9781003410843-14

interference to the urban traffic operation system, the rapid and efficient retrofit of existing bridges will be one of the effective methods and paths to alleviate the urban traffic pressure and promote the development of urban quality improvement.

During the 14th Five-Year Plan period, the State Council, the Ministry of Communications, the Ministry of Construction, and local competent departments issued relevant policies. It proposed to expand and transform urban infrastructure and highways to improve the safety level of infrastructure (Beijing Municipal People's Government 2022; Council 2021; Ministry of Transport 2022; Ministry of Housing and Urban-Rural Development 2022; The State Council 2019, 2021).

2 DOMESTIC TECHNOLOGY APPLICATION AND DEVELOPMENT

Tianjin Shizilin Bridge (Wang & Han 2004): Shizilin Bridge is located on the Haihe River in Tianjin, built in 1954, serving as the earliest prestressed concrete cantilever beam bridge used on Chinese highways. In 2003, the bridge was lifted and retrofitted, and the lifting height reached 1.27 m, which lasted less than one month, creating a miracle in the history of Chinese bridges at that time. The lifting project of Shizilin Bridge is a great breakthrough in bridge retrofit technology in China, which pioneered the field of urban bridge lifting and is the first bridge lifting project in China. In addition, guardrails on both sides of sidewalks were added during the second retrofit of the bridge in 2021, which improved traffic safety.

Beijing Xizhimen North Interchange Retrofit Project (Mu 2018): Xizhimen North Interchange is situated in the northwest corner of the Second Ring Road, the main traffic node in the northwest of the city undertakes important and busy traffic tasks. After being tested by the Highway Science Research Institute of the Ministry of Communications and Municipal Design Research Institute, the bridge was assessed as Class E and in a dangerous state. In order to ensure the safety of citizens' travel, the municipal government decided to repair and retrofit Xizhimen North Interchange by demolishing the old bridge and constructing the new bridge in the way of pushing and shifting from 22:00 on September 8, 2006. The retrofit of the dangerous bridge of Xizhimen North Interchange was completed from 22:00 on September 8th to the morning of September 11th. In less than three days, the bridge was successfully retrofitted, which was also a memorable feat in the history of bridge construction in Beijing at that time.

Sanyuan Bridge Overhaul Project (Mu 2018): The project lasted for 43 hours from 23:00 on November 13, 2015, to 18:00 on November 15, 2015. Beijing Sanyuan Bridge Overhaul Project (across Jingshun Road Bridge) realized the overall replacement of the upper main girder, and completed the cutting of the old bridge, the overall transportation of the old beam and the new beam, the erection, and the deck pavement, and resumed traffic. The retrofit of the Sanyuan Bridge has ushered in a new speed of bridge construction in China, which has been widely concerned by news media and netizens at home and abroad, stirring a sensation in the industry and society.

Reconstruction and Expansion Project of Beijing-Hong Kong-Macau Expressway (Wei 2017): Bridge expansion and reconstruction is also a common form of bridge retrofit. The roads and bridges in the Beijing-Shijiazhuang section of the Beijing-Hong Kong-Macau Expressway completed in 2014 have been widened from two-way four lanes before reconstruction to two-way eight lanes, and the designed traffic capacity will increase from 25,000 to 30,000 vehicles per day before reconstruction to 60,000 to 80,000 vehicles per day.

Reconstruction and Expansion Project of Ji-Qing Expressway (Hou 2016): As the main traffic channel traversing east and west of Shandong Province, the two-way four-lane road condition of the original Ji-Qing Expressway has long been unable to meet the traffic flow demand, and road congestion occurs from time to time. In 2019, the roads and bridges of Ji-Qing Expressway were widened from four lanes in both directions to eight lanes in both directions, which greatly improved the road capacity.

Shanghai Songpu Bridge (Guo 2017): Songpu Bridge is the first bridge on the Huangpu River, which was built in 1976. The main bridge is a double-deck steel truss bridge with a two-lane highway on the upper floor and a single-track railway on the lower floor. In 2019, the retrofit was completed. The upper approach bridge was widened by the structures on both sides, thus realizing the transformation from two lanes in both directions to six lanes in both directions, improving the traffic capacity. The original railway bridge at the lower level can be transformed into a non-motor vehicle lane by lifting and adjusting the slope.

At present, the main forms of bridge retrofit can be divided into beam replacement construction, bridge widening, main girder lifting, bridge parts (bearings, etc.) replacement, local reinforcement (increasing section, sticking steel plates, sticking carbon fiber sheets, planting steel bars, grouting, applying to prestress, etc.), bridge deck pavement and repair, pier column, and foundation reinforcement, etc. (Guo 2017; Hou 2016; Mu 2018; Wang & Han 2004; Wei 2017).

At present, it is a common problem for old urban bridges to incur a large area of main girder deflection overrun and serious cracking. At the same time, the original design bearing capacity and traffic capacity cannot meet the increasing traffic flow and safe operation needs. The beam replacement construction has relative advantages in the economy and construction period, which is one of the mainstream bridge retrofit methods at present. How to adopt advanced construction technology under the condition of limited working space and minimize the impact on the existing traffic in the city is the key to completing the bridge engineering retrofit safely and quickly.

3 ENGINEERING BACKGROUND

A second-ring viaduct in Kunming was completed and opened to traffic in 1998. At the initial stage, the proportion of trucks in transit traffic was high, and the overload phenomenon was serious. Moreover, the original design safety redundancy of the main span beam slab of the viaduct was small, which caused serious damage to the viaduct bridge structure and a severe impact on its bearing capacity after opening to traffic. In the wake of the test in 2006, parts of the bridge span are four types of dangerous bridges. As an urban expressway, with the increase of operating years, although the cargo restriction measures are adopted, the traffic volume of various buses is still much higher than the design traffic volume of the original six-lane highway, resulting in a large number of structural cracks, which seriously affects the service safety and life of the bridge structure and endangers the operation safety. It is exigent to upgrade and retrofit the viaduct in this section to ensure smooth traffic in the core area of Kunming.

The total length of a retrofit project of the Second Ring Road in Kunming is 6.2 kilometers, serving as the largest urban bridge retrofit project in China at that time. The main contents of the project were to replace the bridge slabs of viaducts and some ramps, retain the substructure, and upgrade the load level from the original steam-20 and hanging 100 to city A by replacing concrete beams with I-shaped steel-concrete composite beams. Because the traffic in the city cannot be interrupted for a long time, the total contract period of the project is 7.5 months. The main contents of the retrofit project are as follows: demolish 4685 hollow slab beams, replace 3022 I-beam beams, demolish and replace six cast-in-place concrete box girders with six steel box girders, repair and reinforce the substructure, and overhaul the ground under the bridge.

4 RISKS AND CHALLENGES

4.1 *Large scale, short construction period, and wide influence on existing traffic operation*

The retrofit length of the viaduct is 6.2 kilometers, with a total of 4,685 hollow slab beams replaced and more than 26,000 tons of new steel beams built. It is the largest urban bridge retrofit project in China, with a total construction period of 7.5 months. The construction of this project will affect the traffic of most residents in Kunming, and the traffic guidance is complex. It is critical to rationally organize and deploy the construction and innovate the construction technology on the premise of ensuring safety, speed, and economy.

4.2 *Confined working space*

In the vicinity of the project, there exist substantial private houses, railway fences, traffic ramps, and other buildings, and the area without occupied space on the bridge side accounts for more than 55% of the entire line, with severely confined space and difficult construction operations.

4.3 *Safety assurance of existing structures*

The original bridge cover beam is a prestressed structure, and the cover beam steel bundles are tensioned twice. That is, the hollow slab is erected after the first batch of prestressed steel bundles is tensioned, and the deck ancillary facilities are constructed after the second batch of prestressed steel bundles is tensioned. However, in the process of dismantling the hollow slab beam, the existing cover beam may appear to reverse arch or even crack under the action of prestress due to sudden unloading.

4.4 *Demolition of long-span prestressed concrete box girder with an upper span of the closed-frame passage*

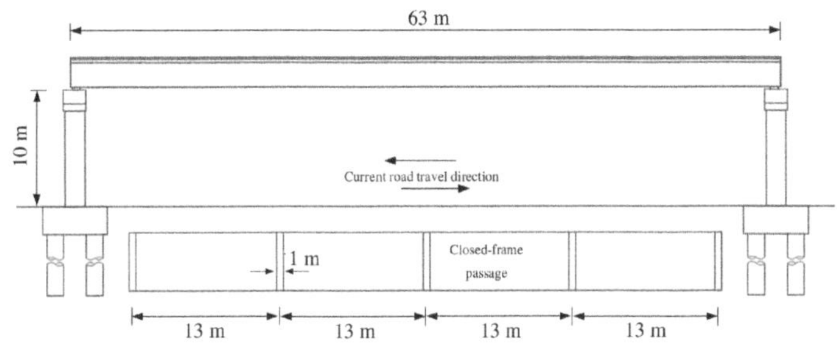

Figure 1. Schematic diagram of long-span concrete box girder with closed-frame passage on the upper span.

As shown in Figure 1, there is a simply supported box girder with a span of 63 m in the project. In the process of box girder disassembly, it is necessary to erect an unloading support system on the culvert of the lower channel with an upper span of 4 × 13 m closed-frame passage. The technical problems encountered are to solve the sudden change of load during the cutting and unloading of box girders by using the rigid-flexible combined support system and to overcome the inability to support the large mechanical equipment in the load-bearing part of the channel under the bridge.

5 KEY CONSTRUCTION COUNTERMEASURES

5.1 *Beam replacement technology by half-range alternate counterweight method*

In the process of dismantling the main girder, the cover beam will appear reverse arch under the action of prestressing, and if the reverse arch passes through, the cover beam will crack. In order to find out the reverse arch situation and internal force change of the cover beam during the demolition of the old beam, the best dismantling sequence of the beam replacement is analyzed to ensure the safety of the cover beam during the demolition of the superstructure of the bridge. A typical span of 20.0 m of the main beam is selected for

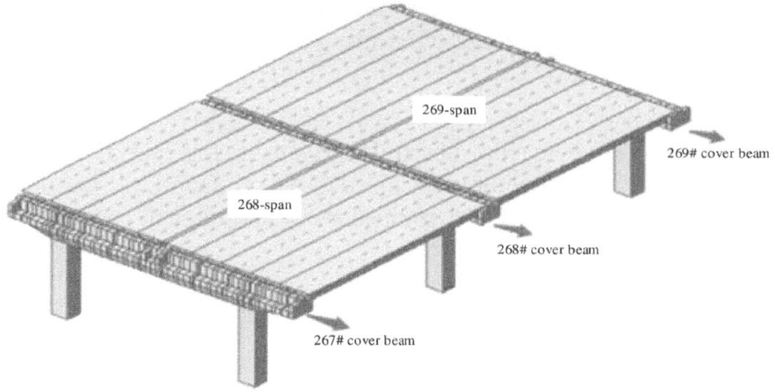

Figure 2. A computational model of the 268# cover beam.

Table 1. Simulated working conditions.

Construction steps	Working condition 1: half-range alternate counterweight method	Working condition 2: replacement before complete dismantlement	Working condition 3: beam replacement method with the whole span at intervals
1	Demolition of 268-span left hollow slab	Demolition of 268-span left hollow slab	Demolition of 268-span left hollow slab
2	Demolition of 269-span left hollow slab	Demolition of 269-span left hollow slab	Demolition of 268-span right hollow slab
3	Installation of 268-span left steel beam	Demolition of 268-span right hollow slab	Installation of 268-span left steel beam
4	Installation of 269-span left steel beam	Demolition of 269-span right hollow slab	Installation of 268-span right steel beam
5	Binding of 268 and 269 spans left steel bars	Installation of 268-span left steel beam	Binding of 268-span steel bar
6	Pouring of 268 and 269-span concrete on the left side	Installation of 269-span left steel beam	Pouring of 268-span concrete
7	Installation of 268 and 269 spans left guardrails	Installation of 268-span right steel beam	Installation of 268-span guardrail
8	The pavement of 268 and 269-span left the asphalt	Installation of 269-span right steel beam	Pavement of 268-span asphalt
9	Demolition of 268-span right hollow slab	Binding of left steel bar	Demolition of 269-span left hollow slab
10	Demolition of 269-span right hollow slab	Pouring of left concrete	Demolition of 269-span right hollow slab
11	Installation of 268-span right steel beam	Installation of left guardrail	Installation of 269-span left steel beam
12	Installation of 269-span right steel beam	The pavement of left asphalt	Installation of 269-span right steel beam
13	Binding of 268 and 269-span right steel bars	Binding of right steel bar	Binding of 269-span steel bar
14	Pouring of 268 and 269- span concrete	Pouring of right concrete	Pouring of 269-span concrete
15	Installation of 268 and 269-span right guardrail	Installation of right guardrail	Installation 269-span guardrail
16	The pavement of 268 and 269-span right asphalt	The pavement of right asphalt	The pavement of 269-span asphalt

analysis. Through MIDAS, the corresponding finite element model is established, as is shown in Figure 2, and the deformation of the cover beam in the whole process of bridge construction under three working conditions is analyzed, as shown in Table 1. Each working condition includes 16 construction steps, among which the first working condition is called the "half-range alternate counterweight method," the second working condition is called "replacement before complete dismantlement," and the third working condition is called "beam replacement method with the whole span at intervals."

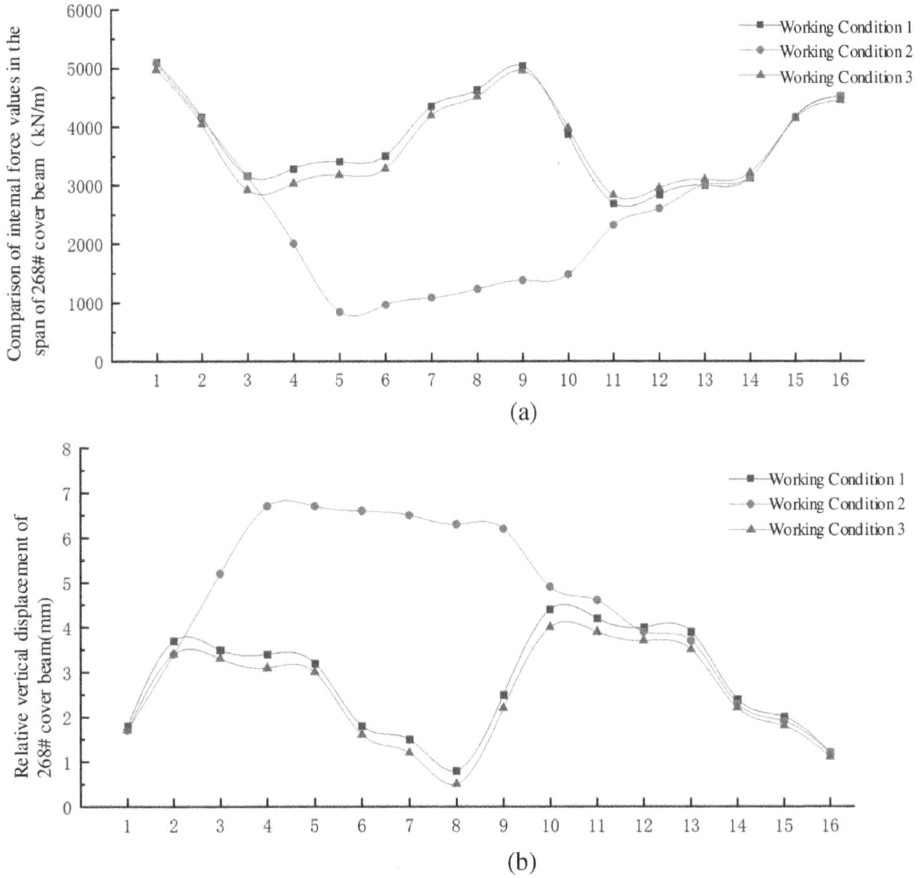

Figure 3. Variation of internal force and displacement in the mid-span of the cover beam. (a) Variation of internal force in mid-span (b) Variation of displacement in mid-span.

The calculation results are shown in Figure 3, from which it can be clearly seen that the deformation of the cover beam is small when using the half-range alternate counterweight method and beam replacement method with the whole span at intervals, but it is very large when using the method of replacement before complete dismantlement. At the same time, the beam replacement method with the whole span at intervals has discontinuous construction organization and great cross-interference in construction. After comparison and selection, the construction scheme of the half-range alternate counterweight method is determined.

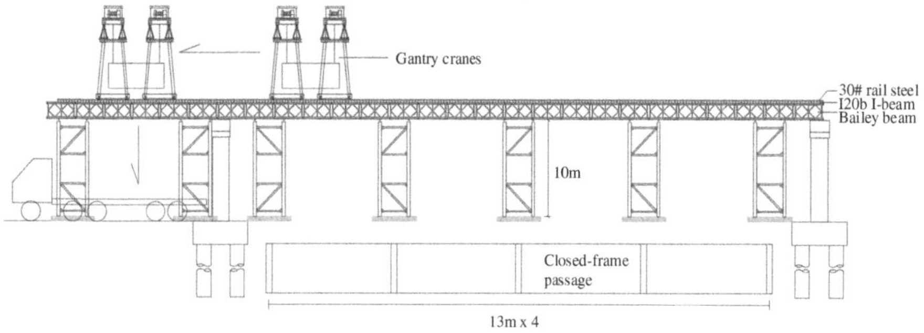

Figure 4. Schematic diagram of demolition technology of long-span concrete box girder with upper span closed-frame passage.

5.2 *Key technologies for demolition of the viaduct in confined space*

Method of demolishing hollow slab beam by gantry crane (Ye 2021)

The method of demolishing the hollow slab girder by gantry crane adopted in this project refers to installing moving tracks on both sides of the bridge, setting up two gantry cranes, and then lowering the hollow slab girder from the inner span through the cooperation of two gantry cranes to realize the demolition. Because the length of the hollow slab beam is larger than the net distance between the two cover beams, it needs to be rotated vertically at a certain angle before it can be lowered smoothly.

Demolition technology of plate beam by span skylight method (Ye *et al.* 2020)

At the same time, the demolition technology of plate beams using the straddle skylight method was developed and applied in this project. In this technology, 2 ~ 3 plate beams are hoisted and placed on the undismantled plate beams by gantry cranes on the bridge, a "skylight" is opened to leave a working face for the operation of automobile crane boom under the bridge, and the hollow plate beams are dismantled by automobile cranes.

Demolition technology of long-span concrete box girder with the upper-span closed-frame passage (Wang 2021)

A gantry crane walking track platform is set up on both sides of the square bridge on the load-bearing wall of the closed-frame passage, which is composed of steel pipe lattice columns and Bailey beams. Then two 40t gantry cranes are elevated to carry out straddle-type beam dismantling operation, and the cut beams are moved outside the bridge span. Based on the dismantling ability of the gantry crane, the beam body is designed in blocks. As for the support problem of the large box girder during demolition, the wing plate is supported by full-frame brackets. On account of the heavy weight of the main girder box room, over 1000t, the "rigid-flexible combined support mode" is adopted. That is, the steel pipe column structure with large stiffness and bearing capacity is used to support at both ends to resist the sudden impact of load when the end of the large box girder is cut off, and a relatively flexible full-frame bracket is arranged in the middle of the bridge span to support when the beam is cut in blocks. Because of the light weight of the wing plate, the wing plate is cut into a comb shape first and then cut longitudinally to fall on the bracket. After the box is stretched at the end, the beam falls on the lower unloading frame. According to the stress condition of the unloading gear under two working conditions of cutting the middle part first and cutting the secondary end first, through calculation and analysis, it is concluded that the buckling load of the unloading gear is greater when cutting the middle part first, so it is determined to adopt the mode of cutting the middle part first, then cutting the secondary end and then cutting into small pieces.

5.3 *Key technologies for efficient installation and construction of steel-concrete composite beam of the viaduct*

For the hoisting of I-beam, the most widely used form in this project is the single machine or double machine 2-1-2 hoisting method, that is, hoisting two pieces, one piece, and two pieces at a time, and installing a total of five steel beams in place. According to the span and construction arrangement of steel beams, the method of hoisting five pieces at a time or one piece at a time can also be adopted.

Aiming at the hoisting of the steel box girder, the steel box girder is prefabricated in pieces in this project. The beam dismantling bracket and gantry crane are used to erect and hoist it in place in pieces, and then the welding method is arranged.

After the I-beam is installed in place, how to pour the concrete bridge deck is also one of the key problems. For the sake of facilitating the construction and reducing interference under the bridge, a portable self-supporting formwork system (Ye 2020) is adopted. The formwork structure includes square steel pipe, square timber, and clear water formwork. By reserving bolt holes on the I-beam and fixing the formwork structure on the I-beam with high-strength bolts and nuts, the bridge deck concrete can be poured. After the concrete curing is completed, the nuts are retained in the concrete, and the bolts and formwork structure can be removed only by manually using mechanical wrenches.

6 CONCLUSION AND PROSPECT

From June 22, 2018, to December 28, 2018, the largest urban bridge retrofit project in China was completed one month ahead of schedule, which played a good demonstration role in bridge retrofit in China.

In 2021, Vice Minister Dongchang Dai of the Ministry of Communications pointed out that China's highway bridges have made remarkable achievements. However, at the same time, there are many dangerous bridges, old bridges, and low-load bridges in China, which are intertwined with the medium-and long-term risks. For example, some bridges are about to reach the design service life, the peak period of bridge maintenance is about to come, and some bridges are undergoing "premature aging." There are complex periodic problems that still need to be solved urgently.

In the future, there will be increasing urban bridge retrofit projects, and the requirements will be expected to rise. Focusing on the goal of double carbon, greening, intelligence, and industrialization are the development directions of urban bridge retrofit in the next step.

REFERENCES

"14th Five-Year Plan" Construction Industry Development Plan [R]. *Beijing: Ministry of Housing and Urban-Rural Development.* 2022.

2021 Statistical Bulletin on the Development of Transport Industry [R]. *Beijing: Ministry of Transport.* 2021.

Guo Jun. Design of Overhaul Project of Songpu Bridge of Huangpu River. *Urban Road, Bridge, and Flood Control.* 2017(11):95–97 + 12.

Hou Jie. *Research on Pavement Widening Technology of Ji-Qing Expressway Reconstruction and Expansion Project [D].* Beijing University of Technology. 2016.

Mu Xiangchun. Practice and Research on Reconstruction of Existing Dangerous Urban Bridges in China. *Urban Road Bridges and Flood Control.* 2018(09):47–50 + 69 + 9.

Outline of National Comprehensive Three-dimensional Transportation Network Planning [R]. *Beijing: The State Council.* 2021.

Outline of Transportation power Construction [R]. *Beijing: The State Council.* 2019.

The 14th Five-Year Plan of Highway Development [R]. *Beijing: Ministry of Transport.* 2022.

The Outline of the 14th Five-Year Plan for National Economic and Social Development of the People's Republic of China and the Long-term Goal for 2035 [R]. *Beijing: The State Council.* 2021.

Urban Renewal Planning of Beijing (Urban Renewal Planning of Beijing during the 14th Five-Year Plan Period) [R]. *Beijing: Beijing Municipal People's Government.* 2022.

Wang Qingsong and Han Zhichen. Control of Horizontal Displacement in the Overall Lifting Project of Shizilin Bridge. *Tianjin Construction Science and Technology*. 2004(06):22–23.

Wang Tao. Research on Removal Technology Of Long-Span Concrete Box Girder With the Upper-span Closed-frame Passage. *Journal of Municipal Technology*. 2021,39(12):152–156.

Wei Yutao. *Research and Comprehensive Evaluation of the Reconstruction and Extension Project of the Zhumadian to Xinyang section of the Beijing-Hong Kong-Macao Expressway* [D]. Hubei University of Technology. 2017.

Ye Jinhua, Cao Yanhui, Chen Yanan. Safe and Efficient Beam Exchange Technology for Urban Viaduct. *Journal of Municipal Technology*. 2020,38(03):75–78 + 89.

Ye Jinhua. Research on the Key Technology of Gantry Cranes to Remove Old Beams Under the Condition of Space Limitation. *Highway*. 2021,66(08):175–180.

Constructional Engineering and Ecological Environment – Chih-Huang Weng (Ed)
© 2024 The Author(s), ISBN 978-1-032-53198-4

Research of smart production and installation planning of prefabricated beams for prefabricated bridges

Jinhua Ye*
Beijing Municipal Road & Bridge Co., Ltd, Beijing, China

Xin Zhao
Beijing Youbi Zhicheng Construction Technology Co., Ltd., Beijing, China

ABSTRACT: The production and installation planning of prefabricated bridge components have many constraints. With few software support in the industry, the planning needs the experience to calculate manually, but errors often affect the construction period. In this paper, the prefabricated beam is selected as an example to analyze the factors affecting the production and installation progress and the balance between supply and demand. It uses Pseudocode to describe the algorithm of experience that developers can refer to and develop the algorithm into a smart management system for the production and planning of prefabricated beam. In this way, the project member can use the tool to automatically calculate the planning quickly and efficiently and make dynamic adjustments in combination with the actual situation on site so as to achieve intelligent control over the production, storage, and installation of prefabricated beams and save a lot of manpower and material resources.

1 INTRODUCTION

With the rapid development of urban infrastructure construction in China, and also with the national requirements for the coordinated development of intelligent construction and building industrialization, prefabricated bridge projects are increasing in China (Li *et al.* 2022).

For the construction of prefabricated bridges, the scientific rationality of the production, storage, transportation, and installation of prefabricated components should be fully considered in advance and should be scheduled properly.

However, due to the prefabricated components containing many types, large quantities, and also the bridge project is usually long and contains many prefabrication yards, there are many factors affecting the actual production and installation schedule, which caused frequent temporary adjustment of production tasks (Sun & Ding 2021). Due to the complexity and intersection of influencing factors, there are few existing planning software for prefabricated components of bridges in the industry.

At present, the planning for the production and installation of prefabricated components mainly depends on traditional experience and manual calculation. However, the workload of manpower calculation is huge, the accuracy is low, and the planning is difficult to keep up with the actual changes on site. These factors often lead to the production and installation cannot meet the requirements of on-site construction, resulting in construction delays and an increase in cost.

Therefore, this paper takes the prefabricated beam as an example, analyzes the factors affecting the production and installation progress, and the balance between supply and demand forms a logical association with the experience expressed by Pseudocode which is used for developers to refer to and develop the algorithm and planning system accordingly. In this way, the project member can use the system to automatically calculate the production

*Corresponding Author: 13488799957@163.com

DOI: 10.1201/9781003410843-15

and installation plan of prefabricated beams quickly and make dynamic adjustments based on the actual situation on site so as to achieve intelligent control over the production, storage, and installation of prefabricated beams.

2 INFLUENCE FACTORS RESEARCH OF PRODUCTION AND INSTALLATION OF PREFABRICATED BEAMS

There are many constraints of prefabricated bridge construction., it is difficult to coordinate the production schedule with the construction schedule, and the construction tasks are frequently adjusted temporarily. In the actual planning process, manual planning is unable to meet the construction requirements of the job site.

The main influencing factors to be considered during the production and installation planning of prefabricated beams are shown in Table 1 (Chen 2020; Jie & Qi 2022; Li 2021; Zhang 2018).

Table 1. Main influencing factors of beam's production and installation planning.

Influence Factor	Influence on Planning	Associate Influence Factors
Classification of Beams	Identification of Production and Installation Planning of Beams	Whole Planning Process
Number of Beams	Influence Number of Beam Yards and Construction Sections	Construction Sections, Number of Beam Yards
Number of Beam Yards	The number of Beams and Construction Sections determine Beam Yard selection	Number of Beams, Construction Sections
Type of Beam	Different Type of Beam needs different Type of Formwork	Type of Formwork
Length of Beam	Different Length of Beam requires different Length of Pedestals and different Type of Formwork	Number of Precast and Storage Pedestals, Type of Formwork
Position of Beam	The same type of Beam, which is at the edge or center uses different formworks	Type of Formwork
Type of Formwork	Based on the Type, Length, and Position of the Beam, to confirm the type of Formwork	Type, Length, Position of Beam
Number of Formwork	Affect Precast Productivity, but limit to the Number of Precast Pedestals and Fabrication Period	Number of Precast Pedestals
Fabrication Period	Related to the Type of Beam, the Length of the Beam, which is a fixed value	Type of Beam, Length of Beam
Precast Productivity	The number of Formwork and Fabrication Periods determine the daily number of Prefabricated beams	Number of Formwork, Type of Beam
Number of Pedestals	The number of Pedestals should align with Storage Ability, Precast, and Installation Productivity	Number of Pedestals, Productivity
Time of Beam Storage	Related to the Completion Time of Beam Fabrication, Productivity	Beam Fabrication Period, Productivity
Stacking of Pedestals	The Storage Pedestal can temporarily store two beams during peak hours	Number of Storage Pedestals, Productivity
Construction Sections	Construction Sections are determined by quantities of beams and the Number of Beam Yards	Number of Beams, Number of Beam Yards
Installation Productivity	Influence the Number of Storage Pedestals and Installation Schedule	Number of Precast and Storage Pedestals

This part of the influencing factors is cross related. In the planning process, project managers need to comprehensively consider the project scale, type, and specification of prefabricated beams, construction sections, etc., and rely on experience to set parameters such as the number of beam yards, pedestals, and formworks, and then calculate the production and installation schedules. The main logic diagram of the thinking process and the relationship between factors considered by project managers in planning is shown in Figure 1.

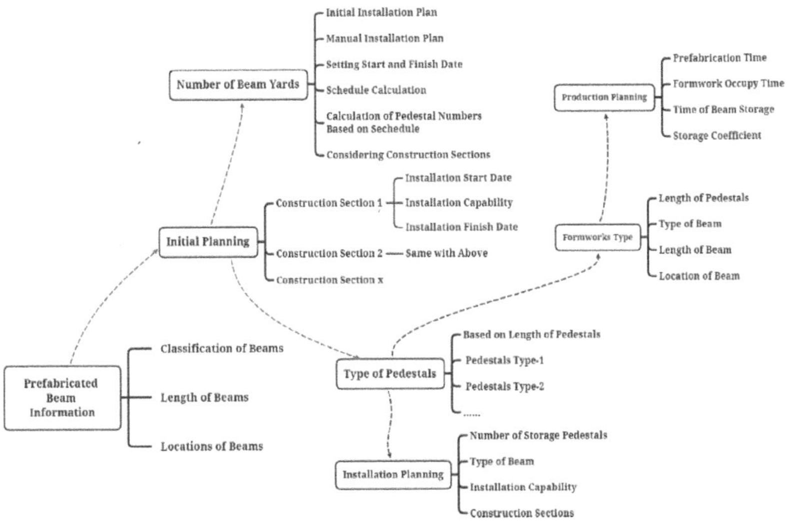

Figure 1. Relationships between influencing factors of prefabricated beam planning.

3 PRODUCTION AND INSTALLATION PLANNING PROCESS OF PREFABRICATED BEAMS

Based on the above factors affecting the production and installation of prefabricated beams, as well as the correlation between them, the project management personnel, according to the work decision-making process in their own production and installation planning of prefabricated beams, explicitly displayed the overall calculation process and the idea of judging rationality based on traditional experience estimation in the form of Pseudocode (China Machine Press 2017; Liang 2018).

```
function Number of Formwork Type
  Formwork Type List
  foreach Prefabricated Beam in Project
    if Formwork TypeList.Contains(x => x.Type of Beam
== Prefabricated Beam.Type of Beam  and x.Length of
Beam  == Prefabricated Beam.Length of Beam  and
x.Location of Beam  == Prefabricated Beam.Location of
Beam )
      continue
    else
      Formwork TypeList.Add(new Formwork
Type(Prefabricated Beam.Type of Beam , Prefabricated
Beam.Length of Beam , Prefabricated Beam.Location of
Beam ))
    end if
  return Count(Formwork TypeList)
end function
```

```
function Average Beam Storage Time
  Total Beam Storage Time
  foreach Prefabricated Beam in Project
    Total Beam Storage Time += Prefabricated
Beam.Installation Date - Prefabricated
Beam.Prefabricate Start Date - Prefabricated
Beam.Fabricated Period of Beams
  Average Beam Storage Time = Total Beam Storage Time
/ Prefabricated BeamNumbers
  return Average Beam Storage Time
end function

function Storage PedestalsNumbers
  Storage PedestalsNumbers = Precast PedestalsOccupy
Time * Average Beam Storage Time * Storage
Coefficient // Storage CoefficientValues0.5 ~ 1
  return Storage PedestalsNumbers
end function
```

Figure 2. Example of using pseudocode to express relationships between influencing factors of production and installation planning of prefabricated beam.

The Pseudocode code prepared by the project manager will be used to provide the logic and process basis for software developers to write the automatic planning system.

4 DEVELOPMENT OF SMART PRODUCTION AND INSTALLATION PLANNING SYSTEMS OF PREFABRICATED BEAMS

4.1 *Introduction of development*

As there are many influencing factors involved in the production and installation planning of prefabricated beams, and they cross influence each other, the traditional calculation cannot accurately express the logical relationship between the influencing factors.

Therefore, the project uses Asp Net Core + MySQL is used as the back end to complete data calculation and storage, and WPF is used to create the client interface. On the one hand, the influencing factors involved in the production and installation planning of prefabricated beams are materialized through programming; on the other hand, the logical relationship and calculation data between the materialized influencing factors are recorded through MySQL.

4.2 *Parameterization of influencing factors*

Based on the logical relationship expressed by Pseudocode code, software developers extract the involved parameter objects and write them into fixed program objects. At the same time, based on the logical relationship between different objects, the algorithm between the fixed objects of precast beam production, storage, transportation and installation is compiled to form a mathematical model with parametric mapping and change synchronization between objects, so as to automatically judge the rationality of the installation plan and achieve intelligent production scheduling.

Based on the calculation logic and mathematical model between entity objects, the project uses MySQL as the database carrier of linkage calculation between objects, which is used to calculate and record the data analysis, algorithm calculation, resource matching requirements, automatic early warning, and other functions of precast beam members in the planning process of production, storage, and installation. It supports the intellectualization of precast beam production and installation.

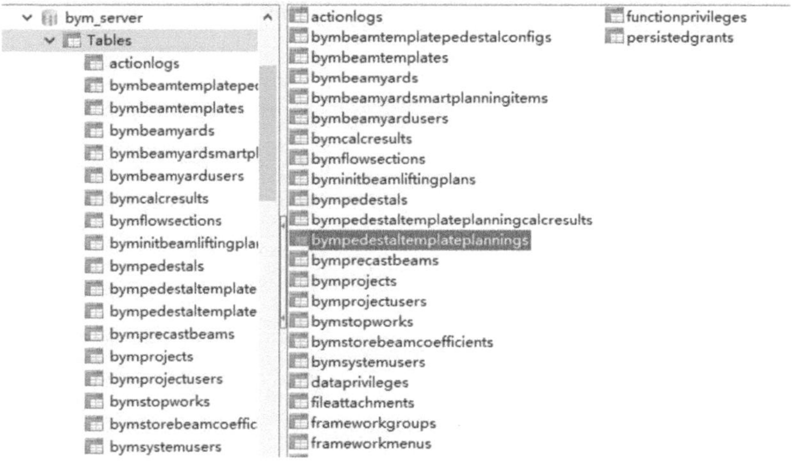

Figure 3. Establish a calculation database of entities.

The SQL database is used to carry the logic and mathematical formulas between entity objects. It is convenient for the linkage calculation between different influencing factors. It can also be used for customizing the pre-warning conditions in the database based on the field's actual production and installation data. Meanwhile, the SQL database can timely back up the calculation results under different preset conditions. The database can provide an interface with other management systems to integrate the planning data with the cost, quality, safety, production, and other information on site and provide a basis for the later project total management.

4.3 *Classification and influencing factors grouping*

A project usually contains thousands of beams. In order to quickly obtain the influencing factor information and data required for planning, the research studied and established a coding standard. The code is the identification of the bridge component, based on the principle of "route-bridge name-work category-axle number-location."

The advantage of this coding method is that it can automatically extract engineering information from drawings or models, establish a mapping relationship between drawings or models with information, and quickly obtain the parameters of prefabricated beams, including type, specification, and location, which can be used for automatic planning.

4.4 *Development of production and installation planning systems*

Based on the classification, the planning system was developed, which can extract the information of beams into a database and classifies and counts them according to the planning influencing factors.

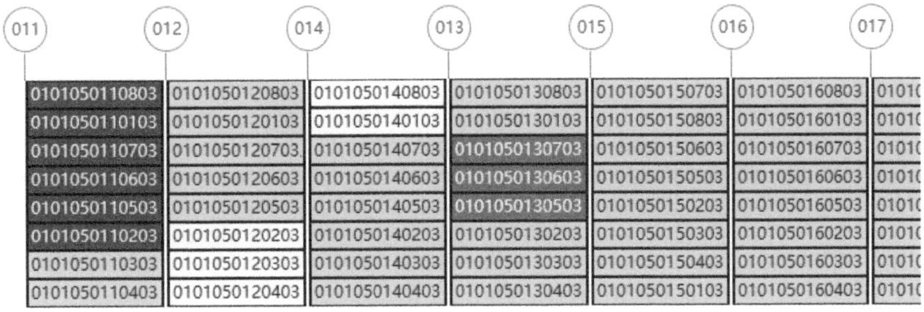

Figure 4. Automatic classification of prefabricated beam.

By imputing the fixed parameters such as the service time of the beam yard, and equipment capacity, the system automatically calculates the relevant parameters based on the logic algorithm, including the type and quantity of pedestals, the type and quantity of formworks, etc.

The database application enables the project member to automatically calculate and adjust the parameters according to the actual situation on site and calculate again according to the adjusted parameters. Based on the final confirmed production parameters of prefabricated beams, the system automatically calculates the production and installation plans in combination with the algorithm.

By using the calculation with SQL, the database can be integrated with other on-site management system data. Based on the actual production and installation progress, the progress trend can be predicted and analyzed according to the algorithm. The project can adjust the production and installation parameters in time, recalculate and correct the deviation according to the calculation method.

The production and installation of prefabricated beams are automatically planned based on the algorithm. And the project can carry out different calculations and analyses of the overall construction process based on different parameters, including 4D progress display, schedule analysis, etc., which can be used as a basis for decision-making.

5 APPLICATION OF SMART PRODUCTION AND INSTALLATION PLANNING SYSTEMS

The smart planning system has been tested and applied in many projects, such as Yuanshan Avenue Project, and G312 Line Project.

In the test project, the project uses traditional planning methods and a smart planning system to work at the same time and compares the work results. By comparison, the efficiency of the smart planning system is more than 95% higher than that of the traditional method, and the error rate is reduced to 0 and provides a scientific basis for rational planning of production capacity and resource allocation (Wang *et al.* 2022).

Through the planning system, the project can reasonably formulate the production, storage, and installation plans, ensuring the project duration performance and efficient use of the beam yard.

Table 2. Comparison of traditional planning with intelligent planning system.

Comparison	Traditional Method	Smart Planning Systems
Initial Planning	Due to a large number of components, the project doesn't have detailed planning and preparation.	Preset various planning parameters to find the most reasonable scheme
Schedule Calculation	There are about 1000 prefabricated beams in the project. It takes about 2 days to calculate a complete planning scheme	The automatic calculation took about 10 minutes, and different planning parameters can be calculated and selected quickly
Planning Adjustment	Many factors changed on site; the planning cannot be adjusted in time according to the actual situation.	Capture the actual on-site situation, real-time prediction, and update based on the algorithm
Error Rate	There are many mistakes in the planning process. The plans are not detailed enough.	Automatically calculate with zero error rate

6 CONCLUSION

In the application of information technology, the human thinking process is represented in the form of data and logic (Fu *et al.* 2022). This study uses Pseudocode and software development to compile logical algorithms that are easy for project managers to carry out heavy planning works with mathematical logic, forming a digital and intelligent work mode, which significantly improves work efficiency. This method is highly referential and promotional to other similar project management objects.

REFERENCES

Chen Jing-peng. *Key Issues and Theoretical Research on Construction Management of Short-line Assemble Precast Beam Yard of Urban Viaduct (Dissertation of Master of Engineering)*. Nanchang; Nanchang University, 2020, in Chinese.

Fu Xu, Zhong Youheng, Zhou Huiwen, SI Wenjing, Zhang Aiqing. Functional Requirement Analysis and System Architecture Research of Digital Twin Intelligent Beam Field. *Journal of Information Technology in Civil Engineering and Architecture*. 2022, in Chinese.

Jie Mao-hai, Qi Shi-chao, Planning and Design of Typical Demonstration Beam Yard of Beijing Qinhuangdao Expressway. *Road and Bridges*, 2022,49(20):152–156, in Chinese.

Li Qing-yang, Guan Tao, Miao Zi-zhen, and Wei Zhi-song, Application of Prefabricated Assembly Technology in Municipal Bridge Engineering. *Construction Technology*. 2022,51(2):66–69, in Chinese.

Li Wei-wei. Formwork Selection and Beam Yard Planning of Expressway Precast Beam Yard. *Proceedings of 2021 National Civil Engineering Construction Technology Exchange Conference (Volume I)*. 2021-06-10:616–618, in Chinese.

Liang Yun-wei. Prefabricated Beam Yard Planning Construction and Construction Management Based on Fuzzy Mathematics Theory. *Journal of Bridge and Tunnel Engineering*. 2018,10:115–120, in Chinese.

Rod Stephens. *Essential Algorithms*. China Machine Press, 2017, Chapter 1.3, in Chinese.

Sun Yong-fang, and Ding Jie. Study on Intelligent Construction of Digital Twinning Technology for Bridge Precast Beam Field. *Fujian Architecture & Construction*, 2021,06(276):109–113, in Chinese.

Wang L. W, Li H. and Ye J. H. Research on Parametric Modeling Technology and Construction Analogue Method of Prefabricated Small Box Girder based on BIM. *Journal of Municipal Technology*. 2022,40 (9):229–233, in Chinese.

Zhang Qing-dong, Innovative Application of Intelligent and Information Technology in High-Speed Railway Precast Beam Yard. *Journal of Shanxi Architecture*, 2018,44(2):130–131, in Chinese.

Constructional Engineering and Ecological Environment – Chih-Huang Weng (Ed)
© 2024 The Author(s), ISBN 978-1-032-53198-4

Research and application of construction method for long span steel-concrete composite beam bridge

Tingting Sun
School of Road Bridge & Harbor Engineering, Nanjing Communication Institute of Technology, Nanjing, Jiangsu, China

Jingnan Yang
CCCC Second Highway Consultants Co. Ltd., Wuhan, China

Zhiliang Jing* & Shengjun Zhou
The Third Construction Co., Ltd. of China Construction Eighth Engineering division, Nanjing, China

ABSTRACT: A long-span steel-concrete composite beam bridge is a continuous beam bridge with 83.5 m + 135 m + 98.5 m. This paper combines the characteristics of site construction, in order to ensure the smooth installation of steel-concrete composite beam section. Based on the actual situation on site, we explored and developed a series of construction technologies such as fabrication, transportation, hoisting, jacking, and installation of composite beams. The temporary support is set up to ensure smooth sliding, and the construction stage is divided to ensure that the lower part can be navigable and the upper part can be hoisted during construction, overcoming the difficulty of smooth hoisting on water. The application results show that the method of splitting, hoisting, installing, and welding the composite beam segments by sliding makes each beam segment smoothly installed in place. Compared with the traditional construction method, it reduces the complex construction process, reduces the difficulty of beam connection, speeds up the construction progress, and also ensures the quality of beam connection.

1 INTRODUCTION

As a steel-concrete composite structure, steel-mixed laminated girders are widely used in various engineering projects, especially in large-span bridges, which can ensure the full utilization of the performance of steel as well as the compressive strength of concrete to achieve the maximum performance of both materials and, at the same time, achieve the maximum industrial production. Yang (Yang & Liu 2022) studied the steel and concrete laminated girders of the 500m span Honghe Bridge and used a new type of bridge deck crane to construct the laminated girders. Shi (Shi et al. 2022) presented an innovative precast composite girder consisting of two full-depth precast concrete (FDPC) slabs with extended reinforcement and reverse notches, an H-beam with welded shear nails, additional transverse reinforcement, and cast-in-place ultra-high performance concrete (UHPC). Berry (Berry & Patrick 2002) discussed the application and practice of combined steel-concrete structural bridges in Australia and described the superior performance of such structures through practical engineering. Tong (Tong et al 2022) investigated the combined structure of

*Corresponding Author: 2284454553@qq.com

 DOI: 10.1201/9781003410843-16

high-strength structural steel (HSS) and ultra-high-performance concrete (UHPC) through experiments and mainly analyzed the load-bearing performance of the steel-concrete combination structure, which showed good mechanical properties. Huang (Huang et al 2022) proposed a steel-concrete combined beam, systematically analyzed its load-bearing capacity performance, and concluded that this structural form has a better seismic and flexural load capacity.

In summary, steel-concrete combination structures in bridges should be very extensive, and some construction methods are not necessarily common due to the characteristics of the location where the respective projects are located (Fang et al 2021; Xiong et al 2022). Based on this, this paper takes the actual bridge under construction as the research background. It systematically introduces the construction process of steel-concrete laminated girders and the installation of girder sections, etc., to provide reference and reference for similar projects.

2 PROJECT

This paper takes a large-span steel-mixed laminated girder bridge as the research background. The project is planned as an urban trunk road, with a span combination of 83.5m +135m+98.5m=317m, the structure form is a steel mixed laminated girder, and the bridge elevation is shown in Figure 1.

Figure 1. Elevation of a steel mixed laminated beam.

3 BRIDGE IMPORTANT STRUCTURE

3.1 *Steel main girder*

The steel main girder adopts a separated steel box structure (Figure 2). The left and right steel boxes are connected by concrete deck slabs and steel cross partitions. The total height of the main girder is 3.5m~6.5m, of which the thickness of the standard section of the bridge deck plate is 0.17m, and the thickness of the axillary part is 0.40m.

There are 7 types of main girders and 56 girder sections, including 14 SA girder sections (SA1-SA14), 11 SB girder sections (SB1-SB11), 11 SC girder sections (SC1-SC11), 17 SD girder sections (SD1-SD17), 1 S0 girder section, 1 S1 girder section, and 1 S2 girder section.

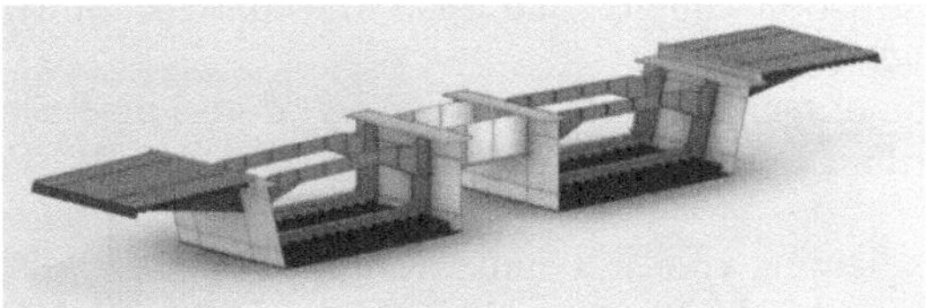

Figure 2. Schematic diagram of steel main beam with separated steel box.

3.2 *Bridge deck slabs*

Precast slabs, the whole bridge is divided into four types of precast slabs, I-s, I-m, II-s, and II-m. The characteristics of each type of precast slab are shown in Table 1 below. Both ends of the transverse joints are set into wedge-shaped joints, and both ends of the longitudinal joints are set into slot joints.

Table 1. Table of classification and characteristics of prefabricated panels.

Type	Dimension (m)	The thickness of the bridge deck plate	Pre-embedded steel plate
I-s	8.7×2.6	1.2 m thick at both ends, 0.17~0.4 m thick, 6.3 thick in the middle, 0.17 m thick	Prefabricated plate I-s along the longitudinal bridgeSet up at both ends: spacerSet flange
I-m	3.8×2.6	1.2 m thick at both ends, 0.17~0.4 m thick, 1.4 thick in the middle, 0.17 m thick	Prefabricated plate I-m along the longitudinal bridgeSet up at both ends: spacerSet flange
II-s	8.7×3.6	1.2 m thick at both ends, 0.17~0.4 m thick, 6.3 thick in the middle, 0.17 m thick	Prefabricated plate II-s along the longitudinal bridgeSet up at both ends: spacerSet flange
II-m	3.8×3.6	1.2 m thick at both ends, 0.17~0.4 m thick, 1.4 thick in the middle, 0.17 m thick	Prefabricated plate II-m along the longitudinal bridgeSet up at both ends: spacerSet flange

3.3 *Joint joints*

Transverse wet joints, the whole bridge has 97 transverse wet joints, of which 67 are cast-in-place at the site, and 30 are cast-in-place at the factory. The size of transverse wet joints is 26m (horizontal bridge direction) × 0.4m (both) (longitudinal bridge direction) × 0.17~0.4m (vertical bridge direction), and the pre-buried steel plates with a thickness of 16mm are laid under the transverse wet joints. The butt welds between the pre-buried steel plates are welded before the transverse wet joints are poured.

4 MAIN BRIDGE CONSTRUCTION TECHNOLOGY

The support platform of the main bridge is designed in three parts: north shore, water, and south shore; north shore, due to the small distance from the original ground to the beam section, the top part of the embankment uses an expanded foundation plus a bery system for support. The rest part uses a steel pipe pile beryl system; the south shore uses a steel pipe pile plus beryl support system, and the cross-road passage gate adopts a steel pipe pile plus steel combination support system; the water span uses steel pipe pile foundation plus a beryl system for support. The temporary support is shown in Figure 3.

Step 1: In the dry period, the organization will enter the site, clean and level the site, review the point construction samples, and apply the construction right-of-way, steel cofferdam, and pier foundation.

Step 2: Construction of temporary supporting piers of steel pipe piles in water.

Step 3: erecting berthing bracket, distribution beam, sliding track beam, and installing the sliding track.

Step 4: Using two floating cranes, symmetrically lifting the sliding, 1# floating crane lifting the north bank SA14~SB2 beam section (17 sections), 2# floating crane lifting the

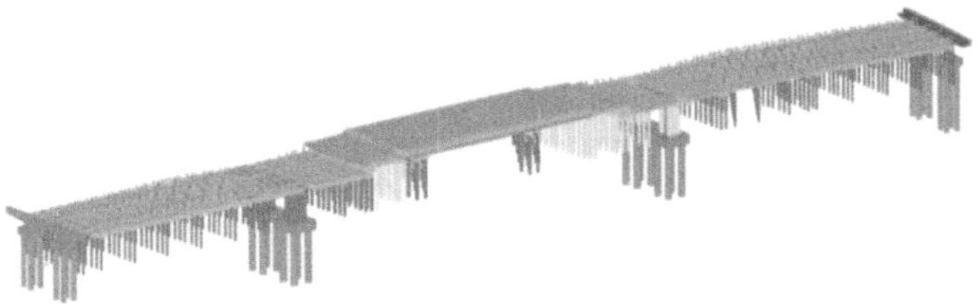

Figure 3. Overall elevation of temporary support.

south bank S D17~SC2 beam section (20 sections), adjusting the elevation line shape, assembling and welding.

Step 5: Symmetrical lifting, 1# floating crane lifting the north bank SB3~SB8 beam section, 2# floating crane lifting the south bank SC3~SC8 beam section, adjusting the elevation line shape, assembling and welding.

Step 6: 1# floating crane lifts SB9~S2~SC9 beam section in the span, adjusts the elevation line type, and assembles and welds.

Step 7: Pour wet joint concrete into the positive moment area, moisturizing and maintaining to design strength.

Step 8: Remove the bailey bracket, then pour the wet joint concrete in the negative moment zone and moisten and maintain it to the design strength.

Step 9: Apply prestressing in the wet joint (internal prestressing).

Step 10: Drop the beam at P8 and P9 piers to the design elevation, and the conversion of the beam system is completed.

Step 11: Apply internal prestressing in the box girder (in vivo prestressing).

Step 12: Construction of painting, sidewalk system, and other ancillary facilities, bridge loading test, and completion acceptance.

4.1 *Analysis of main bridge segmental and lifting working conditions*

According to the characteristics of the bridge structure and the actual site conditions, they are combined with the construction period, quality, cost, and other comprehensive factors. The main bridge steel box girder uses the sliding method of construction, erecting the bery l girder bracket and setting the sliding device on the beryl girder bracket. The processing plant completed the main girder (excluding the pick arm) section loading and waterway transport to the site installation location using a floating crane transfer with the sliding method of installation of the main girder. The steel box girder is divided into three sections in the direction of the bridge width, one section at each end of the pick arm, and one section of the box body, in accordance with the design. The specific segmentation is shown in Figure 4.

Figure 4. Schematic diagram of beam segmentation.

4.2 *Sliding ideas*

The project steel-concrete combination girder bridge structure installation height is high, and the longitudinal and transverse span is large, and the self-weight is large. If the conventional piece-wise aerial bulk assembly scheme is adopted. In that case, a large amount of aerial scaffolding will need to be erected, which will not only result in a huge workload of aerial assembly and welding. It will have a great impact on the construction schedule of the whole project, and the technical and economic index of the scheme is poor.

According to the successful experience of similar projects in the past, after comprehensive consideration (Araujo et al 2022; Zhang et al. 2022; Zou et al 2023), the bridge segments will be assembled in the axes and then installed by "super-large hydraulic synchronous sliding construction technology" (Figure 5). This will greatly reduce the difficulty of the installation construction and is beneficial to quality, safety, and schedule.

Figure 5. Slip construction drawing of part of the beam section.

5 CONCLUSION

Taking a large-span steel-hybrid stacked girder bridge as the research background, combined with the construction characteristics of the site. A new steel-hybrid stacked girder section construction process was explored in order to guarantee the navigation requirements of the waterway and safe and civilized construction, which integrated the construction processes of prefabrication, transportation, lifting, jacking, and installation, and solved all the technical problems faced during the construction of the stacked girder section. The whole temporary support bracket was erected to solve the space limitation and meet the construction requirements. It reduces the weight of the lifting load, ensures the accurate construction of steel-hybrid laminated beams, and improves the safety of construction. The research and application of this project can provide a reference for similar engineering projects and further promote the application and development of assembled buildings.

ACKNOWLEDGMENTS

This research was funded by the Practical Training Program for Young Teachers in higher Vocational Colleges in Jiangsu Province(2022QYSJ036), Teaching Capital Construction Project of Nanjing Transport Vocational and Technical College (JX2149 and JX2148). The authors wish to express their gratitude for this financial support.

REFERENCES

Araujo H. F., Andrade C. M., Basaglia C., *et al.* Lateral-distortional Buckling of Steel-concrete Composite Beams: Kinematics, Constrained-mode GBT and Analytical Formulae. *Journal of Constructional Steel Research*. 192: 107210–107218(2022).

Berry P. A. and Patrick M. Design of Simply-supported and Continuous Beams in Steel-concrete Composite Construction. In: Anson M, Ko JM, Lam ESS, eds. *Advances in Building Technology*. Oxford: Elsevier; 183–190(2002).

Fang J., Bao W., Ren F., *et al.* Behaviour of Composite Beams With Beam-to-girder End-plate Connection Under Hogging Moments. *Engineering Structures*. 235: 112030–112039(2021).

Huang H., Yao Y., Liang C., *et al.* Experimental Study on the Cyclic Performance of Steel-hollow Core Partially Encased Composite Spliced Frame Beam. *Soil Dynamics and Earthquake Engineering*. 163: 107499–107506(2022).

Shi F. W, Sun C. H, Liu X.G, *et al.* Flexural Behavior of Prefabricated Composite Beam With Cast-in-situ UHPC: Experimental and Numerical Studies. *Structures*. 45: 670–684(2022).

Tong L., Chen L., Wang X., *et al.* Experiment and Finite Element Analysis of Bending Behavior of High Strength Steel-UHPC Composite Beams. *Engineering Structures*. 266: 114594–114598(2022).

Xiong G., Li W., Wang X., *et al.* Flexural Behavior of Prefabricated High-strength Steel-concrete Composite Beams With Steel Block Connectors. *Journal of Constructional Steel Research*. 197: 107507–107515(2022).

Yang Yong and Liu Tao. Construction Technology of Steel-concrete Composite Beams of Cable-stayed Bridge in Swamp Area. *Construction Technology*. 51(02): 70–73+109(2022).

Zhang J., Pei Z. and Rong X. Experimental Seismic Study of An Innovative Precast Steel–Concrete Composite Beam–column Joint. *Soil Dynamics and Earthquake Engineering*. 161: 107420–107428(2022).

Zou Y., Yu K., Heng J., *et al.* Feasibility Study of New GFRP Grid Web - Concrete Composite Beam. *Composite Structures*. 305: 116527–116535 (2023).

Constructional Engineering and Ecological Environment – Chih-Huang Weng (Ed)
© 2024 The Author(s), ISBN 978-1-032-53198-4

Ancient buildings and culture empowering Island development

Wenhui Fei*, Jie Yang, Xiaomin Huang, Yun Shen & Waner Li
Zhejiang University of Water Resources and Electric Power, Hangzhou, Zhejiang, China

ABSTRACT: Ancient villages are an important carrier of regional culture, among which the island's ancient villages are facing the situation of becoming barren or over-commercialized development due to the particularity of the geographical region. Therefore, in order to implement the rural revitalization strategy, and inherit and innovate the local traditional culture, it is particularly important to carry out the organic, ecological transformation of the island's ancient buildings. We should base ourselves on national policies, creatively inherit traditional culture, promote the development of island culture and tourism, and combine local actual conditions to achieve sustainable development of island ancient buildings.

1 INSTRUCTION

Zhoushan, Zhejiang Province, located on the shore of the East China Sea, is the largest archipelago in China with a long history, and the ancient villages and buildings on the island are the best testimony of Zhoushan's history over time. Nowadays, the ancient village buildings in Zhoushan are roughly divided into four kinds of situations: the old city has been developed in large numbers, and the original villages have been demolished and changed into today's blocks of high-rise buildings, full of modern commercial atmosphere; some of them have been selectively abandoned by people due to their geographical location or bad development prospect, and gradually deserted and full of vines; some villages have made use of the existing ancient village buildings, and on the basis of protection, they have been reasonably developed and utilized to develop tourism and take the The road of rural revitalization has been well developed; there are also some villages which are more closed and lagging behind in development due to transportation and resource constraints, and their economy stays at the level of a dozen or even decades ago, making it difficult to develop a breakthrough. And how to make the last of these ancient villages in Zhoushan get attention and embark on the road of development has become a topic we need to explore.

2 THE ROAD TO SAVING ANCIENT BUILDINGS IN AN ISLAND VILLAGE

Ancient villages are constructed from vernacular culture and are typical cultural and ecological settlements, a living carrier of the history of vernacular culture. When one hears the description of these images, one can naturally associate them with the entity to which they correspond. For example: when one hears "small bridges, flowing water, homes," one thinks of the water villages in Jiangnan. When one hears "green tiles, white walls, ancestral halls," one thinks of the ancient villages in southern Anhui, etc. When one hears "green tiles, white walls, and ancestral halls," one thinks of the ancient villages of southern Anhui, etc. The

*Corresponding Author: 2019548339@qq.com

 DOI: 10.1201/9781003410843-17

carriers of these images are ancient buildings, and it can be said that the preservation of ancient buildings is the top priority for the preservation of ancient villages.

Zhoushan Islands have a long history, a beautiful and colorful natural landscape, profound marine culture, and a long human history. According to relevant records, as early as the Neolithic Age, more than 5000 years ago, Zhoushan Islands were inhabited. There are 476 ancient buildings, accounting for 56.07% of the registered immovable cultural relics: 232 important historical records and representative buildings in modern times, accounting for 27.33%. Among the traditional villages in Zhoushan, most are traditional villages with ancient buildings, while there are relatively few folk customs villages and natural ecological villages. Specifically, 19 of the 30 villages belong to traditional villages of ancient architecture, 6 belong to folk customs villages, and 5 belong to natural ecological villages. As far as the ancient architectural villages are concerned, although they have typical island architectural features, they are not very old (1 architectural village in the Ming Dynasty, 13 in the Qing Dynasty, and 9 in 1950-1980). There are 2172 existing ancient buildings before 1980, covering an area of 259150 square meters. More than 70% are concentrated in Dongsha Ancient Fishing Town, mainly from the Qing Dynasty to the early Republic of China (Zhang 2021).

The climate of Zhoushan Islands is subtropical maritime monsoon climate, which is significantly influenced by the climate, facing the sea and land breeze, high wind and fog, tide, and disastrous weather at any time (Miao 2017). Therefore, drainage and moisture prevention became the first factor for the people of Zhoushan to consider when building houses. In the era of technology and economic underdevelopment, old houses were generally built with stones as raw materials. The stones used were mainly in the form of blocks, which had a neat appearance and were more convenient for building houses with beautiful shapes. On the one hand, because of the geographical location, the island is rich in stone, which is not only easy to obtain and use, but also economical. On the other hand, the stone is usually a dense structure, which can better improve the solidity of the house, not only it is not easy to collapse, but also it can keep heat and moisture, resist sea wind attack, and be more resistant to flood. The main wall material of island houses is granite, thanks to its high hardness and smooth surface. In ancient times, sand and ash were used to fill the gaps between the stones, making a piece of square stone stacked walls and built. In addition, many parts of the building outside the outer walls are made directly from the stones on the island, such as doors, windows, roofs, etc. This kind of building is known as a stone house. It also adopts a sloping roof to prevent rainwater from accumulating on the roof. The house is built along the mountain with a staggered height, using the sloping terrain to drain the rainwater along the ramp, which is quite an island style.

Only on the islands, do residents use stones to build houses on a large scale, which makes them contrast with inland house styles and shapes the unique style of island houses (Tao 2017). These stone houses have survived the weather and still stand in an ancient village, preserving the historical flavor of Zhoushan, with a special taste of old Zhoushan. Nowadays, with the rapid development of society, these ancient buildings are wrapped up in history, mostly deserted in villages and mountains, recording the loss of culture. At present, the ancient buildings in traditional villages are very fragile. If we do not hurry to repair, the ancient building stock of traditional villages in China will face a major risk.

2.1 *Major risks faced by ancient villages*

With the development and construction of villages, more and more residents have moved out of their ancestral houses and built modern houses in the periphery, resulting in the hollowing out of ancient villages. On the one hand, due to the age of construction, the building structure is near the edge of life, some basic equipment is aging, lack of daily maintenance, which makes the ancient village easy to damage and collapse. On the other hand, the construction of a large number of modern houses makes the pattern and distribution of ancient villages fragmented, which poses a certain threat to maintaining the integrity of the overall style of

Figure 1. A building in which stone is the main material [Photo source: Field shooting].

Figure 2. The ancient village where the stone is the main material [Photo source: Field shooting].

ancient villages (Feng & Fei 2020). In the long run, more and more modern buildings covered the original ancient village landform. The ancient villages that should have developed together with traditional culture were not protected and inherited accordingly and gradually lost their original simplicity and elegance.

2.2 *The significance of saving ancient buildings on islands*

Ancient buildings have always been a controversial issue. As a product of the era of backward productivity, ancient buildings are not enough to protect the value of the hedge against the cost of protection and repair, so it is inevitable that the trend is eliminated. In addition to letting it fall into disuse, most of the time, the choice is to demolish it. However, as a party to the history of the carrier, the protection of ancient buildings may not be unable to play a role in promoting local economic development.

Due to geographical constraints and the scattered distribution of islands, it is difficult for Zhoushan to develop an industrial and large-scale agricultural economy. Therefore, the development of the tourism economy is a good strategy that can drive the rapid development of the economy. As a symbol of island culture, the ancient architecture of the island can be used as a focal point of the tourism economy, adding to the economic development of the island, regional cultural construction, natural ecological protection, and improvement of people's living standards.

2.3 *Measures for the rescue of ancient buildings on islands*

Building on the rural revitalization strategy. The rural revitalization strategy is an important measure to solve the "three rural problems" and improve people's livelihood. The rescue of ancient buildings on the island should rely on the rural revitalization strategy, taking the restoration and transformation of ancient buildings as the entry point. Based on relevant policies, coordinate social development and cultural restoration, take into account the ecological environment and economic growth, and make the rural revitalization strategy an important hand for the rescue of ancient buildings. They achieve the successful rescue of ancient buildings, face the future, develop for a long time, and realize the prosperity of industry, ecological livability, civilized countryside, effective governance, and Wealthy living.

Guided by the "Save Our Homes" initiative. Since its implementation in 2016, the "Save Old House Initiative" in Songyang, Lishui, Zhejiang Province, has completed the repair and acceptance of old houses in the county as of the second half of 2021, while realizing the transformation and upgrading of Songyang's tourism economy and cultural and creative industries. They significantly improve the rural living environment, strengthening the protection and development of traditional culture, and showing the vigorous development of the county.

Song Yang practice for Zhoushan island's ancient building protection and repair work provides excellent guidance, so you can "save the old house action" and Zhoushan island's ancient villages combined with the actual local conditions to promote the island's ancient building repair and protection work. First of all, building a joint effort from all walks of life to ensure the smooth implementation of the rescue of ancient buildings. Promote all walks of life to cooperate with each other and cooperate to promote the rescue of ancient buildings smoothly. Second, improve the technical leadership, highlighting the characteristics of ancient buildings on the island. The main goal of ancient building restoration is to preserve its original values, so if the actual operation of ancient buildings caused damage, the relevant restoration works will also lose their fundamental significance (Zhao 2021). Furthermore, the output of professional technology to achieve the long-term development of ancient building rescue. Ancient building repair experts to the township technicians, craftsmen, villagers, and other related personnel output expertise, so as to create a quality, skilled, professional, and complete team of artisan talent.

With the aim of promoting the development of the island's cultural and tourism economy. The rescue of ancient buildings is not only about preserving the traditional culture of the island villages, but also about using ancient buildings to drive the transformation and upgrading of the cultural and tourism industries on the islands, creating economic benefits and achieving long-term economic development on the islands (Yu 2017). To this end, cultural preservation should be the starting point for the rescue of ancient buildings, while economic development should be the target point for the rescue of ancient buildings, so as to coordinate the development and promote the island villages to keep pace with the times, accelerate economic development, and help to achieve the goal of socialist modernization (Zhang *et al.* 2021).

3 THE CULTURAL TOURISM ECONOMY OF AN ANCIENT VILLAGE ON AN ISLAND THE ROAD

3.1 *Heritage and innovation, culture rushes out of the islands*

The vastness of the ocean and the mountains have made the independence of each ancient village in Zhoushan all the more remarkable. The advantage of this, however, is that the culture has developed steadily and solidly in its unique atmosphere. Although it belongs to the same coastal culture, it has been able to branch out into different strands. Fishing, farming, forestry, and even shipping flourished on this island, and the multiple strands of historical heritage intertwine and run in parallel, giving it more character (Duan 2016).

As a frontier of marine culture, fishing culture is a naturally existing culture in the ancient villages of Zhoushan Island. Maximizing the preservation of the characteristic fishing culture and finding commonalities in individuality is the focus of preserving the local culture of the ancient villages on the islands. Faced with the problem of ancient villages gradually falling into oblivion and fishing culture fading away in the modern machinery industry, diversified conservation can be carried out.

On the one hand, it is important to do a good job of protecting intangible cultural heritage. Ships and the sea are the eternal topics of fishermen, and in the long history, around these two terms, a number of craftsmen closely related to them were born. It is also necessary to preserve local characteristics and promote local economic development.

On the other hand, cultural heritage must also be integrated with the development of the times, keeping up with the times and combining the ancient with the modern, so that the local culture of ancient villages can enter modern society and become better known to more people. In this era of rapid technological development and convenient information dissemination, it is possible to develop products and services with local characteristics by strengthening the capacity for innovation and using the "Internet +" and big data to establish a smart tourism system. And big data establishes a set of intelligent tourism systems to keep pace with the times and improve the competitiveness of the industry. For example, by leveraging the internet and relying on live streaming platforms to tell the stories behind the artisans, you can also create special labels such as "island life," "ocean fishing," and "sea watching" to conduct Live broadcasting can be used to showcase the life of fishermen and satisfy people's curiosity, while attracting interested people to come and experience tourism and increase tourism income. The museum will showcase the unique culture of different islands and form a diversified cultural landscape of the islands.

Only when an island retains and innovates to develop its distinctive culture will it have a soul and be rejuvenated in a new era.

3.2 *Revitalizing ancient buildings to create an island identity*

Zhoushan's unique geographical location and ecological environment as an island region is both a constraint and an advantage. The inconvenience of transportation has allowed the island to retain its relative tranquillity, the ecological conditions are excellent, and the diverse tourism resources add to the local tourism industry (Feng 2015). In a state where primary and secondary industries are difficult to develop on a large scale and as a whole, the development of tertiary industries, mainly tourism, has become the obvious choice for the island region.

At this stage, compared with regions with developed tourism, such as Hainan, Qingdao, and Xiamen, Zhoushan's tourism industry is relatively young. The homogeneity of tourism products is high, so accurate positioning of tourism image and improving popularity have become the primary focus at present. Therefore, based on the unique folk culture, unique architectural form, and unique island food, we should activate the repaired ancient buildings, form a differentiated tourism image and tourism brand, develop an information-based, characteristic, and regional cultural tourism industry, promote the "living protection and organic development" of island villages, and realize the long-term income increase of residents.

Ancient buildings have three main functions in the development of tourism: firstly, the tourist function, and secondly, the accommodation function. And thirdly, the cultural function.

The tourist function relies on the culture of the island, such as the fishermen's paintings painted on the walls, the stone pillars of the rock houses, and the wooden carvings of the platform doors, to show visitors the rich culture of the island.

The accommodation function focuses on the ancient and island character, receiving visitors as a B&B. The inconvenient access to the islands makes it more likely that tourists will stay on the islands, which creates opportunities for the development of ancient B&Bs. The biggest issue facing the development of B&Bs in China's rural areas at present is the operation, i.e., the awareness and ability of the operators and other aspects of the problem. Local people's B&B ventures are generally the first choice for the construction of rural B&Bs, but the management of B&Bs also requires a wealth of operational and management experience as well as professional knowledge reserves. Whether it is the introduction of talent from outside the country to take charge of the management of B&Bs, or the hiring of experienced industry experts by the relevant local authorities to train local B&B operators, the local community needs to adopt certain means to strengthen the overall quality of B&B operators. The local authorities need to take certain measures to strengthen the overall quality of B&B operators (Liu *et al.* 2021).

The cultural function means that the ancient buildings can be used as a venue for cultural activities on the island, such as special cultural performances, theatrical sketches, and cultural lectures. Other recreational activities related to the island culture that can make use of the cultural connotations of the ancient buildings, two complement each other, exporting the distinctive culture with island characteristics to tourists, increasing the stickiness of tourists, and spreading the distinctive culture. In addition, it is also possible to draw on the experience of Dongji Island, where the ancient buildings have been used as a location for filming, and to increase publicity in the clearest possible way, thus attracting people to visit the island and contributing to the development of the cultural tourism economy.

4 CONCLUSIONS

Zhoushan City, Zhejiang Province, was founded as an archipelago, and the ancient villages on Zhoushan Island have obvious island characteristics. In this paper, we selected several sample villages as objects and studied the architectural landscape of ancient villages on Zhoushan Island through field investigation, comparison, and analysis, etc. The layout of the villages is generally built near the water, and most of the sites are among the mountain plain. In order to more effectively promote the common development of ancient architecture and traditional culture, we based on the inheritance of traditional culture, follow the national policy in line with the actual situation of local development, the promotion of island cultural tourism, and promote the sustainable development of island ancient architecture combined.

However, the ancient villages on Zhoushan Island are widely distributed, and their architectural and landscape features cannot be summarized by the few objects selected in this paper, which have certain limitations and are subject to deeper research in the future.

ACKNOWLEDGEMENT

This is a project funded by the Zhejiang University of Water Resources and Electric Power undergraduate innovation training project - Research on the organic and ecological renewal of ancient buildings on Zhejiang islands (S202111481032).

REFERENCES

Beili Duan: *Research on Value Evaluation of Traditional Island Villages: Zhoushan Case [D]*. Zhejiang Ocean University. 2016

Dongfang Zhao: Application of Meticulous Management in Ancient Building Repair Projects. *Engineering Technology Research*. 2021,6(22):208–209. doi: 10.3969/j.issn.1671-3818.2021.22.094.

Mingming Feng: *The Value and Evaluation of Island Traditional Cultural Villages* [D]. Zhejiang Ocean University. 2015

Rui Tao: *On the Protection and Development of Ancient Island Villages* [D]. Tianjin University. 2017

Weiqiu Yu: Protection and Improvement of Historical Buildings in the Urbanization Development of Zhoushan Island. *Journal of Culture*. 2017 (11): 142–144

Yibo Feng and Yue Fei: Research on Disaster Status and Protection and Development of Traditional Villages. *Urban Architecture*. 2020,17 (15): 71–72. DOI: 10.19, 892/j.cnki.csjz.2020.15.026

Yiyun Zhang: *Study on the Landscape Characteristics and Protection Strategies of Zhoushan Island Villages* [D]. Zhejiang Agriculture and Forestry University. 2021. DOI: 10.27, 756/d.cnki.gzjlx.2021.000,104

Yu Liu, Yaling Zhao, Jiajia Zhang and Peng Li: Analysis of the Path of B&B Tourism to Boost Rural Revitalization in Anshun. *Shanxi Agricultural Economics*. 2022(09):60–62. doi: 10.16675/j.cnki.cn14-1065/f.2022.09.020.

Zhe Zhang, Fengying Wang and Zheng Zhu: The Attempts of Architectural Preservation and Restoration for Traditional Chinese Village "Rescuing Old Houses Action "in Songyang. Journal of Architecture, 2021(1): 34–37.DOI: 10.19819/j.cnki.ISSN0529-1399.202101005.

Zhenlong Miao: *Spatial Distribution Characteristics and Cause Analysis of Island Villages* [D]. Zhejiang Ocean University., 2017

Constructional Engineering and Ecological Environment – Chih-Huang Weng (Ed)
© 2024 The Author(s), ISBN 978-1-032-53198-4

Obtaining geometric parameters of complex bridges based on 3D laser scanning

Guilin Zhang, Lijie Meng*, Sheng Yuan, Huipeng Gao & Guangguang Dong
State Grid Hebei Electric Power Construction Co., Ltd., Shijiazhuang, Hebei Province, China
Hebei Electric Power Engineering Supervision Co., Ltd., Shijiazhuang, China

ABSTRACT: According to traditional manual inspection methods, such as steel ruler and total station, the corresponding quality inspection work of bridges is time-consuming, labor-intensive, and error-prone. The actual measurement work is often affected by various conditions, such as large bridge spans and the height of bridge towers. Using three-dimensional laser scanning technology could quickly obtain point cloud data of cable-stayed bridges. Through internal data processing such as splicing, orientation, noise reduction, resampling, packaging, and slice fitting of the point cloud data, the geometric dimensions of the bridge structure can be realized by the rapid extraction of elevation information. This paper could provide technical support for the acceptance measurement of geometric parameters of newly built bridges and subsequent bridge repairs.

1 INTRODUCTION

The size of bridge components can clearly reflect the combination mode and layout of each component of the bridge, and it is also a key link to establish the finite element model and the BIM model (Yang & Lin 2019). The traditional geometric dimension acquisition method requires field measurement (Qiu 2020). But for complex bridges, the actual measurement is time-consuming, laborious, and dangerous. 3D laser scanning technology has the characteristics of high speed, wide range, high precision, and non-contact (Zhao *et al.* 2014). The point cloud model can be used to obtain various geometric parameters of the bridge through convenient office operation (Wang *et al.* 2019). Liang Dong (Liang et al. 2020) discussed the key technical issues of measuring the overall configuration of large and complex Bridges by using 3D laser scanning technology. Liu (Liu & Zhang 2019) designed a set of data acquisition methods based on MS50 three-dimensional laser scanning total instrument free station technology. Sudip (Subedi et al. 2019) analyzed the displacement change of the bridge in the construction stage. Shen (2012) used three-dimensional Lidar to assess bridge damage. In this paper, the point cloud on the surface of a cable-stayed bridge is obtained by 3D laser scanning, and the point cloud model is processed in the office by software to extract the accurate geometric size information of the bridge. Compared with the conventional measurement methods, 3D laser measurement can achieve a comprehensive grasp of the geometric information of the bridge shape, greatly improve the data accuracy, improve the work efficiency, and reduce the field labor.

2 MEASUREMENT OF BRIDGE DIMENSIONS

2.1 *Collection of point cloud data*

In this study, Leica Scan Station P40 ground 3D laser scanner was used to conduct acceptance measurements of the geometric dimensions of a cable-stayed bridge.

*Corresponding Author: 13672067512@163.com

 DOI: 10.1201/9781003410843-18

In order to unify the point cloud to the geodetic coordinate system, it is necessary to arrange the geodetic coordinate system reference points before scanning, which is the key to ensuring the quality of point cloud data acquisition (Chen & Guo 2016). The bridge is arranged with 10 scanning reference points. Obtain the three-dimensional coordinates of the reference point through GPS measurement. Before scanning the bridge with a three-dimensional laser, the target is erected on the control point. By measuring the vertical height from the control point to the target center, the scanning point cloud can be quickly oriented and unified into the geodetic coordinate system.

In this paper, we use the relationship between the scanning resolution, station spacing, and incidence angle of the Leica Scan-station P40 3D laser scanner obtained in the literature (Liang et al. 2020) to set up scanning stations. For some detailed areas, we can set up encrypted stations or improve the resolution to scan. For example, because the reverse modeling of target objects may require extracting boundaries and other features from point cloud data, there should be enough scanning points in the boundary area of the structure, striving to obtain a high-precision point cloud covering the whole area with the minimum number of stations.

When scanning adjacent stations, it is necessary to keep the target of the previous station fixed. The distance between the target and the scanner should not exceed half the distance between adjacent stations. Before scanning the target at each station, it is necessary to number the target. Two identical targets are set between each two measuring stations so that the point clouds are spliced "end-to-end" to facilitate the subsequent point cloud splicing. After each target scanning is completed, check whether the target scanning is qualified, and rescan the target in time if it is unqualified.

2.2 *Data processing*

At present, common splicing methods include target-based splicing, feature point based on splicing, and control-point-based splicing (Ou 2014). This paper adopts the target-based splicing method. For the point cloud data of two stations, we use Cyclone software to automatically splice two or more targets with the same name in the above point cloud data. The maximum splicing error of the two stations obtained by the above method is 3.2 mm, and the splicing error of the whole bridge is 1.5 mm. Therefore, the point cloud data collected by 3D laser surveying and mapping technology has significant practical value in extracting the geometric dimension information of bridges.

In this paper, we scan the target arranged on 6 reference points and add the reference point coordinates to the target fitting center point. The parameter model (Hu 2020) is used to calculate the rotation matrix and translation matrix, which can convert the point cloud data under the temporary coordinate system to the geodetic coordinate system as a whole. Table 1 shows the accuracy of internal coincidence after conversion based on 6 reference points. We select the GPS coordinates of the stations not used in the absolute orientation and compare

Table 1. Internal accuracy calculation.

	Coordinates after conversion			GPS survey coordinates			Deviation		
No	X'	Y'	Z'	X	Y	Z	ΔX	ΔY	ΔZ
X 001	4490180.926	499967.843	1082.547	4490180.931	499967.840	1082.540	–0.005	0.003	0.007
X 002	4490036.609	500039.392	1090.414	4490036.605	500039.395	1090.404	0.004	–0.003	0.010
X 005	4490110.413	500067.137	1082.078	4490110.413	500067.139	1082.080	0.000	–0.002	–0.002
X 006	4490049.466	499914.602	1068.182	4490049.476	499914.607	1068.189	–0.010	–0.005	–0.007
X 009	4490029.991	499937.250	1069.467	4490029.994	499937.256	1069.460	–0.003	–0.006	0.007
X 010	4490062.885	499946.554	1069.981	4490062.883	499946.561	1069.987	0.002	–0.007	–0.006

them with the coordinates obtained from the conversion. Then the external coincidence accuracy calculation of the point cloud results is shown in Table 2.

Table 2. Point cloud accuracy detection.

	Point cloud coordinates			GPS survey coordinates			Deviation		
No	X'	Y'	Z'	X	Y	Z	ΔX	ΔY	ΔZ
X 003	4490067.935	500063.720	1091.040	4490067.924	500063.725	1091.033	0.011	−0.005	0.007
X 004	4490208.757	500074.361	1087.731	4490208.744	500074.376	1087.751	0.013	−0.015	−0.020
X 007	4490090.080	499931.988	1069.945	4490090.094	499931.975	1069.967	−0.014	0.013	−0.022
X 008	4490164.696	500072.404	1082.499	4490164.682	500072.416	1082.490	0.014	−0.012	0.009

In this paper, we use the point cloud processing software Geomagic Studio to reduce the noise of point cloud data. For different types of noise points, we need to use different methods to process them. For some large point clouds and scattered points that are visible to the naked eye and far away from the main body, we use manual interaction, manual framing, and deletion. For points that are difficult to identify and inconvenient to frame manually. It can be deleted through the functions of removing external isolated points and unconnected items, reducing noise, etc., in the Geomagic Studio point cloud processing software, to achieve the effect of simple and efficient removal of noise points.

Excessive point cloud data reduces processing speed and computing efficiency (Ren 2017). In this paper, the unified sampling method is used to control the spacing of point clouds, sample, and the number of points in the high curvature area through curvature priority. We select and maintain the boundary box and reduce the number of point clouds while maintaining the integrity of the model boundary, as well as the shape is not distorted.

Point cloud encapsulation creates tiny triangles by connecting points to fit NURBS surfaces. It uses point models to form surface models, laying a foundation for the subsequent stage of model preservation and extraction of bridge size information. The point cloud data is encapsulated, as shown in Figure 1.

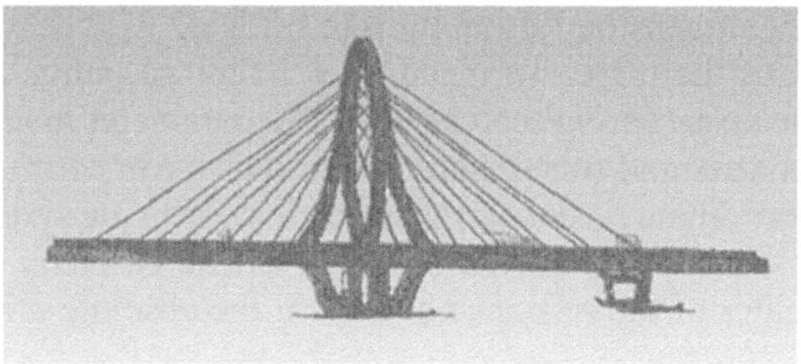

Figure 1. Point cloud package model.

3 MEASUREMENT OF BRIDGE DIMENSIONS

After software point cloud encapsulation, the structural model of the bridge is obtained, and then the verticality of the bridge tower is calculated. In order to ensure the accuracy of fitting straight lines, this paper increases the partition density, as shown in Figure 2a, and the centroid point extraction is shown in Figure 2b. The maximum standard deviation of section fitting is 3.2 mm, and the average standard deviation of all section fitting is 1.9mm. The error is within the allowable range. The direction vector of fitting straight lines through the centroid of the fitting section is 0.0045, 0.0013, 0.9999, the included angle with Z-axis unit

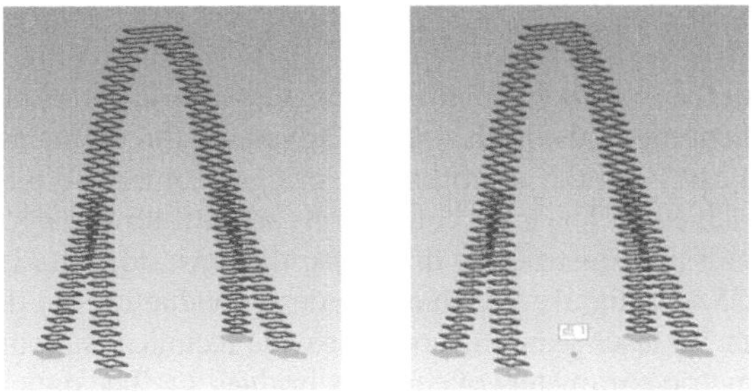

a) Slice Results b) Centroid Extraction

Figure 2. Bridge tower slice fitting.

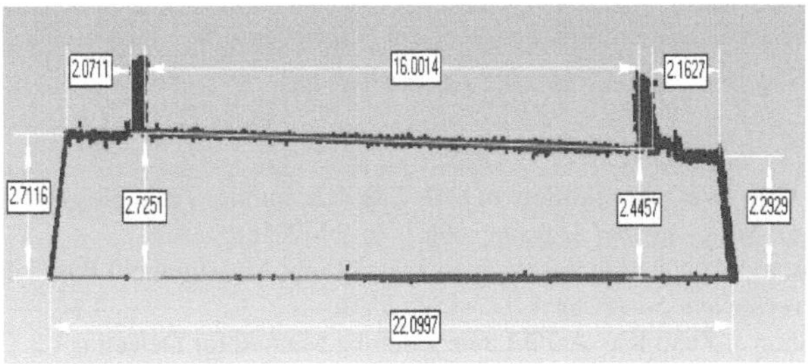

Figure 3. Software measurement.

vector is 0.21°, and the perpendicularity is 3.66 mm, which meet the requirements of specification H/3000 mm.

The designated section of the bridge is intercepted through the Geomagic Studio point cloud processing software through the "Object Section Function," and the section size is measured through the "Size Measurement Tool." The measurement results are shown in Figure 3. Comparing the above-measured section dimensions with the construction drawings, the deviations are less than 10mm, and the cross slope of 1.75% is slightly less than the design value on the drawings by 2%.

To sum up, this paper uses the above two methods and the function of measuring the two-point distance of Geomagic Studio point cloud processing software directly to measure the geometric parameters of different positions of the bridge. The results are shown in Table 3.

Table 3. Measurement results of geometric parameters of the bridge.

Part of bridge	Project	Design value (m)	Point cloud measurements (m)	Deviation (mm)
Main beam	Bridge length	217.00	216.988	27.0
	Bridge width	21.24	21.235	9.0
	Bridge height	2.50	2.530	11
	Cross slope	2.00%	1.75%	0.25%
Main tower	Tower height	38.00	38.008	8.0
	Verticality	0.00	0.004	3.7
	Elevation of the bridge tower top	1136.389	1136.371	18

4 CONCLUSION

It is difficult to obtain the overall geometric parameters of bridges with special structures by traditional measurement methods. In this paper, advanced three-dimensional laser scanning technology is used to carry out the acceptance of bridge geometric dimension measurement, and good results are achieved. The main conclusions are as follows: (1) Comparing measured section dimensions with the construction drawings, the deviations are less than 10 mm, and the cross slope of 1.75% is slightly less than the design value on the drawings by 2%. The obtained method with 3D laser scanning could provide technical support for the acceptance measurement of geometric parameters of complex bridges. (2) The data acquired by 3D laser scanning is rich in quantity and high in accuracy, which has strong applicability in bridge engineering acceptance. (3) The full bridge point cloud data scanned by 3D laser is intuitive and clear, which is convenient for technicians to master the full bridge condition and can provide comprehensive technical support for the later bridge operation and maintenance work.

REFERENCES

Chen Hongquan & Guo Wei. Application of 3D Laser Scanning Technology in Bridge Deformation Monitoring. *Modern Surveying and Mapping*. 39(1), 36–39 (2016).

Hu Yuxiang. Analysis and Application Research on Metro Limit Measurement Based on the 3D Laser Point Cloud. *Urban Surveying and Surveying* 4, 147–152 (2020).

Liang Dong, Zhang Shuo & Zhao Kai. A 3D Laser Scanning Method for Detecting the Overall Configuration of a Pedestrian Bridge. *Highway Traffic Technology*, 37(09), 57–66 (2020).

Liang Dong, Zhang Shuo, Zhao Kai, etc. Three-dimensional Laser Scanning Detection Method for the Overall Configuration of a Pedestrian Bridge. *Highway and Transportation Science and Technology* 37(09), 57–66 (2020).

Liu Yuanxu & Zhang Keqi. Mining Subsidence Monitoring Method Based on 3D Laser Scanning Total Station. *Journal of Qinghai University*. 37(04), 70–76 (2019).

Ou Bin. Research on Field Data Acquisition Method of Terrestrial 3D Laser Scanning Technology. *Surveying and Spatial Geographic Information* 112, 106–108 (2014).

Qiu Jinshun. Planning Acceptance Method and Result Analysis of Municipal Road and Bridge Project Completion. *Geomatics & Spatial Information Technology* 43(9), 193–195 (2020).

Ren Zhiyong. Research on 3D Point Cloud Modeling of Complex Entities Based on Geomagic. *Standardization of Surveying and Mapping*. 33(3), 21–23 (2017).

Shen Enchen, Liu Wanqiu, Bian Haitao, et al. 3D LiDAR Scans for Bridge Damage Evaluations. *Forensic Engineering* 487–495 (2012).

Sudip Subedi, Kalasapudi Vamsi & Pradhananga Nipesh. Spatial Change Tracking of Structural Elements of a Girder Bridge Under Construction Using 3d Point Cloud. *Computing in Civil Engineering* 193–200 (2019).

Wang Fen, Xu Bingqian, Hao Dan. Planning Acceptance Measurement Data Processing and Results Acquisition Under 3D Laser Scanning Technology. *Bulletin of Surveying and Mapping*. 4, 159–161 (2019).

Yang Xubang & Lin Weibin. The Application of Three-dimensional Laser Scanning in Bridge Inspection. *Guangdong Highway Transportation*. 45(5), 139–143 (2019).

Zhao Haiyang, Ge Xiaoping & Du Zhenzhen. Discussion on the Application of Three-Dimensional Laser Scanning Technology in Building Deformation Measurement. *Jiangxi Building Materials* 14, 222–228 (2014).

Constructional Engineering and Ecological Environment – Chih-Huang Weng (Ed)
© 2024 The Author(s), ISBN 978-1-032-53198-4

Study on mechanical property of glass fiber sleeve reinforced bridge pile foundation

Rui Wang & Xinlei Chen*
School of Mechanical Engineering, Tianjin University of Technology and Education, Tianjin, China

Shigang Luo
Carbon Technology Group Co., Ltd, Tianjin, China

Yujun Cai
School of Mechanical Engineering, Tianjin University of Technology and Education, Tianjin, China

Fengning Li & Bo Wu
Carbon Technology Group Co., Ltd, Tianjin, China

ABSTRACT: With the development of steel structure bridges crossing the river and sea, the load-bearing performance of bridge pile foundations is higher. This paper analyzes the characteristics and advantages of glass fiber composite reinforcement for steel pile corrosion. By designing the relevant test schemes, had simulated the axial compression test of steel tubular piles with different corrosion degrees. By observing the experimental phenomenon, the load-displacement curves and load-strain curves, analyzing the mechanical property of samples during the experiment, and comparing the reinforcement effect of glass fiber sleeves on pile foundations with different corrosion degrees.

1 INTRODUCTION

With the improvement of the continuous progress of bridge construction technology, a large number of steel structure trans-river and sea bridges have been greatly developed, and a higher request is also put forward for the bearing performance of bridge pile foundations. However, in the seawater environment, chloride ions in the water will lead to steel corrosion (Zhao 2010), especially the corrosion of steel structure in the splash area is faster, which seriously affects the service life of steel pile foundation and leads to premature failure of pile foundation structure (Wei & Yan 2017).

Currently, the corrosion reinforcement technology of glass fiber composite for steel structure buildings is becoming increasingly mature. It has become the ideal material choice for repairing steel structures due to its lightweight, high strength, and good corrosion resistance (Xiao 2017). In addition, this technology has the advantages of shorting the reinforcement period and large-scale production capacity. In recent years, it has been widely used in the field of restoration and reinforcement of existing buildings.

This paper is to test the bearing capacity of a glass fiber sleeve reinforced pile foundation, simulate two steel pipe piles with different degrees of corrosion reinforced by glass fiber sleeves, and study the mechanical properties of the two steel pipe piles by load-displacement curve and load strain curve.

*Corresponding Author: cxl_0920@126.com

DOI: 10.1201/9781003410843-19

2 VERIFICATION OF REINFORCEMENT PERFORMANCE OF GLASS FIBER SLEEVE

At present, many researchers have studied the axial compressive properties of composite columns such as GFRP tube-concrete-steel tube and CFRP/ FRP-concrete-steel tube (Kiank J. *et al.* 2011; Li *et al* 2022; Qian *et al.* 2008; Tan *et al.* 2017; Wang *et al.* 2008; Zhang *et al.* 2019). Therefore, it is possible to simulate the actual conditions of the site for the axial compression test and study the failure bearing capacity and mechanical properties of samples from the elastic stage, loading to the design load.

2.1 *Experimental material*

The experimental material mainly includes steel pipe pile, glass fiber sleeve, underwater epoxy grouting material, and related accessories. The samples used Φ630 mm*8 mm steel pipe piles of the same material size as those in the engineering.

2.2 *The design of the samples*

Sample 1 simulated a test of the ultimate bearing capacity of the steel pile after glass fiber sleeve reinforcement when the corrosion thinning to 0 mm, and the corrosion thickness of sample 2 is 4 mm. Disconnect two steel pipe piles from the middle position with a distance of 100 mm. Spot weld a steel sheet with a thickness of 0.5 mm in the middle of one steel pipe pile and a steel pipe with a thickness of 4 mm in the middle of the other steel pipe pile. At the same time, the welded temporary support channel steel was at the four corners of the pressing plate above and below the steel pipe pile. The glass fiber sleeve was covered in the middle area. Then the underwater epoxy grouting material was poured. Test sample tooling is shown in Figure 1.

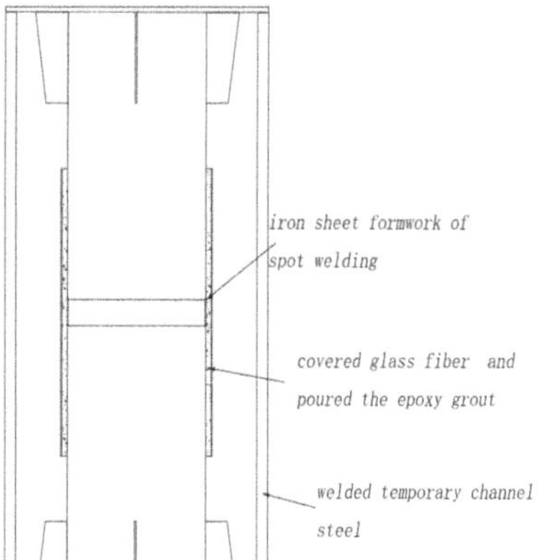

Figure 1. Test sample tooling.

2.3 *Test piece measuring point installation*

In order to measure the strain value of the glass fiber sleeve under static loading, a longitudinal resistive strain gauge is arranged at the measuring points S1 to S3, and a transverse resistive strain gauge S4 is arranged at section S2; In addition, it measures the displacement between the upper and lower steel tubes and the change of the vertical relative displacement between the steel tubes and the glass fiber sleeve during the loading process, 6 displacement gauges(DG) are arranged in the relative positions. The measuring point layout of the sample is shown in Figure 2.

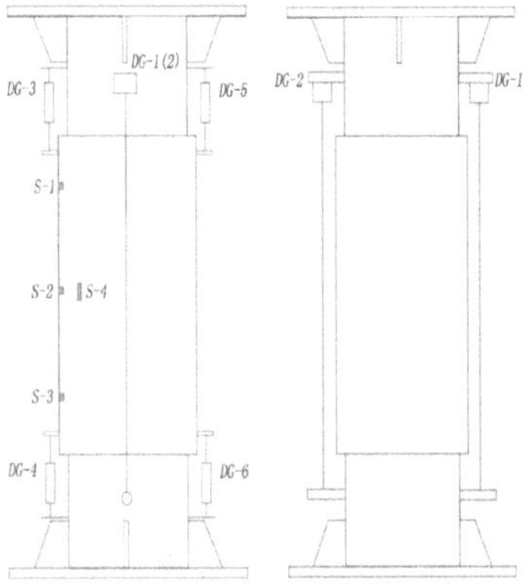

Figure 2. Layouts diagram of measuring points.

3 RESULTS AND DISCUSSION

3.1 *Experimental phenomenon*

When the load reached about 508 kN, the sound of sample 1 appeared. And the sound emitted by sample 1 increased continuously. When the load was loaded to 851.29 kN, the sample made a huge sound, and the load value dropped sharply to 409.28 kN. After the redistribution of internal forces, the bearing capacity began to increase gradually. Only found partial shedding between epoxy mortar and steel pipe, as shown in Figure 3(a). When the load was loaded to 1667 kN, sample 2 made a slight sound, and the load value quickly rose after a small decrease. When the load was loaded to about 1900 kN, the sample made a small breaking sound many times. At the end of the test, there was no obvious damage to the steel tube cylinder and sleeve, as shown in Figure 3(b).

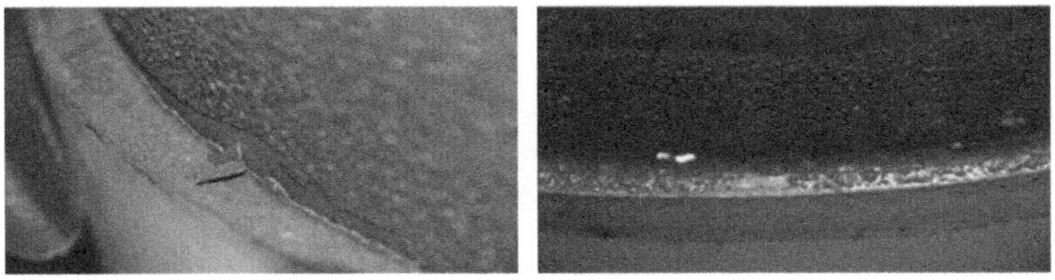

(a) (b)

Figure 3. Failure picture of (a) sample 1 (b) sample 2.

3.2 *The load-displacement curves of the sample*

Figure 4 reflects load-displacement curves in different regions of sample 1. When sample 1 is loaded to the maximum load of 851.29 kN, the relative displacement of upper and lower column segments is the largest, and the maximum displacement value is 0.76 mm, while the relative displacement of other parts has little change, indicating that before reaching the

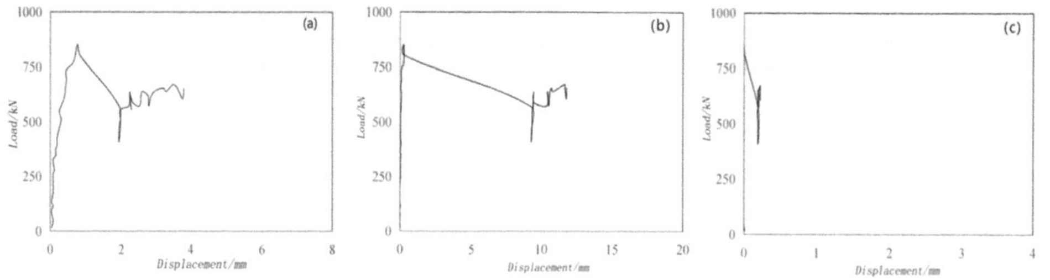

Figure 4. The load-displacement curves of sample 1 (a) Relative displacement of upper and lower steel pipes (b) Relative displacement between the upper steel pipe and the sleeve (c) Relative displacement between the lower steel pipe and the sleeve).

maximum load, The relative displacement of each part is mainly caused by the deformation of the inner structure of the steel pipe. When the sample reached the maximum load, the curve showed a relatively obvious descending process. The load value decreased to 409.28 kN, and the relative displacement between the upper column section and the collar reached the maximum, which increased to 9.27 mm. After the rapid decline of the load curve, there is a short recovery process. When the load recovers to about 640 kN, the load value becomes stable.

Figure 5 reflects load-displacement curves in different regions of sample 2. When sample 2 was loaded to 1701.85 kN, the displacement between the upper column segment and the hoop and the displacement between the lower column segment and the hoop changed slightly. The displacement between the upper and lower column segment increased by 0.8 mm. Still, the load value did not increase or decrease significantly, indicating that the upper and lower column segment had slippage, and sample 2 reached the maximum adhesive failure load. Compared with sample 1, sample 2 has a lower simulated corrosion degree, significantly improved bearing capacity, and smaller deformation.

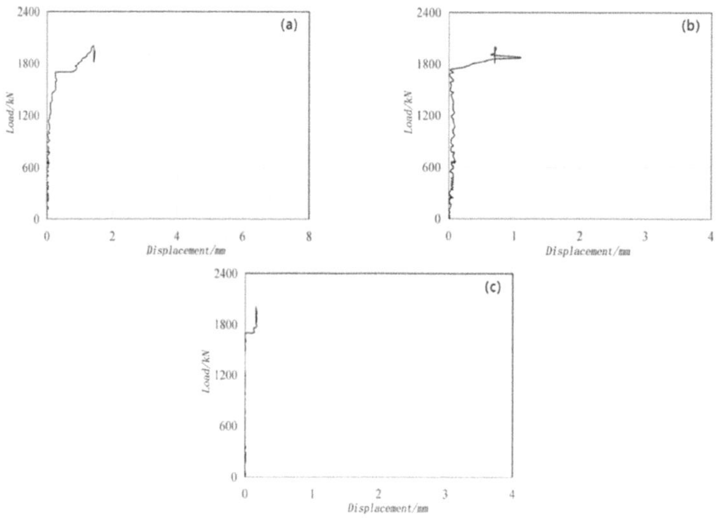

Figure 5. The load-displacement curves of sample 2 (a) Relative displacement of upper and lower steel pipes (b) Relative displacement between the upper steel pipe and the sleeve (c) Relative displacement between the lower steel pipe and the sleeve)

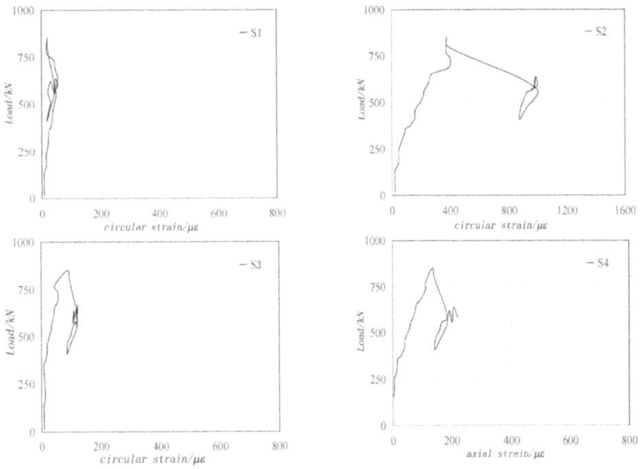

Figure 6. The load-circular strain curves of sample 1 (S1-S3) The load-axial strain curve of sample 1 (S4).

3.3 *The load-strain curves of the sample*

Figure 6 reflects the load-circular strain curves and load-axial strain curves of sample 1. At the maximum bearing capacity of 851.29 kN, the circular strain between the measured heights ranges from 110 με to 380 με, and the axial strain value is 135 με. This indicates that before reaching the peak load, the bond between the steel pipe and the grout material is not damaged, and the circular stress in the hoop is small; When the load reaches the peak value, the circular strain and axial strain of the hoop increase significantly. From figure S1-S4, it can be seen that the strain value (S2) is the largest the closer the area is to the simulated corrosion area, which is because the interface between grouting material and steel pipe is damaged and expands outward, resulting in stress redistribution. At this time, the hoop is mainly subjected to circular tension. The compressive strength of the axial center is significantly increased, and the role of the hoop can be brought into play (Zhang 2008).

Figure 7 reflects the load-circular strain curves and load-axial strain curves of sample 2. Before reaching the maximum adhesive failure load, the load of the hoop increases in a

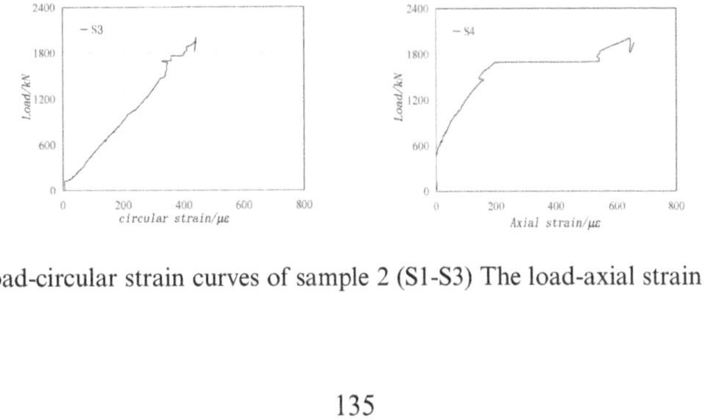

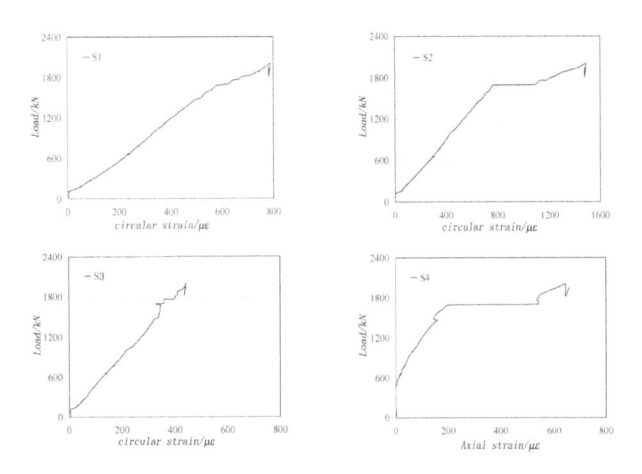

Figure 7. The load-circular strain curves of sample 2 (S1-S3) The load-axial strain curve of sample 2 (S4).

roughly linear relationship with the circular strain and axial strain. When the maximum adhesive failure load is reached, the upper and lower column sections slip, and the annular and axial strains in the zone of S2 increase sharply. After the maximum adhesive failure load, the load of sample 2 still rises. The above shows that sample 2 has a large bearing capacity, and the hoop is involved in the force transfer between the upper and lower steel pipes. With the increase of the load, the restraint effect of the hoop is obvious.

4 CONCLUSIONS

Aiming at the corrosion problem of steel pile foundation, simulated the axial compression tests of steel pipe piles with different corrosion degrees reinforced by glass fiber sleeves, and the following conclusions are drawn:

(1) The maximum bearing capacity of sample 1 reaches 851.29 kN, and the load tends to be stable at 640 kN, so it could well undertake the load.
(2) The simulated corrosion degree of sample 2 is low. After reaching the maximum adhesive failure load of 1701.85 kN, the load of sample 2 still rises, and it can better undertake the load.
(3) The simulated corrosion degree is different. The hoop on sample 1 is subjected to larger circumferential stress, so the bearing capacity is small. The bearing capacity of sample 2 is large because the sleeve is involved in the force transfer between the upper and lower steel pipes.

REFERENCES

Kiank, Michael J. et al.: Testing and modeling of a novel FRP-encased Steel–concrete Composite Column. *Composite Structures*. 93(5),1463–1473(2011).

Li S.B., Pan X.N., Chen Y.N., Qin K.: Experimental Study on Axial Compressive Properties of Concrete-filled Steel Tube Stiffened Mixed Column Restrained by GFRP tube. *Concrete*. (05),31–36(2022).

Qian J.R., Liu M.X. et al.: Experimental Study on Axial Compression of FRP-Concrete-Steel Double-wall Hollow Tube Short Column. *Journal of Building Structures*. 29(2),107–113(2008).

Tan S.Z., Jiao C.J. et al.: RPC Compression Test of CFRP-steel Tube. *Concrete*. (7),146–149(2017).

Wang Y.L., Yu T. et al.: Behavior of FRP-confined Concrete in Annular Section Columns. *Composites: Part B*. 39(3),451–466(2008).

Wei M.H., Yan F.: Application of Glass Fiber Sleeve Reinforcement Technology in High Pile Wharf Maintenance Engineering. Marine Traffic Engineering. (09),202–203+205. (2017)

Xiao Y.H. Research on Strengthening Technology of Underwater Glass Fiber Sleeve. *Urban Roads and Bridges and Flood Control*. (7),146–149(2017).

Zhang C.Y. *Steel-concrete Composite Structure [M]*. China Metrology Press, Beijing. (2008).

Zhang Y.R., Wei Y. et al.: Stress-strain Model of an FRP-confined Concrete-filled Steel Tube Under Axial Compression. *Thin-Walled Structures*. 142(9),149–159(2019).

Zhao S.C. Study on Environmental Action Zoning of Tidal Affected Area Based on Durability of Concrete Structures. *Journal of Highway Communication Technology*. 27(7), 61–64(2010).

Constructional Engineering and Ecological Environment – Chih-Huang Weng (Ed)
© 2024 The Author(s), ISBN 978-1-032-53198-4

Research and analysis of airworthiness of civil and military airport runway pavement and treatment measures

Jixiang Su*
China Airport Construction Group Co., Ltd/ Beijing Super-Creative Technology Co., Ltd, Beijing, China

Gang Yi
Changsha Huanghua International Airport, Changsha, Hunan, China

Fang Lin
Chongqing Jiangbei International Airport Co., Ltd, Chongqing, China

Wei Zhao
Wuhan Tianhe Airport Co., Ltd, Wuhan, Hubei, China

Danze Wu
Northeast Branch Company of China Airport Planning & Design Institute Co., Ltd., Shenyang, Liaoning, China

Shuaituan Tian
China Airport Construction Group Co., Ltd/ Beijing Super-Creative Technology Co., Ltd, Beijing, China

ABSTRACT: Through research and analysis, this paper adopts scientific and reasonable comprehensive evaluation methods to evaluate the status quo and airworthiness of the runway pavement of domestic military-civilian airports and summarizes the typical diseases and targeted airworthiness treatment measures of the runway pavement of military-civilian airports according to the operational requirements of airports and the main airworthiness treatment measures. This research can help to solve airworthiness issues and provide a reference for the safe operation and maintenance construction investment of such airports.

1 INTRODUCTION

1.1 *Domestic civil aviation development*

China's civil airport construction has experienced two main stages of development in the past. The first stage is the slow development stage before reform and opening up, and the second is the rapid development stage after reform and opening up. The number of civil transport airports has increased from 78 in 1978 to 248 in 2021 (Civil Aviation Administration of China 2022), and a national comprehensive airport system has been formed, consisting of four major world-class airport groups, 10 major international air hubs, 29 regional hubs, and non-hub transport airports, and 370 general airports (China Civil Aviation Airport Construction 2022).

1.2 *Introduction of military-civilian airports*

In terms of airport construction, the investment is often huge, and the entire project has a long duration. China has a vast territory with drastically different geographical

*Corresponding Author: 13910977936@163.com

DOI: 10.1201/9781003410843-20

characteristics in different regions. Airport site selection conditions are strict, especially in special regions like highlands, mountainous, and coastal areas. The locations suitable for building airports are scarce resources. Therefore, considering the national conditions, it is of strategic significance and practical benefit to transform existing military airports into military-civilian airports in peacetime. For this reason, China has gradually implemented military-civilian airports since the 1980s. According to statistics, there are more than 60 military airports distributed in various regions of the country, accounting for about 26% of the total number of airports in China (Liu 2018), which is an important part of China's airport system and has played a positive and important role for a long time.

1.3 *Current situation*

With the transformation and optimization of military airports, most of the runway pavements have been in service for 20 or 30 years after they were put into operation in civilian-military airports, and some of them have exceeded their projected lifetime. According to research, the runway pavement of military airports in China was mainly made of hexagonal cement concrete slabs in the early stage (Yang *et al.* 2015). During the transformation from military airports to military-civilian airports, most of them were covered up with an additional layer of cement concrete. Up to now, according to incomplete statistics, the main severe structural damages of the runway pavement of military-civilian airports include corner fractures, broken slabs, voids, and dislocation. The un-airworthiness of the runway pavement has become increasingly prominent, and targeted measures need to be taken to meet its airworthiness.

2 CASE STUDY AND ANALYSIS OF REPRESENTATIVE EXAMPLES

The current military-civilian airports are distributed in various regions of China (Wang *et al.* 2014). The un-airworthiness of the runway pavement put into service is mainly affected by the different geographical and natural climate environments, operating loads, and the structure of the runway pavement itself.

2.1 *Natural climate conditions in different regions*

China has a vast territory, as shown in Figure 1. The geographical and natural climate environments of the regions where airports are located vary greatly. As shown in Table 1, under the influence of temperature, rain and snow, and other external environments, the runway pavement presents different forms of damage.

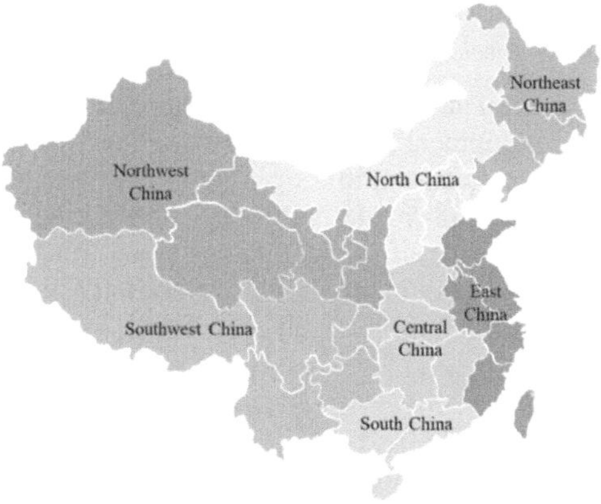

Figure 1. National regional distribution map.

Table 1. Summary of climatic characteristics of different regions in China.

Region	Climatic characteristics
Northeast	The annual temperature difference is large. Summer is short and warm, with an average monthly temperature over 10°C, and the general precipitation is 300 ~ 600mm. Winter is long and cold, with an average temperature of -20°C and frequent snowfall.
North	The annual temperature is from 5 to 20°C. Summer is hot and rainy, and the annual precipitation is from 400 to 1000mm. Winter is cold and dry, with the average temperature below 0°C and snowfall.
Northwest	The daily and annual temperature difference is large, and it's dry throughout the year. The annual precipitation is less than 200mm. The summer is warm and rainy, with an average temperature of 16 to 24°C. Winter is cold and dry as the temperature is below -10°C most of the time and has snowfall.
East	The average annual temperature is from 15 to 18°C, and the annual rainfall is abundant. The annual precipitation is about 1000mm. Summer is hot and rainy, and winter is wet and cold, with a temperature above 0°C most of the time.
Central	The average annual temperature is from 14 to 21°C, and the temperature can get as high as 35 to 40°C. The annual precipitation is from 800 to 2000mm. Summer is hot and rainy, and winter is dry and cold, with a temperature above 0°C most of the time.
South	The temperature difference between day and night is small, the average annual temperature is more than 10°C, and the annual precipitation is from 1400 to 2000mm. Summer is long, and winter is short. Summer is hot and rainy, and winter is warm and dry.
Southwest	The average annual temperature is more than 15°C, and the annual precipitation is mostly about 1000mm. Summer is warm and humid, and winter is warm and dry.

2.2 *Airport basic information*

The structure of the runway pavement, the average annual growth rate of aircraft takeoffs and landings in the past five years, and the main types of aircraft operating at each military-civilian airport in this study are shown in Table 2.

Table 2. Summary of basic conditions at each airport.

Airport	The main structure of the runway pavement	The average annual growth rate of aircraft takeoffs and landings in the past five years	Aircraft type
DB-1	15 cm cement concrete + 20 cm cement concrete + 18 cm block stone + 15 cm cement concrete + 20 cm block stone	27.71%	B737–800
HB-1	27~25 cm cement concrete + 19cm cement concrete + 15 cm cobble stone + 20 cm lime soil + subgrade	11.47%	A320, B737–800
XB-1	25 cm cement concrete + 30 cm sand gravel + subgrade	/	/
HD-1	25 cm cement concrete + 20 cm cement concrete + 22 cm sand gravel base + subgrade	11.40%	A320, B737–800
HZ-1	24 cm cement concrete + 50 cm rubble sand gravel base + subgrade	14.87%	A319, B737–700
HN-1	25 cm cement concrete + 28 cm cement concrete + 18 cm cement concrete + 16 cm sand gravel cushion	23.65%	A320, B737–800
XN-1	25 cm cement concrete + 25 cm cement concrete +20 cm sand gravel base + subgrade	40.73%	A320, A321, B737–800

2.3 *A comprehensive evaluation of runway pavement condition*

In order to scientifically understand the condition of the runway pavement at each of the military-civilian airports studied (Civil Aviation Administration of China 2022) and to comprehensively evaluate the pavement using modern non-destructive testing methods (Civil Aviation Administration of China 2019), the main indicators of the runway pavement condition are summarized in Table 3, typical runway pavement damages are shown in Figure 2 (Federal Aviation Administration 2022), and the results of non-destructive testing are shown in Figures 3 and 4.

Analysis of the research data shows that the annual increase in aircraft takeoffs and landings at each military airport is more than 10%, with some exceeding 20%, and the frequency of load effects is continuously increasing. The voids in the structure of the runway pavement are mostly slight but show a gradually worsening trend from slight to medium.

Table 3. Summary of main indexes of the comprehensive evaluation of runway pavement.

Number	Airport	PCI	Level of voids	PCN	Typical damages
1	Northeast China	80	Slight to medium	55 R/B/W/T	Broken slabs, corner fracture, and freeze-thawing
2	North China	90	None	50 R/B/W/T	Corner flaking and edge flaking
3	Northwest China	56	Slight	25 R/B/W/T	Spalling, broken slabs, corner fracture, joint dislocation, and freeze-thawing
4	East China	90	Slight to medium	46 R/B/W/T	Broken slabs, corner fractures, and cracks
5	Central China	83	Slight	31 R/B/W/T	Broken slabs, corner fractures, and cracks
6	Southwest China	90	Slight to medium	52 R/B/W/T	Broken slabs, corner fractures, and cracks
7	South China	84	Medium	49 R/B/W/T	Broken slabs, corner fractures, and cracks

Figure 2. Typical damages of the runway pavement.

Although the overall PCI condition of the pavement is good, various damages have occurred. Nearly 50% of the airports have PCN values that do not meet the requirements and need targeted airworthiness modification due to the adjustment of aircraft types in operation. The typical types of damages, runway pavement voids condition, and radar detection of pavement structure are shown in Figures 2–4.

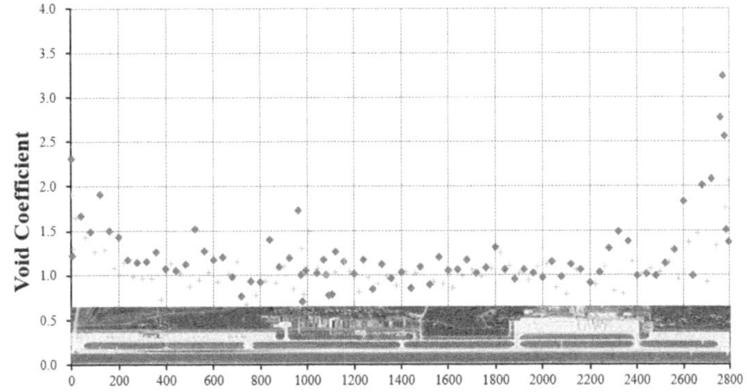

Figure 3. Distribution of voids in runway pavement.

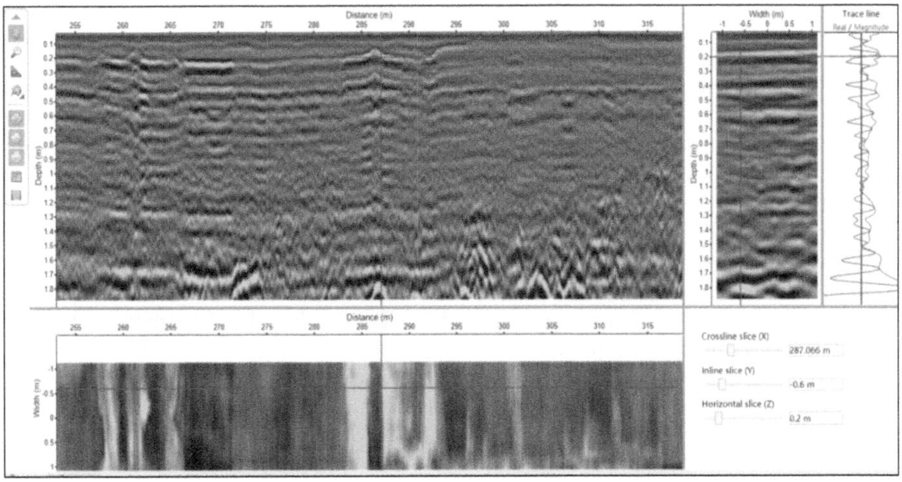

Figure 4. Radar detection of the runway pavement structure.

2.4 *Runway pavement airworthiness issues*

Based on the comprehensive analysis of the research, the runway of a military-civilian airport is typically made of cement concrete covered with cement concrete, as shown in Figure 5. However, the overall pavement structure lacks the strength and rigidity to withstand the adverse effects of the rapid increase in aviation business volume. On the one hand, this is reflected in the fact that airports mainly operate A320 and A738 aircraft. As aircraft takeoffs and landings increase, the runway pavement presents a situation of voids and increased pavement damage. On the other hand, in northern regions, runway pavement is affected by temperature changes, rains, and snow. Thus the typical freeze-thawing would occur at the

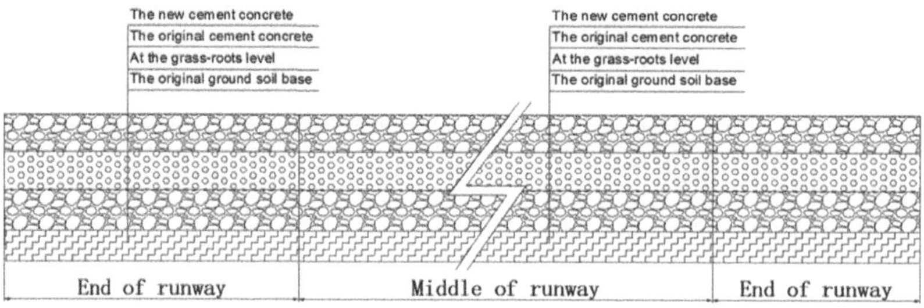

Figure 5. Typical runway pavement structure.

edges and corners of the pavement. In addition, structural damages such as cracks, corner fractures, and broken slabs would appear frequently. Under the dual effects of frequent rain and load in southern regions, the runway pavement would present typical structural damages such as cracks, corner fractures, and broken slabs. Therefore, the existing runway pavement has continuously exhibited serious structural deterioration under the combined effects of internal and external factors. The airworthiness of the runway pavement is the main problem that we need to focus on and urgently address at present and in the future.

3 RESEARCH ON AIRWORTHINESS TREATMENT MEASURES

Currently, there are two main ways to repair the airworthiness of runway pavement structure, including rapid repair with cement concrete and pavement repair with asphalt concrete. According to the severity of the damage, cement concrete repairs can be classified into partial, shallow, half-thickness, full-thickness replacement, and overall covering. Asphalt concrete can be classified into partial repair and overall pavement covering.

This article mainly concerns the serious structural damages of the runway pavement of military-civilian airports and adopts appropriate treatment measures (Su 2020). The differences between the two are analyzed as follows.

Rapid repair with cement concrete can be used generally for cement concrete pavement that is deeply damaged by freeze-thawing and has cracks that do not penetrate the entire panel. After a certain depth of milling, rapid repair with cement concrete can be carried out. As for treating serious structural damages such as corner fractures and broken slabs, a rapid repair can be used after breaking the whole slab.

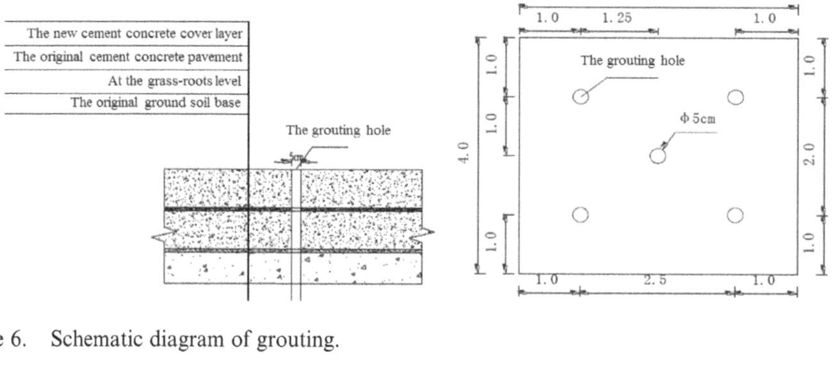

Figure 6. Schematic diagram of grouting.

Currently, repairs and covering with asphalt concrete are common in various civilian airports, but they are not suitable for military-civilian airports. The main reason is that the ground clearance of most mainstream civilian aircraft engines is large, while that of fighter aircraft engines is very small. During operation, the asphalt pavement becomes prone to shedding loss and particle loss with the service time, and the pavement particles can be easily sucked into the aircraft's engines, posing a significant threat to the safety of military fighter aircraft.

In addition, as the runways of military-civilian airports are mostly made of cement concrete covered with cement concrete, the interlayer structure becomes less dense over time due to the combined effects of the environment and load, leading to voids and structural damages. Therefore, grouting is often carried out to stabilize the foundation and non-dense interlayer structure before treating the damages effectively (Liu 2013).

The most appropriate airworthiness treatment process will be decided based on the type and severity of pavement damages, airport operating needs, and available funding. The decision-making process is shown in Figure 7.

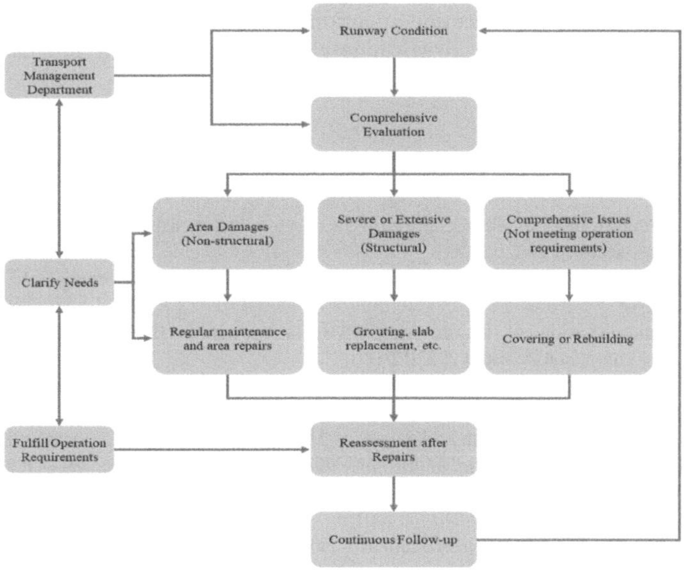

Figure 7. Treatment process.

Table 4. Summary of runway pavement treatments.

Typical damages or not meeting operation needs	Treatment measures	Investment	Suitable for	Investment level (In thousands of RMB)
Freeze thawing and cracks	Area repair or repair with a certain thickness	18000-20000 (RMB/m^3)	Regular maintenance	Millions
Corner fractures and broken slabs	Full-thickness repair and slab replacement	27000-29000 (RMB/m^3)	Security assurance	Millions to tens of millions
Voids	Grouting	80-100 (RMB/m^2)	Security assurance	Millions to tens of millions
Not meeting operation needs	Covering or reconstruction	800-1000 (RMB/m^3)	Upgrade	Billions to tens of billions

4 CONCLUSIONS AND RECOMMENDATIONS

Based on the research and evaluation of the status of cement concrete runway pavement at the military-civilian airports in different areas, this article summarizes the following airworthiness treatment conclusions according to the results of the comprehensive evaluation of the status and the actual needs of airport operation. For areas with non-structural damages, routine maintenance and quick repairs in small areas can be adopted, with a small investment scale to meet daily maintenance needs. For large-scale and severe structural damages, grouting and quick replacement of the pavement slabs can be adopted, with a moderate investment scale to meet operational safety needs. For pavement damages and pavement structures not meeting the airport operation needs, reconstruction of the pavement or pavement covering can be adopted with a large investment scale to enhance the airport's operational security capability.

With the research of the current situation and the comparison of treatment measures, this article can help secure and support the operation and management of the runways of civil-military airports by providing reference.

According to the research and summary of this article, it is recommended that military-civilian airports pay close and scientific attention to the condition of the runway pavement, collect basic data regarding the airworthiness of the runway, and provide data as the decision-making basis for the management department.

REFERENCES

Civil Aviation Administration of China, https://www.mot.gov.cn/tongjishuju/minhang/202204/t20220408_3649981.html, last accessed 2022/12/18.

China Civil Aviation Airport Construction, https://mp.weixin.qq.com/s/RT335inhqyEr4tWedkqwlA, last accessed 2022/12/18.

Civil Aviation Administration of China, http://www.caac.gov.cn/XXGK/XXGK/BZGF/HYBZ/202112/t20211201_210343.html, last accessed 2022/12/18.

Civil Aviation Administration of China: *Specifications for Pavement Evaluation and Management of Civil Airports*. China Civil Aviation Publishing House, Beijing (2019).

Hui, Yang., Jiawei Liu., Zhongling Zhang.: Analysis of Military-civilian Airport Concrete Pavement Surface Characteristics of Consistency. *Journal of Air Force Engineering University*. (5), 39–42 (2015).

Jianbo, Liu.: Study on the Technology of Grouting to Deal with Serious Dislocation of Airports' Cement Concrete Pavements rd. *Trans. Road Traffic*. (2013).

Jianchen, Liu.: Analysis of the Existing Prominent Contradictions of Civil-military Airports. *Science and Technology Research*. (2018).

Jixiang, Su.: Research and Analysis of Civil Airport Cement Concrete Pavement Damage Condition Evaluation and Countermeasure. *IOP Conference Series*. (2020).

Naiyong, Wang.: Discussion on the Layout of Civil-military Airports. *Airport Engineering* (2)., 2–9 (2014).

Ecological environment repair and green development

Constructional Engineering and Ecological Environment – Chih-Huang Weng (Ed)

Does the low-carbon pilot policy promote the building carbon emission reduction: An empirical study based on DID model

Xianghua Wu & Yujiao Li*
School of Economics and Management, Nanjing Tech University, Nanjing, China

ABSTRACT: In order to explore how to promote carbon emission reduction in the construction field to help achieve the "dual carbon" goal, the paper takes the low-carbon pilot policy in China as a quasi-natural experiment to investigate the impact of this policy on building carbon emissions. The empirical results show that the low-carbon pilot policy has a significant inhibitory effect on building carbon emissions, and the policy effect is time-sensitive. The results are still valid after a series of robustness tests. The pilot provinces mainly achieve carbon emission reduction by reducing building energy consumption, while the effect of adjusting the energy structure is not obvious.

1 INTRODUCTION

Global greenhouse gas emissions and climate change are increasing, and low-carbon development has become the consensus of the international community. The development route of low-carbon production aims to reduce greenhouse gas emissions and build a production system based on low energy consumption and low pollution, including a low-carbon energy system, low-carbon technology and a low-carbon industry system. Low-carbon production involves industrial production, agriculture, construction and other industries (Dong & Li 2020). Among them, the huge carbon emissions of the construction industry need more attention. According to the 2021 Global Status Report for Buildings and Construction published by the UN Environment Programme (UNEP), 37% of total carbon emissions were produced by the construction industry. The China Building Energy Consumption Research Report (2021) also shows that the total carbon emissions of the whole building process in 2019 were 4.997 billion tons, accounting for 50.6% of the national carbon emissions. The construction field is an important source of global carbon emissions, which even produced more than half of the total carbon emissions in China. Focusing on carbon emission reduction in the construction field is an urgent task for all countries to explore a green and low-carbon development path.

As a large energy consumer, China has been shouldering the responsibility to actively explore a low-carbon development path to ensure the sustained and healthy development of the economy and society. In 2010, China began to implement the low carbon pilot policy and selected five provinces, including Liaoning, Hubei, Ningxia, Shaanxi and Guangdong and eight other cities, as the first pilot areas. Since then, the second and third batches were launched in 2012 and 2017, respectively. As one of the key aspects of the low-carbon pilot policy, low-carbon buildings are involved in specific work plans issued by the pilot area (Liu et al. 2013). The low-carbon pilot policy is a comprehensive environmental regulation measure. Evaluating the effect of the policy on building carbon emission reduction and

*Corresponding Author: 616827895@qq.com

DOI: 10.1201/9781003410843-21

reflecting on it are of great significance for future energy conservation and emission reduction work (Fan & Liu 2022). Therefore, this paper regards the low-carbon pilot policy as a quasi-natural experiment and uses the difference-in-difference method to scientifically evaluate the effect of this policy on building carbon emissions.

2 THEORETICAL ANALYSIS AND RESEARCH HYPOTHESES

While other countries attach great importance to green and low-carbon development, China is actively exploring ways to achieve low-carbon development, demonstrating its responsibility as a major country. In 2010, implementing the low-carbon pilot policy received extensive attention from the whole society. First, at the level of the central government, low-carbon pilot areas were established by the National Development and Reform Commission (NDRC). The central government attached great importance to the implementation of policies. While monitoring the progress of the policy, it publicized and promoted the low-carbon concept and strived for the understanding and support of all sectors of society. Secondly, all pilot areas immediately responded to the call to prepare for the low-carbon pilot policy. Combining theoretical research and actual situation analysis, they successively issued implementation plans for low-carbon work, established low-carbon goals and tasks of various departments, and actively carried out low-carbon action. Finally, at the level of enterprises and social residents, as the government advocated the concept of low-carbon, they began to integrate low-carbon into production and life. Theoretically, the above support from all sectors of society for policy implementation is strong support for the smooth implementation of low-carbon pilot work. In addition, the implementation plans of all the pilot areas mentioned the construction of low-carbon buildings. Therefore, it is believed that a low-carbon pilot policy can produce certain policy effects and restrain carbon emissions in the construction field.

The relevant literature is analyzed to establish the hypotheses further. Cao *et al.* found through empirical research that the low-carbon pilot policy significantly reduced the carbon emissions of residents through domestic electricity, transportation and heating (Cao & Gao 2021). In addition, considering that the low-carbon pilot policy is a comprehensive environmental policy to promote low-carbon development in many fields, including energy, transportation, construction, production, consumption factors and so on, it is believed that the process of promoting the overall low-carbon level of the city also includes the carbon emission reduction in the construction field. Zhang (2020), Zhou *et al.* (2019), and Dai *et al.* (2015) took different batches of pilot areas as research objects, demonstrating that low-carbon pilot policy has achieved significant carbon emission reduction results, and the dynamic effect is time sensitive.

Based on the above analysis, this paper proposes hypothesis 1.

H1a: Low-carbon pilot policy can promote building carbon emission reduction.

H1b: Low-carbon pilot policy has timeliness in promoting building carbon emission reduction.

Referring to the research of Grossman *et al.* (1995) on how economic development affects the environment, this paper focuses on the intermediary mechanism of scale effect and structure effect in building carbon emission reduction.

Scale effect refers to the need to invest more resources and produce more pollutants in the process of economic development. In the field of construction, which means the continuous investment of energy for buildings will produce a lot of carbon emissions and pollute the environment. In China, resource conservation has always been the core of the energy consumption revolution and reducing building energy consumption is the most direct measure to reduce building carbon emissions. In the process of implementing the low-carbon pilot policy, all pilot areas have set carbon emission reduction targets to save energy use, which is very effective in reducing carbon emissions.

China has always been a coal-based energy structure, and the carbon emission level of coal is much higher than that of clean energy. Reducing the proportion of coal energy consumption and utilizing clean energy is one of the measures to reduce carbon emissions in China. Low-carbon pilot areas plan to reduce the use of high pollution and limited fossil energy and promote the development of new energy technologies to promote building carbon emission reduction.

Based on the above analysis, this paper proposes hypothesis 2.

H2a: Low-carbon pilot policy can promote building carbon emission reduction by saving energy and reducing the scale effect of building energy consumption.

H2b: Low-carbon pilot policy can promote building carbon emission reduction by improving the energy structure and increasing the proportion of clean energy.

3 MODEL SETTING AND DATA DESCRIPTION

3.1 *Model setting*

The difference-in-difference method is a mature method in the field of policy evaluation, which has been widely used in recent years (Li *et al.* 2019). In this paper, the low-carbon pilot policy is regarded as a quasi-natural experiment to assess its impact on building carbon emissions, and the basic equation is designed as follows:

$$EB_{it} = \alpha_0 + \alpha_1 T_{it} \times P_{it} + \sum \alpha_j Control_{jit} + u_i + v_t + \varepsilon_{it} \tag{1}$$

In this formula, subscripts i and t represent provinces and years, respectively. EB_{it} refers to the explained variable. $T_{it} \times P_{it}$ is the double difference estimator to reflect the policy variable. $Control_{jit}$ is the control variable. u_i and v_t represent the fixed effect of province and time, respectively. ε_{it} is disturbance.

3.2 *Variables*

The following is an introduction to the variables involved in this paper.

Explained Variable. This paper calculates the carbon dioxide emissions generated by various types of energy consumed by the construction industry to express the explained variables. With reference to the China Building Energy Consumption Research Report (2018), the specific calculation formula is as follows:

$$EB = \sum e_i f_i \tag{2}$$

In this formula, EB is the building's carbon emissions. e_i is the energy consumption of energy i. f_i is the carbon emission factor of energy i.

As the population base of different provinces varies greatly, in order to better evaluate the effect of policy on building carbon emission reduction, this paper uses two indicators, total building carbon emissions (EB) and per capita building carbon emissions (PEB), as explained variables for research and analysis.

Intermediary Variables. According to the hypotheses set before, the intermediary variables in this paper include building energy consumption and energy structure.

The calculation of total building energy consumption refers to China Building Energy Consumption Research Report 2016, based on the energy balance table, and the specific formula is as follows:

$$BEC = E_l - E_t + E_w + E_o \tag{3}$$

In this formula, BEC is building energy consumption. E_l is the basic energy consumption of “the wholesale and retail industry and accommodation and catering industry”, “others”, and "living consumption". E_t is the traffic energy consumption to be deducted. E_g is the correction of heating energy consumption. E_o is building energy consumption in other sectors.

Energy structure (ES) refers to the proportion of different types of energy consumption in total energy consumption. The carbon emissions generated by fossil energy far exceed that of clean energy. In building energy consumption, fossil energy is mainly coal. The reduction of the proportion of coal energy consumption in China is a direct reflection of the optimization of energy structure. Therefore, the proportion of coal energy consumption in building energy consumption is taken as an indicator to measure the building energy structure.

Control Variables. Combined with previous literature research and the IPAT model, this paper selects urbanization rate (UR) to measure population size, per capita GDP (PGDP) and household consumption level (EX) to characterize the socio-economic situation, and technical level is expressed by energy consumption intensity (EI) and comprehensive carbon emission factor (ICEF).

In this paper, the sample selection considers the availability of building carbon emissions data and the particularity of municipalities directly under the Central Government and ensures that the research results are net policy effects. The final sample excludes Beijing, Shanghai, Tianjin, Chongqing (municipalities directly under the Central Government), Hainan Province (a new pilot province in 2012), and Tibet Province (no data). It selects the remaining 25 provinces for research. The first batch of low-carbon pilot provinces, namely Liaoning, Hubei, Shaanxi, Guangdong and Yunnan, was set as the experimental group, and the other 20 provinces were set as the control group, and 2010 was set as the cut-off point for policy implementation. The research period was from 2006 to 2016, and the panel data with a sample size of 275 was finally composed. The data in this paper are mainly from China Statistical Yearbook China Energy Statistical Yearbook.

The descriptive statistics of each variable are shown in Table 1.

Table 1. Descriptive statistics of variables [self-drawn].

Variables	Observations	Mean	S.D.	Min	Max
EB	275	8.303	0.729	6.332	9.665
PEB	275	1.146	0.334	0.237	4.228
BEC	275	7.524	0.671	5.718	8.919
ES	275	0.219	0.153	0.001	0.691
UR	275	0.491	0.0940	0.275	0.692
PGDP	275	10.30	0.538	8.663	11.48
EX	275	5.272	0.410	4.605	6.069
EI	275	0.171	0.0930	0.0520	0.593
ICEF	275	2.224	0.432	1.019	3.140

4 RESULT ANALYSIS

4.1 *Baseline regression and dynamic effect analysis*

Baseline regression. The average effect results of the low-carbon pilot policy on building carbon emissions are shown in Table 2. All the regression models in this paper control the province-fixed effect and time-fixed effect. Columns (1) and (3) do not add control variables, while columns (2) and (4) are the regression results after the addition of control variables. Column (3) shows that the per capita building carbon emission coefficient is significantly negative. Focusing on columns (2) and (4), after controlling the control variables, DID coefficient is significantly negative at the level of 1%, indicating that the low-carbon pilot policy has a significant inhibitory effect on the total building carbon emissions and per capita

Table 2. Result of baseline regression [self-drawn].

	EB		PEB	
Variables	(1)	(2)	(3)	(4)
DID	–0.0366	–0.0585***	–0.0891*	–0.117***
	(0.0316)	(0.0155)	(0.0505)	(0.0435)
UR		–0.271		–5.399***
		(0.452)		(1.178)
PGDP		0.612***		0.929***
		(0.0935)		(0.322)
EX		0.113		0.832***
		(0.0866)		(0.256)
EI		2.360***		3.046***
		(0.189)		(0.641)
ICEF		0.442***		0.475***
		(0.0468)		(0.0753)
Constant	7.756***	–0.297	1.146***	–11.81***
	(0.0403)	(0.931)	(0.0738)	(3.514)
Province fix	Y	Y	Y	Y
Year fix	Y	Y	Y	Y
N	275	275	275	275
R^2	0.981	0.993	0.908	0.936

carbon emissions in the pilot area, that is, the policy significantly promotes the building carbon emission reduction. Therefore, H1a is valid.

Dynamic effect analysis. The parallel trend test is the premise of DID regression. Referring to Ma *et al.* (2021), taking 2006 as the base period, the first 3 years to the last 6 years of the policy were taken into the explanatory variables as dummy variables for regression, and the dynamic effect trend chart under the 95% confidence interval shown in Figures 1 and 2 was drawn. As shown, before 2010, there was no significant difference in the changing trend between the experimental group and the control group (confidence interval included 0), which

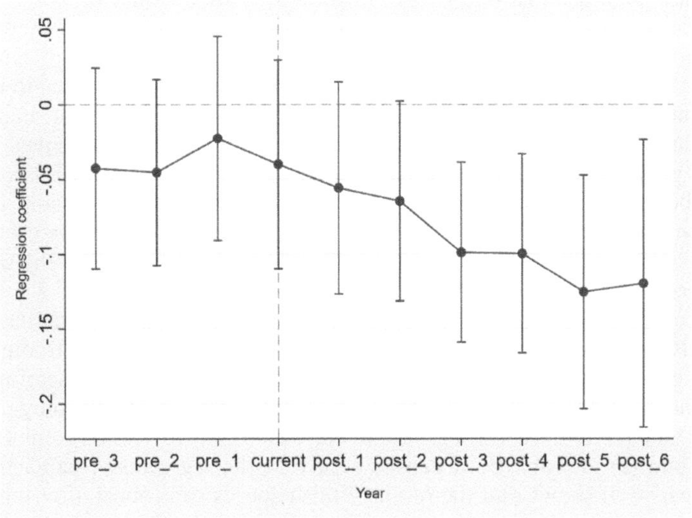

Figure 1. Parallel trend test diagram of EB [self-drawn].

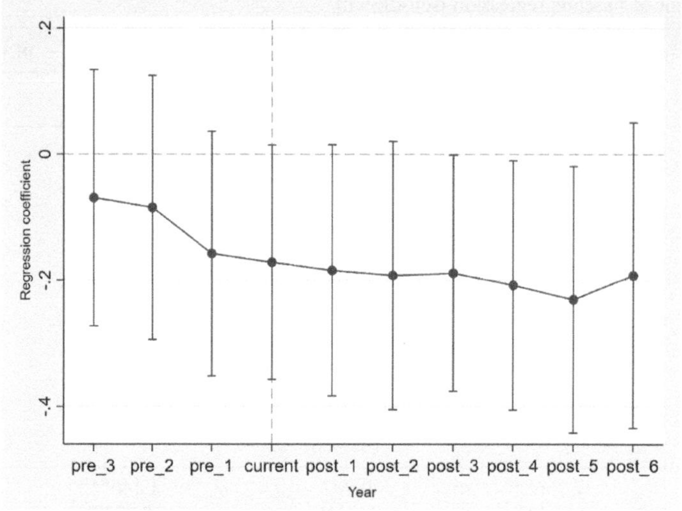

Figure 2. Parallel trend test diagram of PEB [self-drawn].

met the parallel trend. By observing the dynamic effect of the policy, the effect of building carbon emission reduction has been significant since 2013, indicating that the policy has a certain lag. The regression coefficient basically decreased after the implementation of the policy. That is, the effect of carbon emission reduction became more and more obvious, but it began to rise slightly in 2016. Therefore, the building carbon emission reduction effect of low-carbon pilot policies has a dynamic influence, which means H1b is valid.

4.2 *Robustness test*

Two explained variables, total building carbon emissions and per capita building carbon emissions, were chosen above to analyze the effect of policy carbon reduction, which to some extent reflects the stability of the conclusions of this paper. To further ensure the reliability of the research and analysis, the following robustness tests are conducted.

Placebo test. In order to test whether the above results were affected by other random factors or variable omissions, the paper generated an experimental group randomly for a placebo test. Of the 275 samples, 55 were randomly selected to form the pseudo-experimental group for spurious regression. In order to increase the effectiveness of the experiment, the above random process was repeated 500 times. Figures 3 and 4 depict the pseudo-experimental group coefficient distribution of total building carbon emissions and per capita building carbon emissions, respectively. The distribution of the regression coefficient is mostly concentrated around 0, and the p-value is mostly higher than 0.1. In other words, the model setting in this paper did not miss important variables or be affected by other random factors, so the above regression results were robust.

Eliminate other policy distractions. The launch of the first carbon trading platform in Shenzhen in 2013 heralded the official implementation of the pilot carbon trading rights. The second batch of low-carbon pilot areas was unveiled at the end of 2012. Therefore, this paper selects the time sample from 2006 to 2012 and adds Hainan Province, the only pilot province in the second batch, into the control group for regression. The regression results are shown in columns (1) and (2) of Table 3. It can be seen that the regression coefficient is still significantly negative. It shows that the result of this paper is still robust after removing other policy interference.

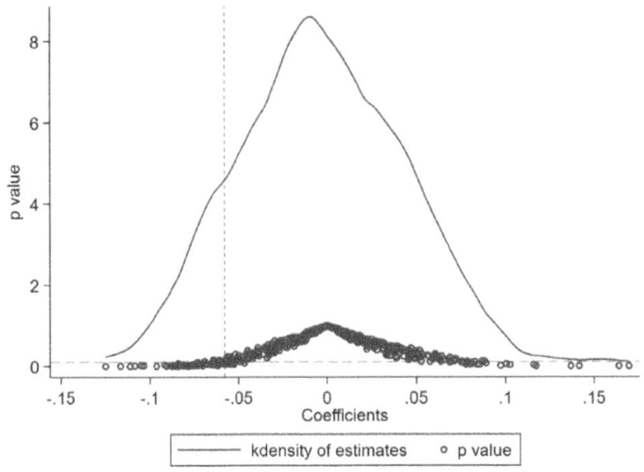

Figure 3. Placebo test of EB [self-drawn].

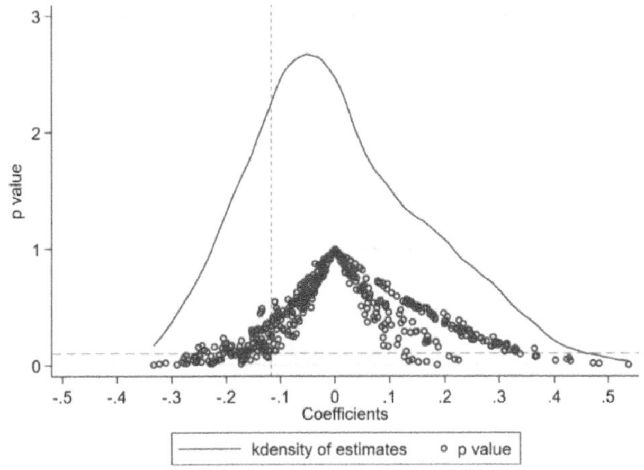

Figure 4. Placebo test of PEB [self-drawn].

Table 3. Result of Robustness test [self-drawn].

Variables	(1) EB	(2) PEB
DID	–0.0247*	–0.107**
	(0.0145)	(0.0528)
Constant	–2.427**	–16.15***
	(1.145)	(5.856)
Control Variables	Y	Y
Province fix	Y	Y
Year fix	Y	Y
N	182	182
R^2	0.997	0.928

4.3 *Mediation mechanism analysis*

For further analysis, this paper referred to Wen *et al.* (2014) to build the mediation model, and the specific equation is as follows:

$$EB_{it} = c_0 + c_1 T_{it} \times P_{it} + \sum c_j Control_{jit} + u_i + v_t + \varepsilon_{it} \quad (4)$$

$$M_{it} = a_0 + a_1 T_{it} \times P_{it} + \sum a_j Control_{jit} + u_i + v_t + \varepsilon_{it} \quad (5)$$

$$EB_{it} = c'_0 + c'_1 T_{it} \times P_{it} + bM_{it} + \sum c'_j Control_{jit} + u_i + v_t + \varepsilon_{it} \quad (6)$$

In order to verify H2a and H2b, building energy consumption and energy structure are taken as intermediary variables to be included in the mediation model. The results are shown in Table 4.

Table 4. Result of mediation model [self-drawn].

Variables	(1) PEB	(2) BEC	(3) PEB	(4) ES	(5) PEB
DID	–0.117***	–0.0421***	–0.0236	0.00135	–0.120***
	(0.0435)	(0.0137)	(0.0345)	(0.0136)	(0.0443)
BEC			2.213***		
ES			(0.223)		1.992***
					(0.285)
Constant	–11.81***	0.0460	–11.91***	–0.471	–10.87***
	(3.514)	(0.859)	(2.112)	(0.695)	(2.690)
Control Variables	Y	Y	Y	Y	Y
Province fix	Y	Y	Y	Y	Y
Year fix	Y	Y	Y	Y	Y
N	275	275	275	275	275
R^2	0.936	0.993	0.972	0.899	0.958
Sobel	–	–0.0932***(z=-2.171)		–0.00269 (z=0.0809)	
Goodman-1	–	–0.0932***(z=-2.167)		–0.00269 (z=0.0806)	
Goodman-2	–	–0.0932***(z=-2.175)		–0.00269 (z=0.0813)	
The proportion of total effect that is mediated	–	0.7977		–0.023	

Column (1) in Table 4 is the test result of the first step. The policy effect is significantly negative, which can be further analyzed. In column (2), the regression coefficient is significant at 1%, indicating that the policy significantly reduces building energy consumption. Meanwhile, in column (3), the regression coefficient of building energy consumption is significantly positive, while the coefficient of DID is not significant. The results of the three-step model show that there is a complete mediation effect, so H2a is valid. Then, from the perspective of energy structure, the regression coefficient of column (4) is not significant, so it is necessary to examine the Sobel model test results. In order to further verify the robustness of the results of the intermediary variables of building energy consumption, a Sobel test is also conducted on it. According to the results, it can be seen that the mediating variable of building energy consumption is significant, while the energy structure variable is not. First, it is verified again that the mediating effect exists in the building energy consumption variable, and the mediating effect accounts for 79.77%. After testing, there is no mediating effect on energy structure. The low-carbon pilot policy does not inhibit building carbon emissions by improving energy structure, so H2b is invalid.

5 CONCLUSION

In this paper, the low-carbon pilot policy in 2010 is regarded as a quasi-natural experiment to evaluate the impact of the policy on building carbon emissions. The main conclusions are as follows: (1) Low-carbon pilot policy can promote building carbon emission reduction, and the policy effect is time-sensitive. (2) Through mediation mechanism analysis, it is found that low-carbon pilot policy mainly promotes building carbon emission reduction by reducing building energy consumption. In contrast, the effect of adjusting energy structure is not obvious. In addition, a series of robustness tests are carried out to ensure the reliability of the research results.

Based on the conclusions, some policy recommendations can be put forward. First, we should further promote a low-carbon pilot policy and expand the scope of the pilot area. Secondly, it is necessary to continuously evaluate the policy effect and establish a long-term development mechanism for building carbon emission reduction to ensure the sustainability of the policy effect. Finally, we should continue to save energy consumption and promote the transformation of the energy structure.

REFERENCES

Dan Fan & Tingting Liu. Research on the Mechanism and Heterogeneity of the Impact of Low-Carbon City Pilot Policies on Total Factor Energy Efficiency. *Industrial Economic Review*. (02), 93–111 (2022)

Di Zhou, Fengnian Zhou & Xueqin Wang. Impact Assessment and Mechanism Analysis of Low-carbon Pilot Policies on Urban Carbon Emissions Performance. Resource Science. 41 (03), 546–556 (2019)

Grossman G.M. et al. Economic Growth and the Environment. The Quarterly Journal of Economics. 110(2), 353–377 (1995)

Hua Zhang. Can the Pilot Policy of Low-carbon Cities Reduce Carbon Emissions–Evidence From Quasi Natural Experiments. Economic Management. 42 (06), 25–41 (2020)

Junqi Ma & Zhang Yue. Water Environment Governance Effect of Ecological Compensation in the Yellow River basin – a Test Based on Double Difference Method. *Resource Science* 43 (11), 2277–2288 (2021)

Linhong Li, Juan Wang & Yanfeng Xu. The Impact of Low Carbon Pilot City Policies on Enterprise Technological Innovation–An Empirical Study Based on DID Double Difference Model. Ecological Economy. 35 (11), 48–54 (2019)

Mei Dong & Cunfang Li. Net Carbon Emission Reduction Effect of Low Carbon Provincial Pilot Policies. China's Population, Resources and Environment. 30 (11), 63–74 (2020)

Rong Dai & Jianhua Cao. Carbon Reduction Effect Evaluation of China's First "Low-carbon Pilot" Policy - DID Estimation Based on Five Provinces and Eight Cities. *Science and Technology Management Research*. 35 (12), 56–61 (2015)

Wei Liu, Liang Liu, Chaofan Chen, Rui Dai & Lei Zang. Planning and Prospect of Low-carbon City Development in China. Modern Urban Research. 28(04), 65–70 (2013)

Xiang Cao & Yu Gao. Does the Pilot Policy of Low-carbon Cities Promote the Formation of Green Lifestyles of Urban Residents? *China's Population, Resources and Environment* 31(12), 93–103 (2021)

Zhonglin Wen & Baojuan Ye. Mediation Effect Analysis: Methods and Model Development. Progress in Psychological Science 22 (05), 731–745 (2014)

Constructional Engineering and Ecological Environment – Chih-Huang Weng (Ed)
© 2024 The Author(s), ISBN 978-1-032-53198-4

Degradation characteristics and mechanism of tetracycline hydrochloride from waste red brick

Shuying Song, Haiyan Sun*, Xue Zhang, Ting He, Yuying Wang, Qian Mo & Chengrui Zhuo
School of Ecological Technology and Engineering, Shanghai Institute of Technology, Shanghai, China

ABSTRACT: For the first time, the red brick obtained from the generation of construction waste was used as raw material to characterize the material by XRD, BET, and SEM, and to study the microstructure of the material. The tetracycline hydrochloride solution was used as simulated wastewater to study the adsorption and degradation ability of the material. The effects of particle size, loading amount, pH, temperature, adsorption kinetics, and isotherm on the removal of tetracycline hydrochloride were studied, and the adsorption mechanism was investigated by Fourier transform infrared (FT-IR) spectroscopy. The results showed that the material had an excellent absorption effect on tetracycline hydrochloride, and the best dosage was 3 g. The material had a strong removal effect for tetracycline hydrochloride in a wide pH range, and the highest absorption rate was 358.3021 $mg{\cdot}g^{-1}$. Combined with the adsorption kinetics results, the pseudo-second-order model can better describe the adsorption process. The adsorption and degradation process of tetracycline hydrochloride is consistent with the Freundlich isotherm model. This study has practical application value for resource utilization of construction waste and pollution control of antibiotics in water.

1 INTRODUCTION

In recent years, with the rapid development of animal husbandry, aquaculture, and other industries, the use of antibiotics has become increasingly common. Tetracycline (TC) has the advantages of low price, a wide application range of antibiotics, high efficacy, and so on. It is widely used in the breeding industry as an antibacterial promoter to accelerate the growth and development of breeding animals (Daghrir & Drogui 2013). However, tetracycline is digested and absorbed only after it enters the human and animal bodies. Most of the tetracycline is released into the animal body in the form of excreta in the form of compounds, resulting in serious pollution of the water and soil environment and posing some danger to human health (Wang et al. 2020; Zhang et al. 2015). At present, the treatment methods for antibiotic contaminants mainly include adsorption, chemical oxidation, membrane separation technology, microbial degradation, and so on. Adsorption is easier to handle and more environmentally friendly than other methods. The selection of a suitable adsorbent is the crucial point in the study of the degradation performance of the adsorption method (Thiele-Bruhn 2003).The materials used as tetracycline adsorbents include polymeric materials (OKOLI, OFOMAJA 2019), montmorillonite (Figueroa et al. 2004), alumina (Chen & Huang 2010), diatomite (Chao et al. 2014), and carbon materials (Ghadim et al. 2013).

In this paper, three types of red brick adsorbent materials with varying particle sizes were prepared by crushing, screening, cleaning, impregnating, and drying typical construction

*Corresponding Author: haiyansun@sit.edu.cn

 DOI: 10.1201/9781003410843-22

waste red brick. The red brick material was characterized using X-ray diffraction (XRD), an automatic specific surface area analyzer (BET), a scanning electron microscope (SEM), and other methods. Meanwhile, the effect of particle size, input amount, solution pH, temperature, adsorption kinetics, and material isotherm on tetracycline degradation was studied. The adsorption mechanism was studied using a Fourier transform infrared absorption spectrometer (FTIR). The findings of the study can be used to provide a theoretical reference value for the reuse of construction waste and antibiotic pollution in water.

2 MATERIALS AND METHODS

2.1 *Chemicals and instruments*

Chemicals: All reagents, such as tetracycline, sodium hydroxide, and hydrochloric acid (HCl) of AR, were purchased from Aladdin Biochemical Co., Ltd. (Shanghai, China). In all experiments, the deionized water was acquired from a Milli-Q system.

Instruments: UNICO UV spectrophotometer; Electric blower drying oven (Shanghai Heng Scientific Instrument Co., Ltd.); Full temperature oscillation incubator (Taicang Chang Cheng Experimental Instrument Manufacturing Co., Ltd. Hcy-123c); pH Meter (PHS-3C); X-ray diffractometer XRD (Brock-D8); BET (MAC ASAP2460); SEM (S-3400N, Nake high-tech Co., Ltd. Japan); Fourier infrared absorption spectrometer FTIR (NICOLET iS10).

2.2 *Synthesis of red brick samples*

The purchased construction waste red brick particles are screened and passed through 1 mm, 2 mm, and 3 mm screens, respectively to obtain 0-1 mm, 1-2 mm, and 2-3 mm particle sizes of discarded red brick particles.

2.3 *Adsorption influence factors*

The adsorption rate and adsorption capacity of tetracycline hydrochloride on red brick particles under the influence of different particle sizes, red brick dosage, initial pH of the solution, temperature, adsorption time, and other factors were studied. A single-factor adsorption experiment was conducted to investigate the optimal parameters of red brick adsorption of tetracycline and explore its adsorption and degradation mechanism.

$100 mg \cdot L^{-1}$ tetracycline hydrochloride (TC solution) storage solution was prepared by dissolving 0.1g solid tetracycline hydrochloride in a beaker and moving it to a 1000mL volumetric flask at constant volume. 100mL tetracycline hydrochloride aqueous solution with a specified mass concentration was added into a 250 mL conical flask. A certain amount of red brick particles was added into the $150 r \cdot min^{-1}$ constant temperature oscillation box for oscillation. There were 3 parallel treatments for each, and the blank control group was set up except for the effects of volatilization, photolysis, and hydrolysis. The content of tetracycline in the residual solvent was determined by a UV spectrophotometer. The adsorption quantity Q_e ($mg \cdot g^{-1}$), and the adsorption rate η (%) of tetracycline hydrochloride for red brick were calculated according to Formulas (1) and (2):

$$Q_e = C_0 - C_e V/m \tag{1}$$

$$\eta = (C_0 - C_t)/C_0 x 100\% \tag{2}$$

Where C_0 ($mg \cdot L^{-1}$) and C_t ($mg \cdot L^{-1}$) are the liquid phase concentrations of the remaining tetracycline hydrochloride solution at the initial tetracycline solution and at time T(min) respectively, V is the volume of TC solution (L), and m is the dry mass of adsorbent red brick (g).

The adsorption effect mainly includes the following factors:

1) Influence of particle size on TC adsorption. The initial mass concentration of tetracycline hydrochloride was 10 $mg \cdot L^{-1}$, the temperature was 25°C, and the oscillation time was 2h. The supernatant was centrifuged at 3000 $r \cdot min^{-1}$ for 10 min (the same as below). The solution was filtered through a 0.45μm filter membrane (the same as below) and tested.
2) Influence of the amount of red brick on TC adsorption. The initial mass concentration of tetracycline was 10$mg \cdot L^{-1}$, and the temperature was 25°C. The particle sizes of red bricks were weighed to optimize the particle sizes. The additional amount of red bricks was 1, 3, 5, 10, and 15 g, respectively, and the shaking time was 2h.
3) Influence of pH on TC adsorption. The initial mass concentration of tetracycline hydrochloride was 10 $mg \cdot l^{-1}$, and the pH of the initial solution was adjusted to 3, 5, 7, 9, and 11 by 0.1 $mg \cdot L^{-1}$ sodium hydroxide solution and hydrochloric acid solution, respectively, at 25°C. The dosage of red brick was the optimal dose, and the oscillation time was 2h. The solution was oscillated, centrifugated, and filtered to be tested.
4) Influence of temperature on TC adsorption. The initial mass concentration of tetracycline hydrochloride was $mg \cdot L^{-1}$, and the temperature was 15, 25, and 35°C, respectively. The dosage of red brick was the optimal dose, and the solution was tested after centrifugation and filtration.
5) Influence of adsorption time on TC adsorption. The initial mass concentration of tetracycline hydrochloride was $mg \cdot L^{-1}$, the temperature was 25°C, and the dosage of red brick was the optimal dose. The solution oscillated for 10min, 30min, 1, 2, 4, 8, 12, 24, 48, and 72h, and the solution was oscillated and centrifugated for filtration.
6) Influence of cycles on TC adsorption. The mixed solution of red brick and tetracycline hydrochloride was filtered through a 0.45 μm filter membrane to obtain the adsorbed red brick particles. After drying in an oven at 80°C, it was mixed with ethanol solution and oscillated for 6 hours. The temperature was set at 25°C, and the speed was 150 $r \cdot min^{-1}$. After oscillation analysis, the mixture of red brick and ethanol was filtered through a 0.45μm membrane and washed with distilled water three times. The cleaned red brick particles were dried in an oven at 80°C, the mass concentration of tetracycline hydrochloride was set at 10 $mg \cdot L^{-1}$, the ambient temperature was set at 25°C, the dose of red brick was analyzed at 3g, the oscillation time was 2h, and the solution was oscillated, centrifuged and filtered for the test. The experimental conditions of each recovery are the above conditions.

2.4 *Adsorption kinetics experiment*

Particle size when the environment temperature is constant, red brick, red brick, the solution of the initial pH value, initial temperature and mass concentration of the tetracycline solution as the best conditions, adsorption by sampling and analysis of different points in time, the dynamic model test data for pseudo-first-order and pseudo secondary linear fitting, fitting coefficient R^2 more in accord with a dynamic model, which is used to determine the adsorption process (Rajapaksha et al. 2016).

2.5 *Adsorption isotherm experiment*

At 25°C, the adsorption process of tetracycline hydrochloride by red brick was optimized by particle size, dosage, and pH value. The 48-hour mixed oscillation experiment was carried out with a tetracycline initial solution of different mass concentrations. Freundlich and Langmuir's isothermal models were used to fit the adsorption process, and the fitting effect of the two isothermal adsorption models was judged according to the correlation coefficient R^2 (Chung et al. 2015).

3 RESULTS AND DISCUSSION

3.1 *Analysis and characterization of red brick materials*

SEM characterization. Figure 1 is the scanning electron microscope of red brick particles at three different magnifications. It can be observed that the surface of red brick particles is rough and has developed a pore structure, which provides more adsorption sites and thus has a strong adsorption capacity.

Figure 1. SEM images of red brick.

Specific surface area and pore volume. According to the analysis of images 2 and 3, the average pore size is 15.7632 nm, belonging to the mesopores (also called mesopores) with pore sizes ranging from 2-50 nm, and the average pore size of mesopores is 32.6350 nm. Studies have shown that mesoporous materials are common characteristics, such as large specific surface area, regular pore structure, narrow pore size distribution, and continuously adjustable pore size, and are good adsorption materials (Zhang et al. 2017).

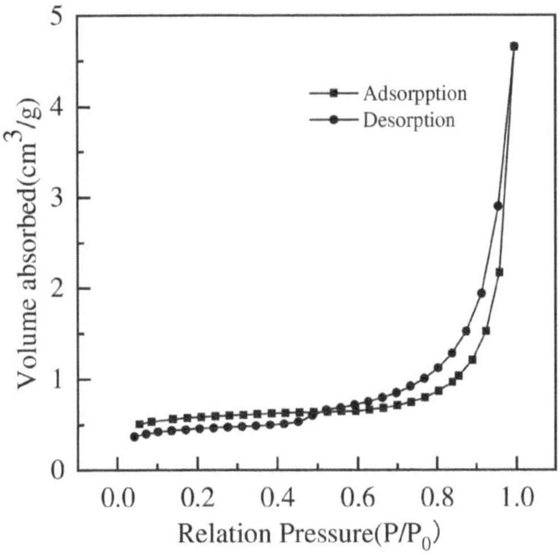

Figure 2. BET isothermal adsorption diagram.

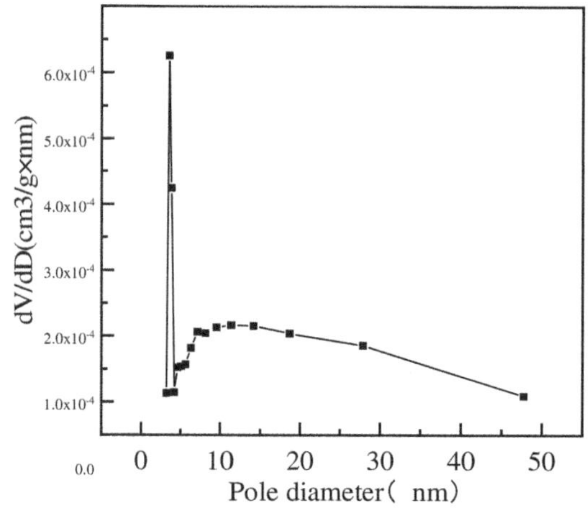

Figure 3. Aperture distribution of red brick.

3.2 *Influencing factors on TC adsorption*

Redbrick diameter. As shown in Figure 4, with the gradual increase of particle size, the adsorption capacity and the removal rate of tetracycline hydrochloride decreased gradually, and red brick particles with particle size 0-1mm had a better adsorption effect on tetracycline. With the increase in particle size, the adsorption capacity decreased from 212.9838mg·g^{-1} to 136.829 mg·g^{-1}, and the removal rate decreased from 86.3% to 55.45%. It can be seen that red brick particles with small particle sizes have better adsorption performance than larger particle sizes. The main reason for this phenomenon is that the smaller the particle size is, the larger the corresponding specific surface area is, and the more adsorption sites are provided (Cepuritis et al. 2017). Subsequent experiments were carried out with 0-1mm particle size.

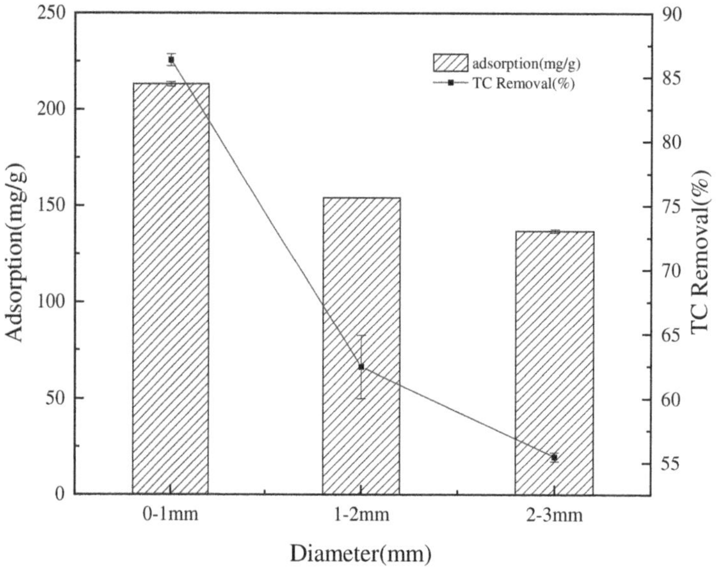

Figure 4. Influence of red brick diameter adsorption of tetracycline.

Amount of red brick. According to the results in Figure 5, with the increase of the additional amount, the removal rate gradually increases and tends to be stable. In the beginning, the adsorption amount decreases rapidly and then gradually tends to be stable. This phenomenon may be caused by the rough surface of red brick particles and a large number of pore structures conducive to the adsorption of tetracycline (Moreno-Maroto & Alonso-Azcárate 2018). Therefore, the redder brick is added to the solution, and more tetracycline attachment sites are exposed to the solution. However, when the amount of red brick is added to a certain extent, tetracycline in the solution will be insufficient, and the excess adsorption sites on the surface of the red brick cannot be fully utilized. In the reaction system, the intersection of adsorption capacity and removal rate is the best amount of red brick addition, which can be reflected in Figure 6. In the follow-up experiment, the amount of red brick was 3 g.

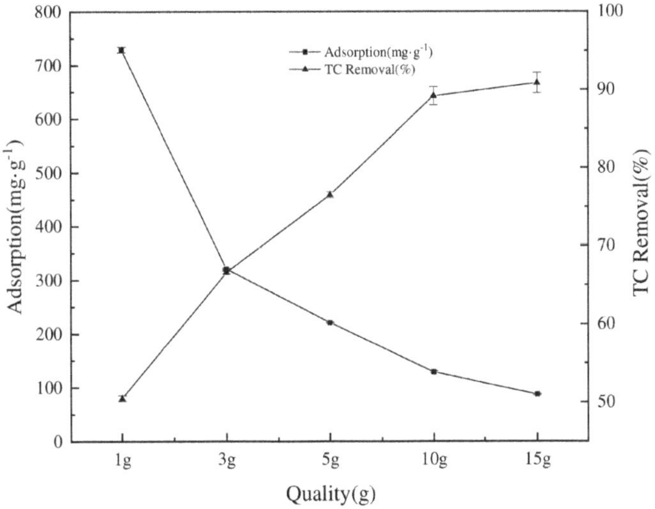

Figure 5. Influence of amount of red brick on adsorption of tetracycline.

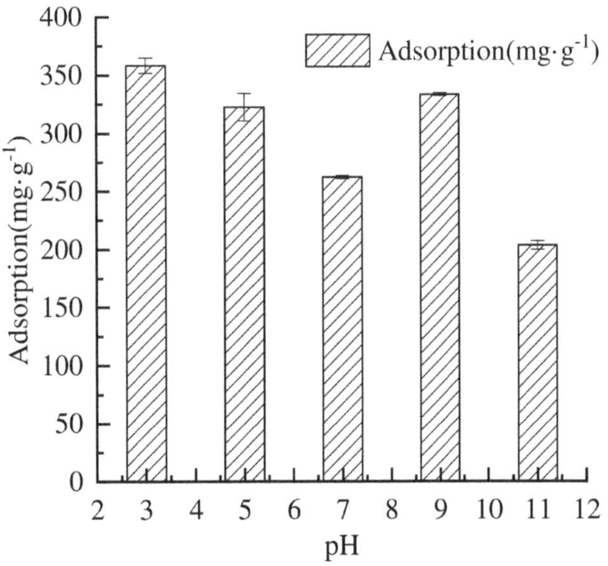

Figure 6. Effect of pH on tetracycline removal.

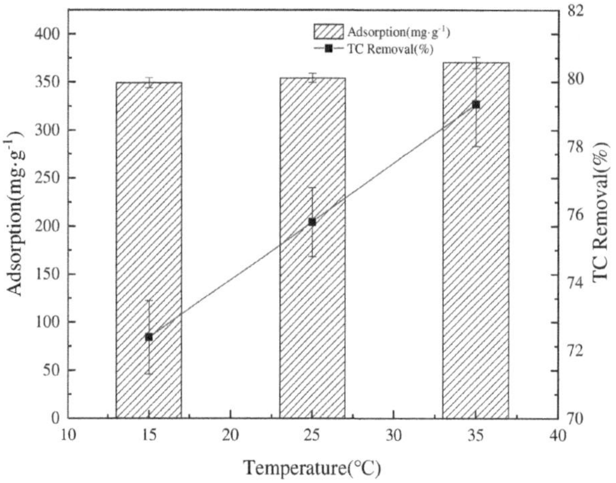

Figure 7. Influence of temperature.

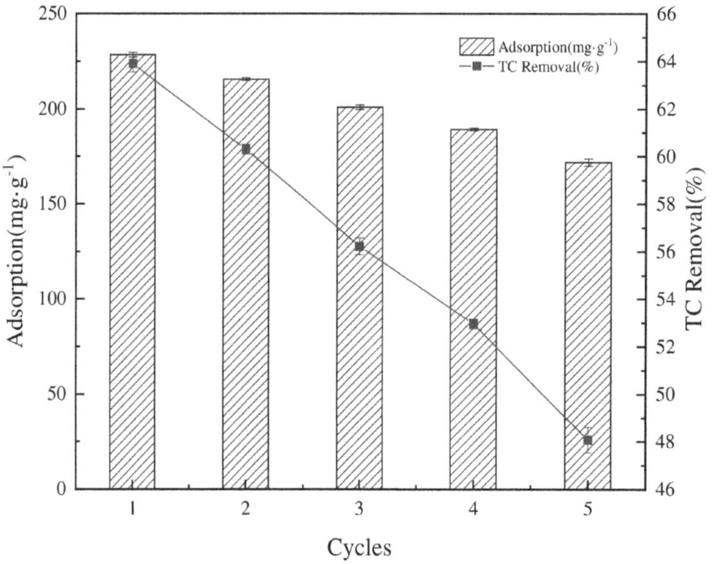

Figure 8. Influence of red brick cycles.

Influence of pH on TC adsorption. The adsorption effect of red brick on tetracycline in water is greatly affected by pH, and the overall adsorption effect under acidic conditions is better than that under alkaline conditions. Under acidic conditions, the adsorption capacity of tetracycline decreased with the increase in pH, and there was a turning point at pH 9. The ionic morphology at pH = 3 is conducive to hydrogen bond formation, so the adsorption capacity is the largest, which is related to -OH can promote adsorption. However, when pH = 9, a new isoelectric point appears, and ionic morphology changes. Adsorption capacity increases slightly, but it is smaller than when pH = 3, indicating that ionic morphology can also form hydrogen bonds at this time, but the number is not large.

Influence of temperature on adsorption effect. When the temperature gradually increased from 15°C to 35°C, the removal rate of tetracycline from red brick increased from 72.38% to 79.24%. The higher the temperature, the faster the molecular movement rate. Temperature

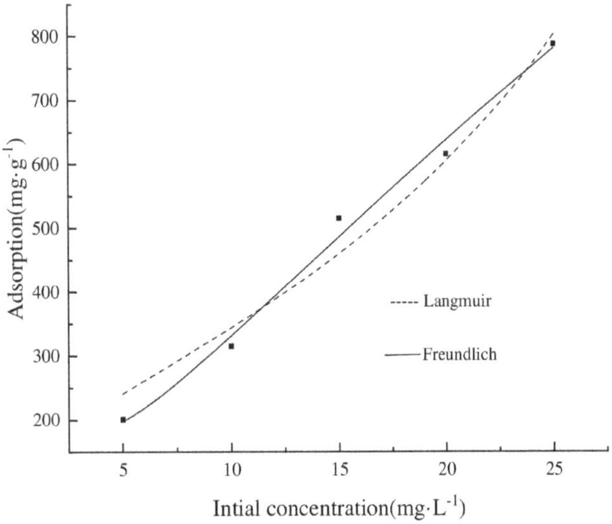

Figure 9. Adsorption isotherm model of tetracycline.

change does not change the interaction force between adsorbent and adsorbent. In general, a low temperature (15°C) will increase the physical adsorption between adsorbents and adsorbents, while a high temperature (35°C) will increase the chemical interaction between them. Combined with the results of this experiment, the temperature is one of the factors affecting the adsorption performance, and higher temperature conditions are conducive to the adsorption process. The adsorption and degradation process of tetracycline hydrochloride includes physical adsorption and chemical adsorption, and the influence of chemical adsorption on the experimental process is greater than that of physical adsorption.

Cycles. Due to the good adsorption capacity of 0-1 mm red brick particles, they can be recycled after analytical desorption. Specific experimental results are shown in Figure 10. According to the results in Figure 10, in the tetracycline solution system with a mass concentration of $10mg{\cdot}L^{-1}$, the removal rate and adsorption capacity of red brick to tetracycline

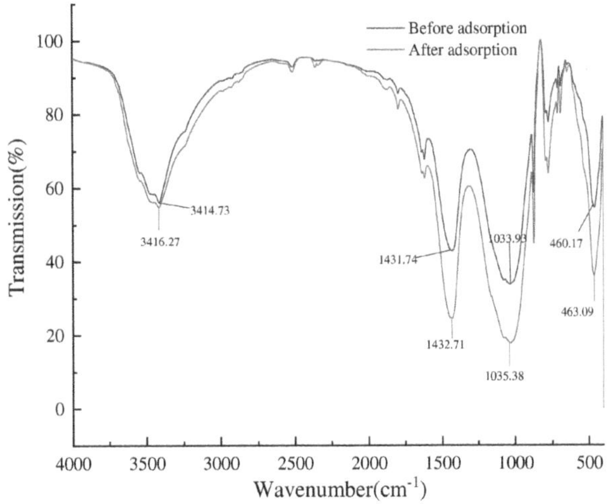

Figure 10. FTIR spectra of red brick.

hydrochloride gradually decreased with the increase in recycling times. In the fifth recovery, the removal rate is still 48.07%, and the adsorption capacity is 171.87 mg·g^{-1}. It can be seen that the red brick particles still have the ability to re-adsorb tetracycline after analytical desorption, indicating that red brick has a good recovery effect as a tetracycline hydrochloride adsorbent.

3.3 *Adsorption kinetics*

To explore the adsorption and degradation mechanism of tetracycline hydrochloride by red brick, pseudo-first order, and pseudo-second order kinetic models were selected to fit and analyze the experimental data (Nguyen et al. 2021).

$$\text{Pseudo-first order kinetic model equation:} \quad q_t = q_e\left(1 - e^{-k_1 t}\right) \tag{3}$$

$$\text{Pseudo-secondary order kinetic model equation:} \quad q_t = q_e{}^2 k_2 t/(1 + k_2 q_e t) \tag{4}$$

In Equations (3) and (4), q_t and q_e represent the adsorption capacity (mg·g^{-1}) at time t (min) and equilibrium respectively; k_1 and k_2 are rate constants of pseudo-first-order and pseudo-second-order kinetic models, respectively.

Table 1 shows the experimental data fitting results of pseudo-first order and pseudo-second order dynamics models. By calculating the slope and intercept of the image, the theoretical adsorption capacity q_e calculated in Table 1 is 90.46 mg, and that calculated in Table 1 is 429.19 mg, which is closer to the actual equilibrium adsorption capacity of 426.967 mg. Among them, the correlation coefficient R^2 of the pseudo-first-order dynamics model, is 0.96625, and that of the second-order dynamics model is 0.99986. The pseudo-second-order kinetic model is more consistent with the adsorption process of tetracycline hydrochloride by red brick than the pseudo-first-order kinetic model, indicating that chemical adsorption is the limiting factor in the adsorption process of tetracycline by red brick (Yan et al. 2020). This process may involve a series of chemical reactions, such as complexation and ion exchange.

Table 1. Kinetic model parameters for the adsorption of tetracycline onto the red brick.

Pseudo-first order			Pseudo-second order		
k_1	q_e	R^2	k_2	q_e	R^2
0.00136	90.464	0.96625	8.278×10^{-5}	429.185	0.99986

3.4 *Adsorption isotherms*

To further explore the adsorption characteristics of tetracycline hydrochloride on red bricks, the effects of different initial concentrations of tetracycline on the adsorption and degradation effects of tetracycline were studied, and Freundlich and Langmuir's isothermal models were used to simulate the adsorption experiment to fit the data (Budhiary & Sumantri 2021). The fitting results are shown in Figure 9.

$$\text{Freundlich isotherm model equation:} \quad q_e = K_F C_e{}^{-1/n} \tag{5}$$

$$\text{Langmuir isotherm model equation:} \quad q_e = Q_m K_L C_e/(1 + K_L C_e) \tag{6}$$

In Equations (5) and (6), K_F and K_L represent the adsorption constants of Freundlich and Langmuir isotherm models, respectively. $-1/n$ represents the surface non-uniformity and adsorption strength; C_e is the concentration at adsorption equilibrium (mg·L^{-1}); Q_m is the maximum adsorption capacity of red brick (mg· g^{-1}).

According to the results in Table 2, the Freundlich model has a better fitting effect on the adsorption of tetracycline hydrochloride than the Langmuir model ($R^2 = 0.99072$), indicating that the adsorption of tetracycline on red brick tends to be nonlinear and the adsorption sites are not uniform (Jang et al. 2018; Yu et al. 2021). According to the fitting results of the Langmuir model, the maximum adsorption capacity of red brick to tetracycline $Q_m = 1094.39$ mg·g^{-1} was calculated, indicating that red brick particles have good adsorption performance and are good adsorbent materials.

Table 2. Isotherm model parameters for adsorption of tetracycline onto the red brick.

Freundlich model			Langmuir model		
K_F	$1/n$	R^2	K_L	Q_m	R^2
5.39834	1.775	0.99072	0.227	1094.39	0.96458

3.5 *Mechanism of degradation*

The experimental results showed that the degradation of tetracycline hydrochloride by red brick involved electrostatic action. To further reveal the mechanism of absorption and degradation of tetracycline by red brick, an infrared spectrum analysis was carried out on red brick particles before and after the absorption of tetracycline. As shown in Figure 10: the FT-IR spectrum shows that the infrared spectrum characteristics of red brick particles are significantly different before and after adsorption. Red brick mainly shows a strong and wide wave peak between 1200-950 cm^{-1}, which is caused by the antisymmetric contraction of the Si-O-Si bond and the symmetric contraction vibration of the O-Si-O bond (Ali et al. 2016). The maximum absorption peak before adsorption is 1033.93 cm^{-1}, and the maximum absorption peak after adsorption is 1035.38 cm^{-1}. The bending vibration of the Si-O-Si bond is reflected in the 460 cm^{-1} wave peak, and the position of the maximum absorption peak and the characteristic peak of the Si-O-Si bond change before and after adsorption, indicating that the chemical bond forces of the Si-O-Si bond and O-Si-O bond occur during the adsorption of tetracycline hydrochloride. Red brick adsorption and degradation of tetracycline in water is a chemical adsorption process.

4 CONCLUSION

1) This study shows that according to the characterization results of red brick particles before and after modification, red brick particles with small particle size have a richer specific surface area and total pore volume pores, thus proving that the existence of these mesoporous pores is conducive to the attachment of tetracycline hydrochloride. The optimal conditions for the adsorption of tetracycline by red brick were as follows: the particle size was 0-1mm, the amount of red brick was 3 g, the initial concentration of tetracycline hydrochloride was 10 mg·L^{-1}, the solution pH was 3-7, the ambient temperature was 25°C and the adsorption time was 48h.
2) The results of recycling showed that with the increase in recycling times, the removal rate and adsorption capacity of tetracycline hydrochloride by red brick gradually decreased, and the red brick still had a high removal rate and adsorption capacity of tetracycline hydrochloride in solution after the fifth recycling.

3) Adsorption kinetics studies showed that the pseudo-second-order kinetics model was more consistent with the red brick adsorption process of tetracycline. Combined with the Fourier transform infrared spectroscopy functional group analysis, the red brick adsorption process of tetracycline hydrochloride involves chemical reactions such as complexation and static electricity, and chemical adsorption plays a dominant role.
4) The adsorption isotherm study showed that with the increase of tetracycline hydrochloride concentration, the adsorption capacity of red brick to tetracycline hydrochloride also increased. This is because the higher the concentration of tetracycline hydrochloride is, the higher the collision contact frequency of the tetracycline molecule, and the more effective the adsorption site on the red brick in the diffusion process is, the more conducive to the adsorption process. The adsorption capacity of tetracycline was not balanced during the whole experiment, indicating that there were many unutilized adsorption sites on red bricks. The theoretical maximum adsorption capacity of tetracycline hydrochloride for red brick was Q_m = 1094.39 mg·g^{-1}, and the fitting effect of the Freundlich model on the adsorption of tetracycline hydrochloride was better than that of the Langmuir model, indicating that the adsorption of tetracycline hydrochloride for red brick had nonlinear adsorption tendency.

REFERENCES

Ali R. M., Hamad H. A., Hussein M. M., *et al.* Potential of Using Green Adsorbent of Heavy Metal Removal From Aqueous Solutions: Adsorption Kinetics, Isotherm, Thermodynamic, Mechanism and Economic Analysis. *Ecological Engineering*, 2016, 91: 317–332.

Budhiary K. N. S. and Sumantri I. Langmuir and Freundlich Isotherm Adsorption Using Activated Charcoal From Banana Peel to Reduce Total Suspended Solid (TSS) Levels in Tofu Industry Liquid Waste. *IOP Conference Series: Materials Science and Engineering*, 2021, 1053(1): 012113.

Chen W.R and Huang C.H. Adsorption and Transformation of Tetracycline Antibiotics With Aluminum Oxide. *Chemosphere*, 2010, 79(8): 779–785.

Chao Y., Zhu W., Chen F., *et al.* Commercial Diatomite for Adsorption of Tetracycline Antibiotic from Aqueous Solution. *Separation Science and Technology*, 2014, 49(14): 2221–2227.

Cepuritis R., Garboczi E. J., Ferraris C. F., *et al.* Measurement of Particle Size Distribution and Specific Surface Area for Crushed Concrete Aggregate Fines. *Advanced Powder Technology*, 2017, 28(3): 706–720.

Chung H-K., Kim W-H., Park J., *et al.* Application of Langmuir and Freundlich Isotherms to Predict Adsorbate Removal Efficiency or Required Amount of Adsorbent. *Journal of Industrial and Engineering Chemistry*, 2015, 28: 241–246.

Daghrir R. and Drogui P. Tetracycline Antibiotics in the Environment: A Review. *Environmental Chemistry Letters*, 2013, 11(3): 209–227.

Figueroa R. A., Leonard A. & Mackay A. A. Modeling Tetracycline Antibiotic Sorption to Clays. *Environmental Science & Technology*, 2004, 38(2): 476–483.

Ghadim E. E., Manouchehri F., Soleimani G., *et al.* Adsorption Properties of Tetracycline onto Graphene Oxide: Equilibrium, Kinetic and Thermodynamic Studies. *PLoS ONE*, 2013, 8(11): e79254.

Jang H. M., Yoo S., Choi Y-K., *et al.* Adsorption Isotherm, Kinetic Modeling and Mechanism of Tetracycline on Pinus Taeda-derived Activated Biochar. *Bioresource Technology*, 2018, 259: 24–31.

Moreno-Maroto J. M. and Alonso-Azcárate J. What is Clay? A New Definition of "Clay" Based on Plasticity and its Impact on the Most Widespread Soil Classification Systems. *Applied Clay Science*, 2018, 161: 57–63.

Nguyen T-B., HO T-B-C., Huang C-P., *et al.* Adsorption Characteristics of Tetracycline Onto Particulate Polyethylene in Dilute Aqueous Solutions. *Environmental Pollution*, 2021, 285: 117398.

Okoli C. P. and Ofomaja A. E. Development of Sustainable Magnetic Polyurethane Polymer Nanocomposite For Abatement of Tetracycline Antibiotics Aqueous Pollution: Response Surface Methodology and Adsorption Dynamics. *Journal of Cleaner Production*, 2019, 217: 42–55.

Rajapaksha A. U., Vithanage M., Lee S. S., *et al.* Steam Activation of Biochars Facilitates Kinetics and pH-Resilience of Sulfamethazine Sorption. *Journal of Soils and Sediments*, 2016, 16(3): 889–895.

Thiele-Bruhn S. Pharmaceutical Antibiotic Compounds in Soils – a Review. *Journal of Plant Nutrition and Soil Science*, 2003, 166(2): 145–167.

Wang H., Chen T., Chen D., *et al.* Sulfurized Oolitic Hematite as a Heterogeneous Fenton-like Catalyst for Tetracycline Antibiotic Degradation. *Applied Catalysis B: Environmental*, 2020, 260: 118203.

Yan L., Liu Y., Zhang Y., *et al.* ZnCl2 Modified Biochar Derived From Aerobic Granular Sludge for Developed Microporosity and Enhanced Adsorption to Tetracycline. *Bioresource Technology*, 2020, 297: 122381.

Yu J., Feng H., Tang L., *et al.* Insight Into the Key Factors in Fast Adsorption of Organic Pollutants by Hierarchical Porous Biochar. *Journal of Hazardous Materials*, 2021, 403: 123610.

Zhang Q-Q., Ying G-G., Pan C-G., *et al.* A Comprehensive Evaluation of Antibiotics Emission and Fate in the River Basins of China: Source Analysis, Multimedia Modeling, and Linkage to Bacterial Resistance. *Environmental Science*, 2015: 41.

Zhang B., Luan L., Gao R., *et al.* Rapid and Effective Removal of Cr(VI) From Aqueous Solution Using Exfoliated LDH Nanosheets. *Colloids and Surfaces A: Physicochemical and Engineering Aspects*, 2017, 520: 399–408.

Constructional Engineering and Ecological Environment – Chih-Huang Weng (Ed)
© 2024 The Author(s), ISBN 978-1-032-53198-4

Regional disparities and evolution trends of maize green total factor productivity in China

Yakun Wang & Jingli Jiang
School of Economics and Management, Hebei University of Science and Technology, Shijiazhuang, China

Dongqing Wang
Anhui Electric Power Co., Ltd. Bozhou Power Supply Company, Bozhou, China

Xinshang You*
School of Economics and Management, Hebei University of Science and Technology, Shijiazhuang, China

ABSTRACT: The study uses the super-efficiency SBM model to explore the GTFP of main maize-producing regions from 2001 to 2020 and applies the Malmquist-Luenberger index and kernel density estimation to analyze the change trend and evolution mechanism of GTFP. The result shows that the GTFP of maize has not reached an effective level as a whole, and the maize output and ecological environment management are in an imbalanced state. Benefiting from the "dual driving" effect of technical efficiency and technological progress, the GTFP in main maize-producing regions has shown an upward trend. There may be two types of "club convergence." The GTFP of maize in each production region changed from a multi-polar trend to a polarized trend, and the regional concentration of high-efficiency values gradually increased.

1 INTRODUCTION

Maize has important strategic significance for China's food security. The continuous use of pesticides and fertilizers in maize production has resulted in ecological damage. The continuous expansion of arable land resources leads to the over-exploitation of resources. These problems seriously endanger the sustainable development of China's maize production. Therefore, improving the production efficiency of maize without exceeding the carrying capacity of resources and the environment is the key to realizing the healthy development of maize in China.

This paper summarizes the research on the production efficiency of maize as follows: Firstly, measure the growth rate and composition of efficiency. The measurement methods adopted by scholars mainly include parametric methods represented by SFA (Xu & Wang 2021) and non-parametric methods represented by DEA (Chen & Zhang 2014; Zhang *et al.* 2021). DEA has relatively loose requirements on production forms and is widely used. The second is the horizontal and vertical analysis of the production efficiency of maize. Scholars mainly study the trend characteristics, spatial distribution, and evolution of production efficiency of maize (Chen *et al.* 2022; Wang *et al.* 2017, 2019; Wang 2018; Zhang *et al.* 2018). Scholars have reached a consensus on the existence of regional differences in the production efficiency of maize in China (Kuang *et al.* 2021; Wang 2018; Zhang *et al.* 2018). Domestic research on the production efficiency of maize mainly focuses on production technology

*Corresponding Author: youxinshang@hebust.edu.cn

DOI: 10.1201/9781003410843-23

efficiency and total factor productivity, and there is less analysis on green total factor productivity (GTFP). Scholars' analysis of GTFP mainly uses carbon emissions as an environmental pollution variable (Kuang *et al.* 2021; Wang 2018).

Green agriculture is the trend of agricultural development in the future. GTFP is the concentrated embodiment of green agricultural production. Therefore, this paper analyzes the GTFP of maize production. This study provides a reference for improving the overall efficiency of maize production.

2 MATERIALS AND METHODS

2.1 *Model construction*

SBM-ML Model. This paper uses the non-radial and non-angular super-efficiency SBM model (Tone 2003) to measure the efficiency value to improve the comparability between effective decision-making units. Each province serves as a decision-making unit (DMU). Suppose that each DMU has L inputs $X = \{x_1, x_2, x_3, \cdots x_L\} \in R_L^+$ and produces M desired outputs $Y = \{y_1, y_2, y_3, \ldots y_M\} \in R_M^+$ and produces N undesired outputs $Z = \{z_1, z_2, z_3, \ldots z_N\} \in R_N^+$. Finally, the super-efficient SBM model that incorporates the undesired output into the decision-making unit is:

$$D_v^t(x_i^t, y_i^t, z_i^t) = \widehat{p} = \min \frac{1 - \left[\frac{1}{L}\sum_{l=1}^{L}\frac{s_l^x}{x_l^t}\right]}{1 + \left[\frac{1}{M+N} + \left(\sum_{m=1}^{M}\frac{s_m^y}{y_m^t} + \sum_{n=1}^{N}\frac{s_n^z}{z_n^t}\right)\right]} \tag{1}$$

$$s.t.\begin{cases} \sum_{i=1}^{I} u_i^t y_{i,m}^t - s_m^y = y_{i,m}^t = 1, 2, \ldots M; \\ \sum_{i=1}^{I} u_i^t z_{i,l}^t - s_l^x = x_{i,l}^t = 1, 2, \ldots L; \\ \sum_{i=1}^{I} u_i^t z_{i,n}^t - s_n^z = z_{i,n}^t = 1, 2, \ldots N; \\ \sum_{i=1}^{I} u_i^t = 1, u_i^t \geq 0, s_m^y \geq 0, s_l^x \geq 0, s_n^z \geq 0, i = 1, 2, \ldots, I \end{cases} \tag{2}$$

At the same time, this study measures GTFP, including undesired output, by constructing the ML index (Chung 1997). The ML index is decomposed into the product of the technical efficiency change index (EC) and the technological progress rate index (TC), and the specific formula is as follows:

$$(SBM - ML)_t^{t+1} = EC_t^{t+1} \times TC_t^{t+1} \tag{3}$$

Kernel density estimation (KDE). KDE builds a continuous probability density function by linearly adding discrete sample points, resulting in a smooth sample distribution. The paper chooses the Gaussian kernel function, which is a scalar function that is symmetrical along the radial direction. The formula is as follows:

$$K(u) = \frac{1}{\sqrt{2\pi}}\exp\left(-\frac{u^2}{2}\right), \quad K_h(u) = \frac{1}{h}K\left(\frac{u}{h}\right) \tag{4}$$

Table 1. Carbon emission coefficients and reference sources of various carbon sources.

Source	Emission factor	Reference source
Fertilizer	0.8956kg /kg	Oak Ridge National Laboratory (2009)
Pesticide	4.9314kg /kg	School of Biotechnology, China Agricultural University
Diesel fuel	0.5927kg /kg	Inter-Political Committee of Experts on Climate Change

u represents the euclidean distance from any point x in the space to the center point. The bandwidth h controls the relative weight of each sample by affecting the value of the independent variable in the kernel function.

2.2 *Definition of variables*

This paper takes 20 major maize-producing provinces in China as the research object. According to the characteristics of natural resources in different regions and relevant documents (Wang *et al.* 2019), the ecological areas for maize planting are defined as follows: the northern spring sowing region, the Huang-Huai-Hai summer sowing region, the southwest mountain sowing region, the northwest irrigation sowing region. The data used come from the National Compilation of Agricultural Products Cost and Benefit Data, China Statistical Yearbook, China Rural Statistical Yearbook, and provincial statistical yearbooks.

The study comprehensively considers the inputs of material, economy, resources, and environment. The final selected input indicators include the number of working days, the amount of seeds, the pure amount of chemical fertilizers, the cost of pesticides, and the cost of mechanical action. The main product output is the expected output. Environmental pollution is undesired output. In order to reduce measurement error, this paper uses the unit input and output factors as indicators. In addition, in order to eliminate the impact of price fluctuations, according to the price index of agricultural production materials in various provinces over the years, the pesticide cost and the mechanical action cost are adjusted based on the price in 2001.

Greenhouse gas emissions are an important source of agricultural pollution. This paper mainly measures the CO_2 emissions from chemical fertilizers, pesticides, and diesel. The emission of N_2O comes from nitrogen fertilizer and maize straw returning to the field. The variables are as follows:

CO_2 emissions are measured by carbon emissions. The calculation formula is: $C = \sum E_n \times \theta_n$. C is the total carbon emission. E_n is the number of the n carbon emission source. θ_n is the coefficient of the n carbon emission source.

N_2O from agricultural land mainly comes from nitrogenous fertilizer and straw returning to the field. This paper refers to Zhang (Zhang et al. 2021) to calculate the direct N_2O emission from maize cultivation:

$$N_2O_{direct} = (N_{staw} + N_{fertilizer}) \times EF \tag{5}$$

$$N_{straw} = (M/L - M) \times \beta \times K + M/L \times \alpha \times K \tag{6}$$

EF is the direct emission factor of N_2O. The specific values are shown in Table 2.

Table 2. Values of each parameter.

Parameter	Symbolic	Numerical value
Grain yield per hectare of maize field (%)	M	
Nitrogen content of straw (%)	K	0.00580
Economic coefficient (%)	L	0.438
Root-to-shoot ratio (%)	α	0.170
Straw returning rate (%)	β	9.30

3 EMPIRICAL RESULTS

3.1 *Change trend of maize GTFP in regions*

From Figure 1, the GTFP of maize has not reached an effective level as a whole, and the maize output and ecological environment management are in an imbalanced state. The main maize-producing regions benefited from the "dual driving" effect of technical efficiency and technological progress, and the overall GTFP showed an upward trend. In order to more accurately analyze the changing trend of maize GTFP in each region, the study period is divided into 4 stages: T1 (2001-2005), T2(2006) -2010), T3(2011-2015), and T4(2016-2020). The growth rate of maize GTFP in each region showed a "U" shaped fluctuation trend during the sample period. During T2, due to climate change and the continuous occurrence of low-temperature rain and snow freezing, the maize-growing environment was poor. The growth rate of maize GTFP in each region decreased significantly in T2. The trend of growth rates in all regions is basically the same, and the growth gap is gradually narrowing. The northern spring sowing region and the Huang-huai-hai summer sowing region grew steadily in each period. The northwest irrigated sowing area maintained a leading position in T1. The yield fluctuated greatly in each period in the southwest mountainous planting area.

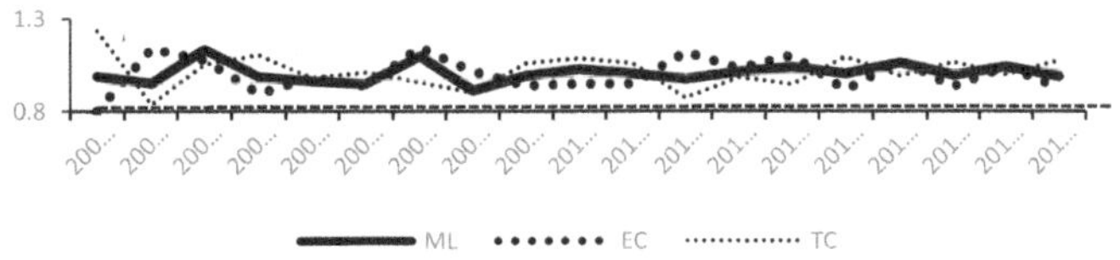

Figure 1. Growth rate and decomposition trend of GTFP from 2001 to 2020.

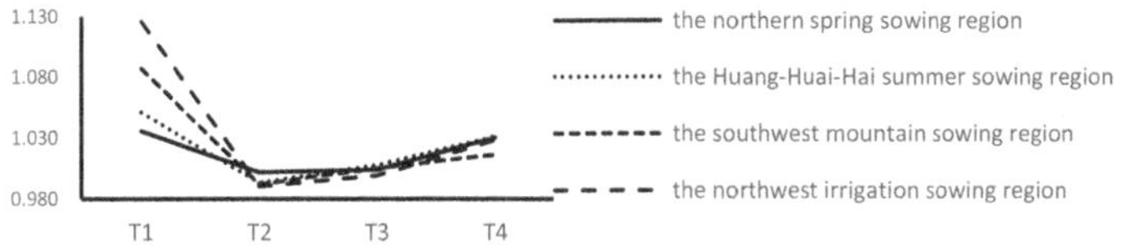

Figure 2. Changes of regional maize GTFP growth rate in Periods.

3.2 *Evolution mechanism of maize GTFP in regions*

This study selected five representative years of 2001, 2005, 2010, 2015, and 2020 to more intuitively examine the distribution shape and evolution mechanism of regional maize GTFP. From Figure 3, the inter-provincial maize GTFP nuclear density curve in China presents an obvious "double-peak" shape in the sample year. The GTFP among provinces shows a trend of polarization. The "double peak" pattern persists, and there may be two types of "club convergence." There is an obvious tailing phenomenon between the sample years, indicating that the distribution of GTFP in each province is not balanced. The "left peak" is always lower than the "right peak," indicating that the number of provinces with high-efficiency values is more than the number of provinces with low-efficiency values. The GTFP of each province is clustered towards high values. According to Figure 4, from the overall shape, the nuclear density curves of GTFP in the four regions gradually transformed from a "multimodal" shape to a "bimodal" shape. The inter-regional GTFP values changed from a multi-polar trend to a polarized trend. From the perspective of kurtosis, it showed a trend of developing from a "spiky peak" to a "broad peak." The absolute difference of GTFP in each region expanded.

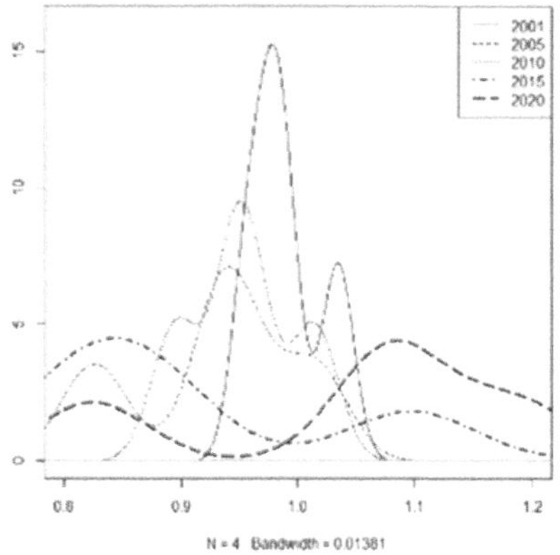

Figure 3. KDE curve of GTFP in provinces.

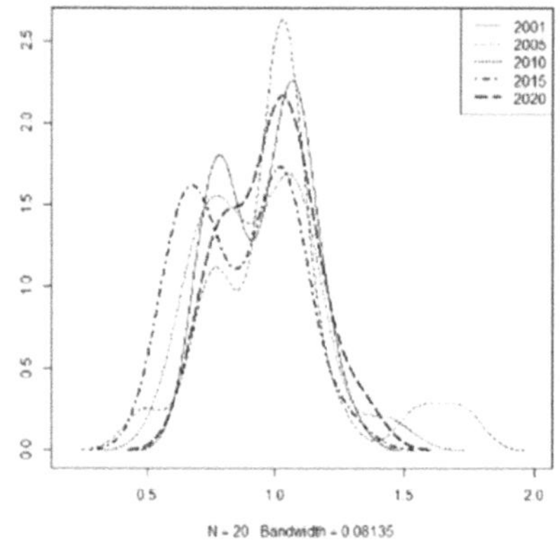

Figure 4. KDE curve of GTFP in regions.

4 CONCLUSIONS

This paper measures China's maize GTFP and reveals the time and regional variation characteristics of maize GTFP. GTFP in main maize-producing regions has not yet reached an effective level. The relationship between output and ecological environment governance is in an unbalanced state. There are significant regional differences in the GTFP of maize. GTFP in the southwestern mountainous maize region has generally reached an effective level. The technical efficiency and technological progress (5.4%) have increased significantly, which has greatly promoted the growth of GTFP. However, the efficiency fluctuates in various periods due to natural disasters. The northwest irrigation sowing region has the lowest level of efficiency. But it has accelerated the introduction of advanced agricultural technology in recent years, and technological progress and technical efficiency have increased rapidly. The GTFP of maize in the Huang-Huai-Hai summer sowing region and the northern spring sowing region maintains steady growth. Shanxi, Ningxia, and Inner Mongolia have relatively high levels of GTFP. Finally, the GTFP of maize in China is gradually concentrated in high-efficiency regions. The GTFP in various provinces presents a trend of polarization, and there may be two types of "club convergence." GTFP of maize in each production region has changed from a multi-polar trend to a polarized trend.

ACKNOWLEDGMENTS

This research was funded by the Social Science Development Research Project of Hebei Province, grant number 20210501002; the Social Science Foundation of Hebei Province, grant number HB22YJ045, HB21MZ005.

REFERENCES

Chen S.Q. and Zhang G.S.: An Empirical Study on the Technical Efficiency of Different Types of Labor Transfer Farmers' Maize Production-Taking Liaoning Province as an Example. *Jiangsu Agricultural Sciences*, 2014, 42(09): 416–419.

Chen H.W., Mu Y.Y. and Hou L.L.: Study On The Production Efficiency And Its Spatial-Temporal Variation of Winter Wheat and Summer Maize Multiple Cropping. *Chinese Journal of Agricultural Resources and Regional Planning*. 2022, 43(12): 9–17

Chung Y.H., Färe R. and Grosskopf S.: Productivity and Undesirable Outputs: A Directional Distance Function Approach. *Journal of Environmental Management*. 1997, 51(3).
Kuang A.P., Hu C. & Han D.D.: China's Corn Total Factor Productivity Measurement and Its Temporal and Spatial Differentiation Under Carbon Emission Constraints: Based on the Empirical Evidence of China's 20 Main Producing Areas. *Areal Research and Development*. 2021, 40(03): 115–120.
Tone K., Biresh K. and Sahoo.: Scale, Indivisibilities and Production Function in Data Envelopment Analysis [J]. *International Journal of Production Economics*. 2003, 84(2).
Wang H., Mu Y.Y. and Hou L.L.: Research on Temporal and Spatial Evolutions of Environmental Cost and Total Factor Productivity of Maize Production in China [J]. *Journal of Natural Resources*, 2017, 32 (07):1204–1216.
Wang M.X., Zhu Y.Y. and Wang D.: Maize Production Efficiency and Its Spatio-temporal Difference Based on Non-point Source Pollution Constraints. *Scientia Geographica Sinica*, 2019, 39(05):857–864.
Wang H.N.: Study on Carbon Emission Efficiency of Maize Production in China [D]. Jilin University. 2018.
Xu J.B. and Wang Y.: Study on the Influence of Agricultural Productive Services on Corn Production Technical Efficiency – Empirical Analysis Based on Micro Data. *Chinese Journal of Agricultural Resources and Regional Planning*., 2021, 42 (07): 27–36.
Zhang L.N., Chen Z. and Yang M.L.: Spatio-temporal Feature of Maize Production Efficiency in Main Producing Provinces of China. *Transactions of the Chinese Society for Agricultural Machinery*, 2018, 49 (01):183–193.
Zhang M., Dong S.Y., Zhu J.Y. and Zhao H.X.: Research on the Influence of Socialized Service of Mid-production Technology of Corn on Production Efficiency. *Journal of Maize Sciences*, 2021, 29 (06):175–183.
Zhang X.Z., Wang J.Y., Zhang T.I., Li B.l. and Yan L.: Assessment of nitrous oxide emissions from Chinese agricultural system and low-carbon measures. *Jiangsu Journal of Agricultural Sciences*. 2021, 37(05), 1215–1223.

Constructional Engineering and Ecological Environment – Chih-Huang Weng (Ed)
© 2024 The Author(s), ISBN 978-1-032-53198-4

Research status of low concentration fluoride wastewater treatment process

Jiewen Zheng & Ying Fu*
School of Civil Engineering and Architecture, University of Jinan, Jinan, Shandong Province, China

Yanting Yang
Jiangsu Hehai Environmental Science Research Institute Co, Ltd., Nanjing, Jiangsu Province, China

Xi Tian
Shuifa Technology Group Co., Ltd., Jinan, Shandong Province, China

ABSTRACT: Fluoride is one of the essential trace elements in the human body. China is a typical area with large areas of high fluoride, so it is very necessary to study the removal of fluoride ions in water. At present, the main methods of fluoride removal in water include the ion exchange method, adsorption method, and chemical precipitation method, among which the coagulation precipitation method is the most widely used method of fluoride removal. This paper mainly reviews the current methods of fluoride removal in water and the latest research progress. It is pointed out that in the actual treatment process of fluorine-containing wastewater, appropriate methods should be adopted for different kinds of wastewater.

1 INTRODUCTION

Fluorine pollution has attracted more and more attention. The treatment technology of fluorine-containing wastewater has always been an important topic in the field of environmental protection at home and abroad. In recent years, a lot of work has been done on the treatment of fluorine-containing wastewater at home and abroad, and great progress has been made in the theoretical knowledge, method technology, and fluorine removal process. At present, the composition of fluorine-containing industrial wastewater treated at home and abroad is complex and diverse, and there are many treatment methods. The commonly used methods are the precipitation method and the adsorption method. In addition, there are induced crystallization method and electrodialysis.

2 CHEMICAL PRECIPITATION METHOD

Chemical precipitation is the most commonly used method for fluoride wastewater treatment, especially for high concentration fluoride wastewater. The most commonly used is the lime precipitation method, which is to add lime to the fluorine-containing wastewater, fully react to form a precipitate, and then remove it. The method is simple, easy to operate, and low cost.

*Corresponding Author: cea_fuy@ujn.edu.cn

DOI: 10.1201/9781003410843-24

The solubility of calcium fluoride in water at 18°C is 16.3 mg/L, and the concentration of fluoride ion is 7.9 mg/L. Calcium fluoride above this concentration will form precipitates (Gu *et al.* 1984). However, the solubility of lime is small, which can only be added in the form of lime milk or lime powder, and the generated CaF_2 precipitate will wrap the surface of the particles in the solution so that it cannot fully contact and react with the solution, so the dosage is large. When fluorine concentration is low, the rate at which precipitate is formed slows down. The fluorine content in the treated wastewater can only be reduced to 20~30 mg/L, which is difficult to meet the national first-level emission standards. In addition, the chemical precipitation method also has the disadvantages of slow sedimentation, difficult dehydration, and inadaptability to continuous treatment and discharge (Wang 2000).

Yan Xiuzhi *et al.* (1998) combined the use of calcium chloride and phosphate treatment of fluorine containing 38 mg/L of electronic components, cleaning wastewater generated by fluoride calcium phosphate method to remove fluoride. By controlling the appropriate PH value, the dosage of calcium salt and poor acid salt, reaction time, and clarification time, the concentration of fluorine can be reduced to 5 mg/L after the reaction.

3 COAGULATION SEDIMENTATION METHOD

Coagulation sedimentation is generally only suitable for low-fluorine wastewater treatment. It can be used in combination with chemical sedimentation to treat high-fluorine wastewater. Coagulation precipitation and chemical precipitation method is similar. It has the advantages of easy operation, simple equipment, economical and practical. There is a large amount of coagulant, resulting in more difficult dealing with the waste residue. The fluoride removal effect is not stable. After fluoride sulfate ions, there is an increasing trend. The treated water contains a large number of dissolved aluminum and other issues.

3.1 *Aluminum salt coagulant defluorination*

The commonly used aluminum salt coagulants mainly include aluminum sulfate, polyaluminum chloride, polyaluminum sulfate, etc. All of which have good coagulation and fluoride removal effects. The mechanism of fluoride removal by aluminum salt coagulation is complex, mainly including adsorption, ion exchange, and complexation sedimentation. The defluorination mechanism of AI salt coagulant is shown in Figure 1.

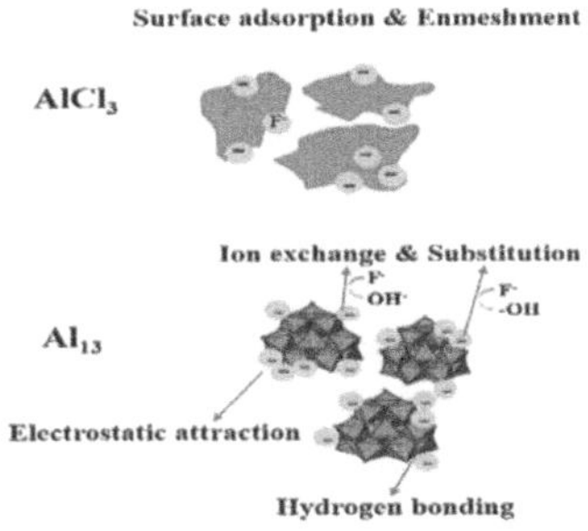

Figure 1. The defluorination mechanism of AI salt coagulant.

Liu Shanshan (2021) used aluminum salts (aluminum sulfate, PAFC, PAC) as coagulants in the optimization of the deep defluorination process of copper smelting wastewater, which can achieve the purpose of defluorination. When $Al_2 (SO_4)_3 + CaCl_2$ was used, the removal rate of F^- was more than 50%, and the defluorination effect was better. With the increase in aluminum salt dosage, the defluorination effect was better and better. When the pH is adjusted by lime, the dosage of aluminum sulfate is 170 times the mass concentration of F^-, and the mass concentration of F^- in treated water is less than 2 mg/L; the pH value was adjusted by alkaline, the mass concentration of F^- in the treated water was less than 2 mg/L when the dosage of aluminum sulfate was 300 times the mass concentration of F^-. Chen Wei *et al.* (2014) used a multi-stage coagulation precipitation method to treat acidic fluorine-containing wastewater produced by photovoltaic enterprises. Using calcium chloride as a precipitant, polyaluminium chloride as a coagulant, and polyacrylamide as a coagulant aid, the fluorine-containing wastewater with a pH of about 3 and F^- concentration of 1.0 g/L can be reduced to 10 mg/L by two-stage precipitation method, which meets the requirements of industrial first-level discharge standard. The fluoride removal process by the secondary precipitation method is shown in Figure 2.

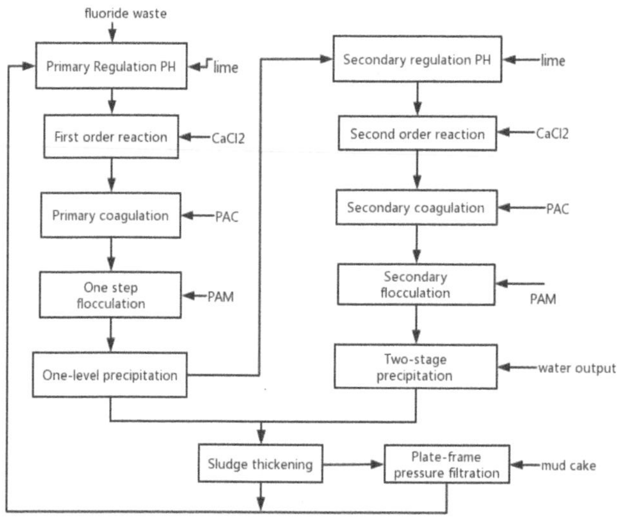

Figure 2. Schematic diagram of fluorine removal process by secondary precipitation.

3.2 *Fluoride removal with iron coagulant*

The iron salts used in coagulation precipitation are mainly modified poly iron, ferric chloride, ferrous sulfate, etc. Similar to aluminum coagulants, when iron coagulants are added to the wastewater, Fe^{3+} is complexed with F^-, and the intermediate products of iron salt hydrolysis and the final generated $Fe (OH)_3$ (am) flocs adsorb, ion exchange, and network capture and sweep $F^-$to remove F^-in the water.

Adsorption. The intermediate products and amorphous $Fe (OH)_3$(am) flocs produced in the process of fluoride removal by coagulation have a large specific surface area. The F^- strong electronegativity, ion radius is small, Fe as the active center, it is easy to F^-hydrogen bond adsorption (Han & He 2004).

Ion exchange. The radius and charge of F^- and OH^- are close. In the process of fluoride removal by adding iron salt coagulant, F^- can exchange ions with OH^- on the intermediate product of iron salt hydrolysis. Similar to aluminum-based coagulants, F^- and OH^- are equal charge exchange, so the charge carried by the floc after exchange does not change, and the Zeta potential does not change. Generally, the optimum pH for ion exchange between F^- and OH^- is 6~7 in a neutral and weakly acidic environment.

Complexation precipitation. The F^- in the aqueous solution can form an iron-fluorine complex $FeFx^{(3-X)+}$ with the added Fe^{3+}. The Fe^{3+} and F^- will undergo the following complexation reactions:

$$Fe^{3+} + 6F^- \rightarrow FeF_6^{3-}$$
$$Fe^{3+} + 5F^- \rightarrow FeF_5^{2-}$$
$$Fe^{3+} + 4F^- \rightarrow FeF_4^-$$
$$Fe^{3+} + 3F^- \rightarrow FeF_3$$

In the subsequent stage, the addition of slaked lime makes the wastewater alkaline, and some ferric ions and hydroxide ions form iron hydroxide precipitates.

4 ADSORPTION METHOD

Adsorption refers to the fluorine-containing wastewater through a device equipped with a fluorine adsorbent, fluorine and the adsorbent in the effective ions or groups in the ion exchange or chemical reaction to remove fluorine ions, adsorbent by regeneration to restore the exchange capacity. The adsorption method is simple to operate and has a stable fluoride removal effect. The commonly used filler adsorbents have dense pore structures and a large specific surface area. The surface has groups suitable for forming chemical bonds with fluoride ions. There are mainly aluminum-containing adsorbents (activated alumina, poly aluminum salt, molecular sieve, etc.), rare earth adsorbents, modified fly ash, modified zeolite, resin, etc. (Li 2018). The adsorption capacities of various adsorbents at optimum pH are shown in Table 1.

Table 1. The adsorption capacities of various adsorbents at optimum pH

Adsorption	The adsorption capacities. (mg/g)	The optimum pH
Activated alumina	0.8~2.0	4.5~6.0
Modified fly ash	0.01~0.03	3.0~5.0
Modified zeolite	0.06~0.3	7.3~7.9
Active magnesium oxide	6~14	6.0~7.0
Zirconia resin	30	3.5~7.0

It is mainly used to treat low-concentration fluoride wastewater or as an advanced treatment of fluoride wastewater. However, due to the problems of bed loss, low adsorption capacity, bed regeneration, and complex treatment of regenerated liquid, the treatment cost is high, which limits its practicability.

5 INDUCED CRYSTALLIZATION METHOD

The induced crystallization method is an improvement of the traditional precipitation method. At present, the induced crystallization method is mainly applied to the softening of water and the removal and recovery of heavy metals, phosphates, and fluorine. The seeds used are often calcium fluoride (CaF_2), cryolite (Na_3AlF_6), and fluorapatite ($Ca_5(PO_4)_3F$). Huang Tinglin (Huang *et al.* 2014) used the induced crystallization method to remove fluorine. The optimum process conditions were obtained by single factor experiment: 8 g/L fluorapatite was added, and NaH_2PO_4 and $CaCl_2$ were added. The molar ratio of calcium ion, phosphate ion, and fluoride ion was 10: 5: 1, the stirring speed was 100r/min, and the reaction time was 1h. The utilization of phosphate ions and calcium ions in the reaction reached more than 98% and 25%, respectively. Scanning electron microscopy (SEM) showed that there was crystallization on the surface of the seed after participating in the reaction. The results show that the fluorine ion concentration in water can be reduced from 5~10 mg/L to less than 1 mg/L by the induced crystallization method, which meets the drinking water quality standard. Oujieli (2021) designed calcium fluorophosphate coating modified calcite and used it as an induced defluorination crystal seed to carry out defluorination experiments in the reactor. The effects of coating ratio, dosing ratio, phosphate type, hydraulic retention time, and other conditions were investigated. During the operation of the reactor to 696 BV, the effluent fluoride ion concentration was less than 1.5mg/L.

6 ELECTRODIALYSIS

Electrodialysis is a kind of membrane separation technology. Its principle is to use the selective permeability of the ion exchange membrane to make the anions and cations in water migrate directionally under the action of an external DC electric field. An ion exchange membrane is formed by ion exchange resin, so electrodialysis is actually another application form of the ion exchange resin method. Kunming Smelter uses a lime-basic aluminum chloride coagulation method to treat fluorine-containing electrolytic workshops and flue water, and the fluorine removal effect is poor. Further electrodialysis of the water treated by the coagulation method can not only effectively remove fluorine ions, but also remove calcium, magnesium, sodium, chlorine, sulfate, and other heavy metal ions (Wang et al. 1982). The electrodialysis device is complex, the power consumption is large, the maintenance intensity is high, and the technical requirements for the operator are strict. If there are high-valent metal ions in the water, it is easy to cause membrane poisoning and damage to the electrode.

7 CONCLUSION

There are many treatment methods for fluorine-containing wastewater. Among them, the chemical precipitation process is the simplest and most convenient to operate, but the dosage of the agent is large, and the effect of fluoride removal is not good, which will bring secondary pollution. The coagulation precipitation method is generally suitable for the treatment of wastewater with low fluorine content. It has the advantages of less dosage, large treatment capacity, and reaching the national discharge standard after one treatment, which is suitable for the treatment of industrial fluorine-containing wastewater. With the emergence of new technologies and new methods, the future research direction should be to maximize the optimization of the feasibility of various methods, and economy and enhance the performance of materials and other aspects of development.

FUND PROJECTS

Shandong Province enterprise technology innovation project from Department of Industry and Information Technology of Shandong Province (202150100867), Research Center of Shuifa Technology Group, University of Jinan (W2021009), and University-enterprise joint project (w2021125; w2021356).

REFERENCES

Chen Wei & Shi Wubin. Experimental Study on the Treatment of Photovoltaic Solar Energy High F^-Content Wastewater by Two-stage Chemical Coagulation Sedimentation Method. *Water Treatment Technology*. 2014,40 (3): 103–105.

Gu Huiming, Guo Shuangling, Jiang Daqiang & Qi Chunxiu. Study on the Treatment of Calcium Fluoride Wastewater by the Lime Method. *Chemical Environmental Protection*, 1984 (04): 200–204 + 260.
Han Jianxun & He Aiguo.Fluorine-containing Wastewater Treatment Method. *Organic Fluorine Industry*, 2004, (03), 27–36.
Huang Tinglin, Sun Tian, Deng Linyu, *et al.* Removal of Fluoride From Groundwater by Induced Crystallization. *Journal of Environmental Engineering*. 2014,8 (1), 1–5.
Li Runbo, Yu Huarong & Li Guibai. Study on Adsorption Fluoride Removal Performance of Activated Alumina. *Water Supply Technology*, 2018, 12 (1):21–24.
Liu Shanshan. Practical Study on Optimization of the Deep Defluorination Process of Copper Smelting Wastewater. *Industrial Water Treatment*, 2021,41 (1), 118–121.
Ogerie.Application of Calcium Fluorophosphate Coating Modified Calcite in Induced Crystallization Defluorination. Xi'an: Xi'an University of Architecture and Technology, 2021.
Wang Dacui. *Handbook of Special Pollutants Treatment in Industrial Wastewater*. Beijing: Chemical Industry Publisher, 2000: 49–53.
Wang Rongjiu, Zheng Honggen, Shi Jixing, *et al.* Study on Fluoride Removal From Wastewater by Electrodialysis. *Environmental*, 1982, 3 (5): 52–56.
Yan Xiu-zhi & Wang Shu-fen. Treatment of Fluorine-containing Wastewater by $CaCl_2$ + Phosphate Method. *Environmental Protection Science*, 1998, 24 (2): 12–14.

Constructional Engineering and Ecological Environment – Chih-Huang Weng (Ed)

Effect of different conditions on mechanical properties of geopolymer mortar at low temperature

Lv Yan
Aviation Engineering School, Air Force Engineering University, Xi'an, Shaanxi, China

Jiang Bo*
Air Force Logistics Engineering Agent Construction Office Construction Management Office, Beijing, China

Wang Zhihang, Wang Tengjiao, Xia Wei & Qin Lijun
Aviation Engineering School, Air Force Engineering University, Xi'an, Shaanxi, China

ABSTRACT: Aiming at the problems of insufficient comprehensive performance, heavy pollution, and poor low-temperature applicability of pavement repair materials, a kind of geopolymer pavement repair material suitable for the low-temperature environment was prepared. Compressive and flexural tests were carried out on geopolymer mortar to explore the early strength performance of different geopolymer mortar under a low-temperature curing environment and the effects of solution cement ratio and cement sand ratio on the basic working performance and mechanical properties of geopolymer mortar at low-temperature environment were analyzed in depth. The results show that the early mechanical properties of slag-based geopolymer mortar are better than those of fly ash-based geopolymer mortar at a low-temperature environment; With the increase of cement-sand ratio, the compressive strength and flexural strength of geopolymer mortar increase continuously; With the increase of solution-cement ratio, the strength first increases and then decreases. Finally, the mechanical properties of geopolymer mortar are the best when the cement-sand ratio is 1:2, and the solution-cement ratio is 0.55. The test can provide a reference for further research and the application of geopolymer mortar.

1 INTRODUCTION

Currently, the selection of pavement repair materials at home and abroad is mostly focused on cement-based materials (Lee *et al.* 2017; Zhao *et al.* 2019). The research is relatively mature, but there are also many problems. For example, ordinary Portland cement repair material is not suitable for rapid repair because of its inherent defects such as low early strength, slow setting and hardening, inability to use in low-temperature environments, and high production energy consumption; Although the special cement can achieve early strength and quick hardening, it is easy to produce strength reversion, set too fast, be greatly affected by temperature, and has high energy consumption, material pollution, and high cost in the production process, which does not conform to the development strategy of environmental protection in China, and is difficult to put into use in a large scale (Li *et al.* 2018; Shen *et al.* 2019). In view of the defects of current pavement repair materials, it is urgent to develop a green, high-quality, and efficient cement concrete pavement repair material suitable for low-temperature environments.

As a green material, geopolymer was first proposed by French scientist Davidovits (Davidovits 1989). It is mainly a kind of amorphous to semi-crystalline three-dimensional

*Corresponding Author: bwxkgy@163.com

 DOI: 10.1201/9781003410843-25

aluminosilicate structure cementitious material, which is generated by the excitation of alkaline activator with industrial solid wastes as raw materials (fly ash, blast furnace slag, etc.). As the preparation of geopolymer materials can consume a large amount of excess blast furnace slag and other silicon aluminum industrial wastes, it has low production energy consumption, low carbon dioxide emissions, good environmental friendliness, and can be widely used (Liu *et al.* 2019a, 2019b). On the other hand, because of its unique three-dimensional reticulated framework structure performance, it has the characteristics of early strength and rapid hardening, good stability, strong corrosion resistance, and durability (Li F. *et al.* 2022; Serhan & Ahmet 2022). Compared with cement-based materials, it has more excellent performance and broad application prospects in the field of road repair.

At present, domestic and foreign scholars have carried out relevant research on the mechanical properties of geopolymer materials. The results show that the mechanical properties of geopolymer materials are affected by many factors, such as the selection of raw materials, types and dosage of activator, solution cement ratio, types of additives, etc. (Huang *et al.* 2022; Sun *et al.* 2019). On the other hand, geopolymer materials react rapidly in normal and high-temperature environments, with fast setting speed and high strength. Therefore, there are many studies on material properties in normal and high-temperature environments (Çelikten *et al.* 2020; Sun *et al.* 2018). However, there are relatively few studies on the polymerization process, reaction product types, and strength development process of geopolymers in low-temperature environments. It is worth noting that temperature, as the most common change factor, has a great impact on mortar quality. The average temperature in most regions of China in spring and autumn and cold regions in summer is 0~10°C. Low temperature will affect the reaction process of geopolymer materials, thus hindering their early strength development. Therefore, a low-temperature environment (0~10°C) has more stringent performance requirements for rush repair materials, which needs further research and verification.

In view of this, under the condition of low-temperature curing, this paper selects fly ash and slag as two kinds of geopolymer to prepare geopolymer mortar and conducts a contrast test at 10°C. Through the early mechanical strength, appropriate geopolymer is selected as the main raw material; Secondly, study the influence of solution cement ratio and cement sand ratio on the basic working performance and mechanical properties of the geopolymer mortar at low-temperature environment, analyze and determine the best value of each factor, and finally determine the basic mix ratio of the geopolymer mortar, so as to provide a reference for further research and application of geopolymer mortar in the field of rush repair and construction projects in the next step.

2 TEST MATERIALS AND METHODS

2.1 *Test materials*

Slag: S95-type slag produced by Chengzun Mineral Products Processing Plant in Lingshou County, Hebei Province, with a density of 2.89 g/cm3. The main chemical components are shown in Table 1; Fly ash: Grade I fly ash produced by Henan Hengyuan New Materials Co., Ltd., the main chemical composition is shown in Table 1; Alkaline activator: 3401 sodium silicate prepared by Gulf Group is used, with the modulus of 3.3 and Baume degree of 40, in which Na2O content is 8.3%, and SiO2 content is 26.5%. Sodium hydroxide is flake sodium hydroxide provided by Baoshun Chemical Technology Co., Ltd., with a

Table 1. Chemical composition of Slag, metakaolin, and silica fume (wt/%).

Composition	SiO_2	Al_2O_3	Fe_2O_3	CaO	MgO	Na_2O	K_2O	SO_3	P_2O_5	Others
Slag	26.96	12.36	1.07	39.87	14.64	0.68	0.89	1.57	—	1.96
Fly ash	63.73	20.64	5.73	3.67	3.25	0.52	—	1.67	0.35	0.44

concentration of 99%; Water and sand: Baqiao tap water and Bahe medium sand are used, with a bulk density of 1510 kg/m3, fineness modulus of 2.7 and silt content of 1.5%.

2.2 *Preparation and testing of test pieces*

The preparation of test pieces requires five steps: preparation of the alkaline activator, weighing, mixing, molding, and curing. The specific preparation process of geopolymer mortar is shown in Figure 1.

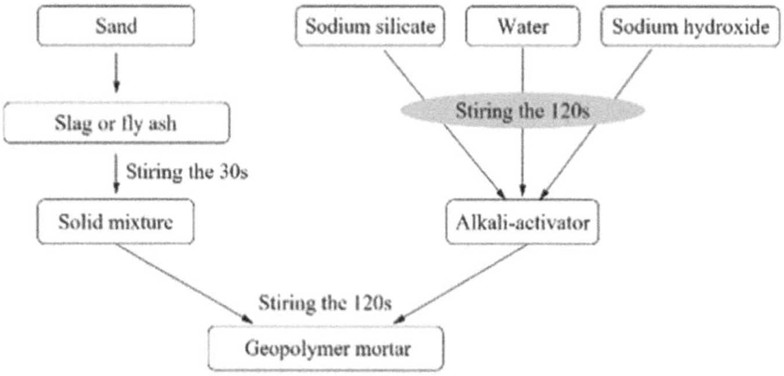

Figure 1. The preparation process of geopolymer mortar.

The specimen shall be prepared at an ambient temperature lower than 10°C, and then placed in a 10°C curing box for curing. Among them, the test piece for the fluidity test does not need to be formed and cured, and the mixed mortar can be directly used for the fluidity test; The test pieces for the compression and bending resistance test shall be poured with triple moulds, and each group shall prepare three 40 mm × 40 mm × 160 mm test piece.

In this paper, the fluidity of geopolymer mortar is measured by using an NLD-3 cement mortar fluidity tester produced by Beijing Zhongkeluda Test Instrument Co., Ltd. The test method is referred to as the Method for Determining the Fluidity of Cement Mortar (GB/T2419-2005); the compressive strength and flexural strength are measured using DYE-300-10S full-automatic bending and compression test machine. The test method is referred to as the Method for Testing the Strength of Cement Mortar (ISO Method) (GB/T 17671-1999).

3 RESULTS AND DISCUSSION

3.1 *Mechanical properties of different geopolymer raw materials*

Fly ash and slag are used as raw materials. The cement sand ratio is 1:2, the solution cement ratio is 0.55, the concentration of sodium hydroxide solution is 40%, and the mass ratio of sodium silicate to sodium hydroxide solution is 1:1. Fly ash-based geopolymer mortar and slag-based geopolymer mortar are prepared respectively for the test. A suitable geopolymer is selected as the main raw material. The two kinds of geopolymers are mixed and shaped according to the mix proportion. The fly ash-based geopolymer mortar and slag-based geopolymer mortar are put into the curing box with the temperature set at 10°C for curing, and their compressive strength and flexural strength are recorded for 3 hours, 1 day, and 3 days. The compressive and flexural strength of two kinds of geopolymer mortar is shown in Figure 2.

It can be seen from Figure 2: (1) With the increase in curing time, the compressive strength and flexural strength of fly ash-based geopolymer mortar gradually increase. The growth rates of compressive strength for 3 h~1 d and 1 d~3 d are 2.75 and 1.40, respectively, and the growth rates of flexural strength are 0.50 and 1.16, respectively. It can be seen that the early

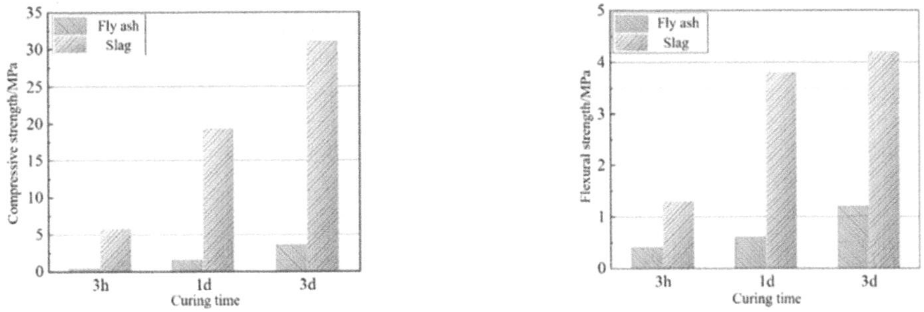

Figure 2. Compressive and flexural strength of two kinds of geopolymer mortar.

compressive strength of fly ash-based geopolymer mortar increases rapidly, while the later compressive strength increases slowly; On the contrary, its early flexural strength increased rapidly, and its later growth rate was slower, but the overall trend was increasing; (2) With the increase of curing time, the mechanical strength of slag-based geopolymer mortar shows an increasing trend. The compressive strength growth rates of 3h~1d and 1d~3d are 2.34 and 0.60, respectively, and the flexural strength growth rates are 1.92 and 0.11, respectively. It can be seen that the early strength growth rate of slag-based geopolymer mortar is fast, and the late growth rate is slow. The flexural strength and compressive strength have the same growth trend; (3) Under the same curing environment, the mechanical strength of fly ash-based geopolymer mortar and slag-based geopolymer mortar gradually increases with the increase of curing time. The strength of slag-based geopolymer mortar is much higher than that of fly ash-based geopolymer mortar. It can be seen that slag is more suitable as the raw material of geopolymer mortar than fly ash under low-temperature environments.

Therefore, slag is selected as the raw material for preparing geopolymer mortar. In the following tests, slag is used as raw material to determine the sol ratio, glue sand ratio, and sodium hydroxide concentration through comparative tests.

3.2 *Effect of cement sand ratio on geopolymer mortar*

The cement sand ratio is the mass ratio of geopolymer raw materials to standard sand. Select 1:1, 1:1.5, and 1:2 as three different mortar ratios, keep the solution cement ratio 0.55, the sodium hydroxide concentration 40%, and the mass ratio of sodium silicate to the sodium hydroxide solution 1:1 for the test. The influence of different cement sand ratios on the compressive and flexural strength of slag-based geopolymer mortar is shown in Figure 3.

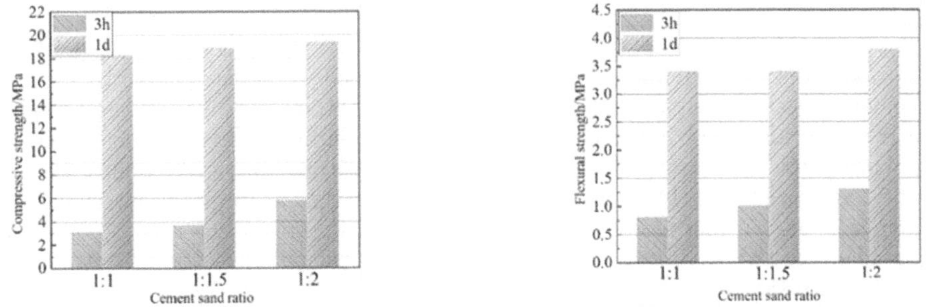

Figure 3. Influence of different cement sand ratios on mechanical properties of slag-based geopolymer mortar.

It can be seen from Figure 3 that: (1) With the increase of cement sand ratio, the compressive strength of slag-based geopolymer mortar shows a trend of gradual increase. When the curing time is 3h, the trend is most obvious. When the cement sand ratio is 1:2, the compressive strength of slag-based geopolymer mortar is the largest; (2) With the increase of cement sand ratio, the flexural strength of slag-based geopolymer mortar gradually increases in a small range, and the flexural strength reaches the maximum when the cement sand ratio is 1:2. Therefore, the best cement sand ratio in this test is 1:2.

3.3 *Influence of solution cement ratio on geopolymer mortar*

The solution cement ratio is the mass ratio of alkaline activator and geopolymer raw materials. Four different solution cement ratios of 0.5, 0.55, 0.6, and 0.65 are selected for the test. The cement sand ratio is 1:2, the sodium hydroxide concentration is 40%, and the mass ratio of sodium silicate to sodium hydroxide solution is 1:1. The influence of different solution cement ratios on the compressive and flexural strength of slag-based geopolymer mortar is shown in Figure 4.

It can be seen from Figure 4 that: (1) With the increase of solution cement ratio, the compressive strength of slag-based geopolymer mortar increases first and then decreases. With the increase of curing time, the trend is more and more obvious; (2) When the solution cement ratio is 0.55, the compressive strength of slag-based geopolymer mortar is the largest; (3) With the increase of solution cement ratio, the flexural strength of slag-based geopolymer mortar does not increase obviously. When the curing time is 1 d, the flexural strength almost has no obvious change. Therefore, the best solution cement ratio in this test is 0.55.

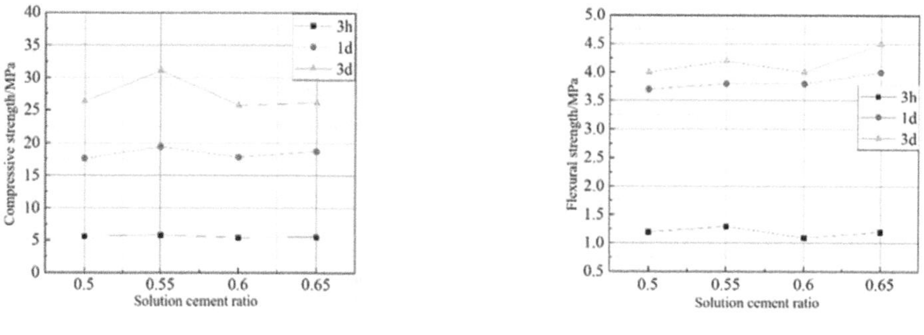

Figure 4. Influence of different solution cement ratios on mechanical properties of slag-based.

4 CONCLUSION

(1) Two kinds of raw materials, fly ash and slag, are used to prepare geopolymer mortar. Under the same curing environment, the mechanical strength of fly ash-based geopolymer mortar and slag-based geopolymer mortar will gradually increase with the increase of curing time. The strength of slag-based geopolymer mortar is much higher than that of fly ash-based geopolymer mortar, so slag is determined as the main raw material for preparing geopolymer mortar at low temperatures.

(2) Based on the early compressive strength and flexural strength of the geopolymer mortar, the best cement sand ratio and the best solution cement ratio of the geopolymer mortar are determined through comparative tests. The best cement sand ratio is 1:2, and the best solution cement ratio is 0.55, which lays the foundation for the next step of the slag base polymer mortar mix proportion optimization test.

REFERENCES

Çelikten S., Saidemir M. and Akçaözoğlu K. Effect of Calcined Perlite Content on Elevated Temperature Behavior of Alkali Activated Slag Mortars. *Journal of Building Engineering*. 2020, 32.

Davidovits J. Geopolymers and Geopolymeric Materials. *Journal of Thermal Analysis*. 1989, 35(2).

Huang Hua, Guo Mengxue, Zhang Wei, et al. Mechanical Properties and Microstructure of Fly Ash - Slag Based Polymer Concrete. *Journal of Harbin Institute of Technology*. 2022, 54(3):74–84.

Lee S. K, Jeon M. J, Cha S. S et al. Mechanical and Permeability Characteristics of Latex-Modified Fiber-Reinforced Roller-Compacted Rapid-Hardening-Cement Concrete for Pavement Repair. *Applied Sciences*. 2017, 7(7).

Li F. P., Chen D. F., Lu Y. Y. et al. Influence of Mixed Fibers on Fly Ash Based Geopolymer Resistance Against Freeze-thaw Cycles. *Journal of Non-crystalline Solids*. 2022, 584.

Li Haibin, Yang Tiejun, Li Tao,. et al. Main Factors Affecting Properties of Magnesium Phosphate Cement and its Development Prospect. *Construction Science and Technology*. 2018(19):55–60.

Liu Jizhong, Zhao Qingxin, Zhang Jinrui, et al. Microstructure and Composition of the Hardened Body of Alkali Slag - Slag Composite Cementitious Material. *Journal of Building Materials*., 2019a, 22(06):872–877. (In Chinese)

Liu Qing, Zang Haoyu, Wang Junxiang, et al. Review on Preparation and Properties of Slag-based Polymers. *Journal of Shandong University of Science and Technology (Natural Science)*. 2019b, 38(03):43–49.

Serhan İ., Ahmet Ö. Investigation of Mechanical Properties, High-temperature Resistance and Microstructural Properties of Diatomite-containing Geopolymer Mortars. *Arabian Journal of Geosciences*., 2022, 15(6).

Shen Yan, Zhang Wei, Chen Xi, et al. Research Progress of Sulphoaluminate Cement Modification. *Silicate Bulletin*., 2019, 38(3):683–687.

Sun J. W., Wang Z., Chen Z. H. *Hydration Mechanism of Composite Binders Containing Blast Furnace Ferronickel Slag at Different Curing Temperatures*. Springer Netherlands, 2018, 131(3).

Sun Jia, Lv Xuesen, He Yan, et al. Study on Migration and Distribution of Sodium Ion in Slag Aggregates Stimulated by Alkali. *Journal of Guangxi University (Natural Science Edition)*. 2019, 44(2):561–569. (In Chinese)

Zhao Z. Q., Ma Q. N., Xu Q., et al. A Review: Fast Repair Technology of Cement Concrete Pavement. *E3s Web of Conferences*, 2019, 136.

Constructional Engineering and Ecological Environment – Chih-Huang Weng (Ed)
© 2024 The Author(s), ISBN 978-1-032-53198-4

Research on the resilient and high-quality development of small towns based on AHP and entropy weight method: Take Wuhan as an example

Sirun Guo, Xiang Duan* & Haoni Zhang
School of Engineering and Architecture, Wuhan Polytechnic University, Wuhan, China

ABSTRACT: This study applies resilience to urban systems, constructs a small town resilience evaluation system, uses AHP and entropy weight to establish weights from four dimensions of economy, society, ecology and infrastructure, and calculates the resilience of small towns. It shows that economic and social are the main factors affecting the resilient development of small towns, and exploring the resilient development of small towns has important guiding significance for the high-quality development of small towns in Wuhan.

1 INTRODUCTION

As an important catalyst connecting cities and villages, small towns play a key role in the high-quality overall planning and construction of urban and rural areas. However, in the process of development, the impact of natural disasters and the lack of talent and technology have disturbed small towns. Building an urban resilience system has become an effective measure to resist uncertain risks. Scholars have introduced the concept of resilience into the field of urban disaster prevention and mitigation, which is helpful in addressing the impact of disasters.

Resilience is used to express the tendency of something to respond positively to difficulties and disturbances and to recover from damage (Manyena 2011). Foreign scholars' research on resilient cities focuses on their composition, ranging from the threat of disasters to the city to the micro-level, such as the ecological environment. Domestic scholars' research on urban resilience is in the stage of theoretical exploration. Chen constructed a framework for analysis of the urban industrial decline and transformation based on path dependence from the perspective of economic resilience (Chen 2018); Fan Le studied the safety resilience of southwest mountain towns based on crowd evacuation behaviour (Fan 2020).

2 CONSTRUCTION OF URBAN RESILIENCE EVALUATION INDEX SYSTEM IN WUHAN

2.1 *Selection of resilience indicators*

The selection method of community vulnerability indicators is based on foreign scholars Cova, Fernandez (Cova 1997; Cutter 2003; Fernandez 2002). We determine the indicators by reflecting the city's ability to resist risks and the ability to recover after the occurrence of risks and calculate the weights by using the analytic hierarchy process and the entropy weight method and using the resilience of cities and communities at home and abroad as a reference to build an urban resilience indicator system.

*Corresponding Author: 287164@qq.com

 DOI: 10.1201/9781003410843-26

2.2 *Building resilience systems*

2.2.1 *Data sources and normalization*

Taking small towns in Wuhan as the research object, an in-depth investigation of the factors affecting the development of resilience in small towns was conducted. Standardizing data of different dimensions enables data to be compared at the same level, condition, and system. The positive index is that the larger the index, the better the development of small towns. The negative index is that the smaller the index, the lower the development of the small town. The standardized calculation is shown in mathematical formulas (1) and (2).

$$a_{ij} = \frac{x_{ij} - \min(x_{ij})}{\max(x_{ij}) - \min(x_{ij})} \tag{1}$$

$$a_{ij} = \frac{\max(x_{ij}) - x_{ij}}{\max(x_{ij}) - \min(x_{ij})} \tag{2}$$

2.2.2 *Build a resilience indicator system*

Set four dimensions and 20 indicators. AHP determines the indicator weights as follows:

Construct a judgment matrix and calculate weights.

Let the judgment matrix be F. If F is a non-singular matrix, then take the vector corresponding to the largest eigenvalue of F as the weight vector. We calculate the index weight of the criterion layer (as shown in Table 1).

Table 1. Original judgment matrix of criterion level indicators.

Criterion layer	Economic dimension	Social dimension	Ecological environment dimension	Infrastructure dimension
Economic dimension	1	1/4	4	3
Social dimension	4	1	6	6
Ecological environment dimension	1/4	1/6	1	1/3
Infrastructure dimension	1/3	1/6	3	1

1) Do a consistency check

$$CI = \frac{\lambda_{\max} - n}{n - 1} = \frac{4.2153 - 4}{4 - 1} = 0.0718$$

2) Look up the table properly when n=4, the average consistency index value of the matrix RI=0.89.

$$CR = \frac{CI}{RI} = \frac{0.0718}{0.89} = 0.0808 < 0.1$$

When CR<0.1, the criterion-level judgment matrix passed the consistency check. The weights derived from this matrix are reasonable. In the same way, the above steps are also performed for the weight of the indicator layer.

The entropy weight method determines the index weight method as follows.

1) Build the initial matrix, Calculate information entropy

Build the initial matrix A= (X_{ij}) m×n, carrying out the standardized calculation of formulas (1) and (2). Normalizing each row and column, we get the matrix B= (Y_{ij}) m×n.

$$Y_{ij} = \frac{X_{ij}}{\sum_{i=1}^{m} X_{ij}} \tag{3}$$

$$e_j = \frac{1}{\ln m}\sum_{i=1}^{m}\left(Y_{ij} \times \ln Y_{ij}\right) \tag{4}$$

2) Calculate the weight of each indicator

Coefficient of Difference g_j=1-e_j, then the weight of the j indicator is calculated as follows.

$$\omega_j = \frac{g_j}{\sum_{j=1}^{m} g_j} \tag{5}$$

In order to avoid errors caused by different experts scoring, taking the arithmetic mean of the two methods as the comprehensive weight of each indicator (as shown in Table 2).

Table 2. Small town resilience evaluation index system.

Target layer	Criterion layer	Indicator layer	Indicator meaning	AHP weights	Entropy weight	Weight
Small town resilience evaluation index system	Economic Dimension Resilience	Local financial expenditure	Government Disaster Relief Capability	0.0629	0.0900	0.0765
		Urban GDP per capita	The economic strength of the residents	0.1034	0.0880	0.0957
		The proportion of primary industry	The proportion of urban disaster-vulnerable industries	0.0270	0.0081	0.0176
		Urban unemployment	The proportion of actual unemployed	0.0218	0.0291	0.0255
	Social Dimension Resilience	The proportion of the population over 60 years old	Aging population	0.0392	0.0034	0.0213
		The proportion of the population with a minimum living guarantee	The development level of urban relief work	0.0387	0.0134	0.0261
		The number of patients in the health center	Social security capacity	0.1528	0.2268	0.1898
		The proportion of employees in the tertiary industry	Urban Employment Structure	0.0587	0.0626	0.0607
		Annual Savings Balance of Urban Residents Per Capita	The degree of economic development and agglomeration of cities and towns	0.0552	0.0297	0.0425
		Population density	Population distribution	0.1639	0.1581	0.1610
		Health insurance rate	The overall disaster relief capacity of society	0.0959	0.1104	0.1032
	Ecological environment dimension Resilience	Green coverage in built-up areas	Urban ecological greening level	0.0058	0.0026	0.0042
		The green area per capita	Residential ecological greening level	0.0035	0.0005	0.0020
		Centralized treatment rate of urban domestic sewage	Ecological Environment Governance	0.0106	0.0105	0.0106

(*continued*)

Table 2. Continued

Target layer	Criterion layer	Indicator layer	Indicator meaning	AHP weights	Entropy weight	Weight
			Response			
		The comprehensivelization rate of industrial solid waste	Comprehensive utilization level of waste	0.0179	0.0223	0.0201
		Discharge of industrial wastewater per 10,000 yuan of GDP	Pollution intensity of urban rivers	0.0278	0.0299	0.0289
	Infrastructure dimension Resilience	Road area per capita	Urban traffic accessibility	0.0314	0.0338	0.0326
		Drain pipe length	Urban drainage capacity	0.0520	0.0556	0.0538
		Number of internet users	Status of urban communication equipment	0.0104	0.0182	0.0143
		Number of beds in medical institutions	Urban medical level	0.0211	0.0236	0.0224

3 EVALUATION OF HIGH-QUALITY RESILIENT DEVELOPMENT OF SMALL TOWNS IN WUHAN

3.1 *Resilience measurement of small towns*

We selected the data from 2007–2020 as the research interval for a comprehensive evaluation (See Table 3).

Table 3. Comprehensive assessment of resilience measurement in small towns.

Years	Economic Resilience	Social Resilience	Ecological Resilience	Infrastructure Resilience	Comprehensive Resilience
2020	0.1854	0.5148	0.0653	0.0552	0.3617
2019	0.1977	0.5229	0.0494	0.0896	0.3721
2018	0.1803	0.4683	0.0423	0.0393	0.3291
2017	0.1579	0.4307	0.0548	0.0401	0.3025
2016	0.1392	0.4370	0.0514	0.0472	0.3029
2015	0.1266	0.3733	0.0494	0.0483	0.2616
2014	0.1177	0.3099	0.0451	0.0248	0.2184
2013	0.1018	0.2308	0.0387	0.0376	0.1683
2012	0.0923	0.1461	0.0263	0.0394	0.1144
2011	0.0571	0.0925	0.0233	0.0232	0.0724
2010	0.0406	0.0807	0.0300	0.0168	0.0614
2009	0.0288	0.0870	0.0109	0.0126	0.0609
2008	0.0241	0.0513	0.0085	0.0045	0.0372
2007	0.0209	0.0332	0.0014	0.0012	0.0247

3.2 *Resilience outcome evaluation*

The comprehensive resilience of cities and towns in Wuhan has gradually increased from 2007 to 2020. The rate of increase has significantly accelerated since 2012-2016. COVID-19 has hit 2020. The level of resilience development has dropped, and all indicators show a downward trend. It makes the medical system out of step with residents' needs. The medical supply and the external environment cause an imbalance of supply and demand in cities and towns (Guo 2022). This imbalance affects social and infrastructural resilience.

From the perspective of economic resilience, the global economic crisis in 2008 had a certain impact on the urban system. The data showed that the investment in urban development decreased and the number of urban unemployed increased. After 2008, the external environment of the economy gradually recovered. The ability of urban systems to resist risks continues to grow. From the overall trend of changes in urban resilience (as shown in Figure 1), the economic resilience of small towns in Wuhan is relatively small, and the growth is relatively slow.

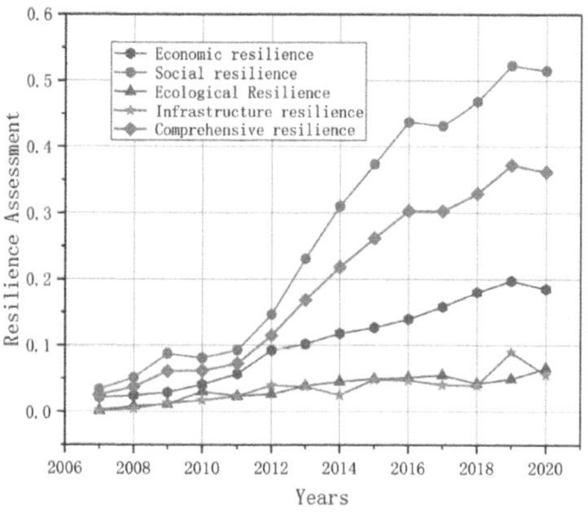

Figure 1. Tenacity development trend of small towns in Wuhan.

From the perspective of social resilience, the overall social resilience of small towns in Wuhan increased rapidly from 2013 to 2016 and above the overall resilience level. In 2017, there was a certain decline in social resilience. From the statistical results, it is due to the substantial decline in the number of urban medical insurance participants in 2017 that urban residents have difficulties in seeking medical care.

Infrastructure and ecological resilience are at a lower level than other dimensions. This could provide the conditions for towns to recover after facing a disaster. Buying critical recovery time for small towns and increasing redundancies for town resilience. At the same time, promote the improvement of the comprehensive resilience of cities and towns, and improve the adaptability and resilience of cities and towns to disasters.

3.3 *A comprehensive evaluation of resilience development level*

The small towns in Wuhan are mainly divided into three categories (See Table 4). The results show that urban resilience in Wuhan is at a medium level. With the economic development of Wuhan and the entry of new industries, the pace of industrial upgrading and transformation will be accelerated, and the development of related towns will be driven.

Table 4. Classification of urban development resilience levels in Wuhan.

Town Resilience Rating	High level of resilience ($r>0.5$)	Moderate level of resilience ($0.2<r\leq 0.5$)	Low level of resilience ($r\leq 0.2$)
Years	no	2014–2020	2007–2013

3.4 *Main research conclusions*

According to the resilience evaluation, the economic and social dimensions are the main factors. From the index level, the number of hospital visits, population density, and medical insurance rate in the social dimension is ranked first, second, and third. The economic dimension of urban per capita GDP and local fiscal expenditure ranks fourth and fifth. In the event of sudden disasters, the social affordability and economic level of small towns play an important role. On the one hand, it can resist and absorb the invasion and damage of external disasters; On the other hand, it can promote the rapid recovery of towns after a disaster. The length of drainage pipes, per capita road area, and industrial wastewater discharge of urban GDP in terms of ecological environment and infrastructure are ranked seventh, ninth, and tenth. This reflects that the defensive nature of infrastructure is a key factor in resisting risks and securing towns.

4 RESEARCH ON THE IMPROVEMENT STRATEGY OF URBAN RESILIENCE AND HIGH QUALITY IN WUHAN

4.1 *The level of urban resilience development needs to be improved*

The economic stability of small towns determines the degree of adaptation and coordination between systems. The development of small towns should fully combine the current level of economic development, and increase policy support for emerging parks, high-tech industries, etc. It is suggested to increase employment opportunities in small towns, reduce unemployment, and achieve stable economic growth. Focus on building an economic resilience evaluation system to promote the economic development of small towns in Wuhan. Second, the level of social development is an important measure of the ability of small towns to cope with changes in risk. It is necessary to increase the urban medical insurance rate, enhance the overall disaster relief capacity of cities and towns, and ensure that the urban system can respond quickly after a disaster occurs.

4.2 *Strengthen infrastructure construction and attach importance to ecological environment protection*

To improve urban traffic accessibility, it is necessary to strengthen the construction of smart towns. big data, Internet+, etc. are used to analyze the communication and traffic construction of small towns, set up an early warning mechanism, and carry out effective identification and investigation. It is necessary to improve the medical and health level of small towns. Especially in the post-epidemic era, the normalization of epidemic prevention has become the general trend. In the era of "carbon neutrality and carbon peaking", improving the comprehensive utilization rate of solid waste and the level of ecological greening, increasing urban pocket parks, green space coverage and river greening efforts. It can guide the public to protect the ecological environment, effectively manage urban water systems and rivers, and enhance the resilience of the urban ecological environment.

4.3 *Coordinate the urban system and promote balanced regional development*

There are large spatial and level differences in the development of small towns in Wuhan, which are divided into regional centre type, urban-suburban integration type, etc. Balance the layout of different types of towns to avoid the rapid development of the central urban area and the pressure of high-density population due to agglomeration of population. Formulate policy guidance for small towns in the far urban areas, drive the development of surrounding areas, and form a balanced urban development pattern dominated by central urban areas and guided by surrounding towns.

4.4 *Focus on building a resilience evaluation system and establishing an effective mechanism for resilience development*

We must give full play to the role of the market in resource allocation, focus on building an urban resilience evaluation system and promote urban economic and social development. Regularly evaluate the resilience development of small towns in Wuhan with the evaluation index system, examine the trend of resilience development in small towns, and formulate a scientific and reasonable resilience development mechanism to promote the rational use of resources and help the resilience development of small towns.

5 CONCLUSION

On the basis of analyzing and studying the resilience development of urban communities at home and abroad, this research constructs an evaluation system for the resilience development of small towns in Wuhan and measures the resilience of small towns from 2007 to 2020. The results show that the resilience development of small towns is at a moderate level, and the overall resilience of small towns shows a steady growth trend. The resilience development level of each sub-unit varies. In this regard, this paper puts forward countermeasures and suggestions for the resilient development of small towns in Wuhan from the aspects of urban economy, social development level, infrastructure construction, ecological, environmental protection, urban system coordination, and resilience system construction. This provides a reference for the resilient development of small towns in Wuhan.

ACKNOWLEDGMENTS

I would like to express my gratitude to all those who have helped me during the writing of this thesis. I gratefully acknowledge the help of my supervisor Professor Xiang Duan. I appreciate his patience, encouragement, and professional instructions during my thesis.

REFERENCES

Cova T. J. *et al.* Modelling Community Evacuation Vulnerability Using GIS. *Int. J Geographical Information Science.* 1997(11): 763–784.

Cutter S. L. Social Vulnerability to Environmental Hazards. *Social Science Quarterly.* 2003, 84(2):242–261.

Chen Zuoren, Li Xun. Path Dependence and Path Creation of Industrial Evolution About Towns From The Perspective of Economic Resilience: Based on the Comparative Analysis of Zhang Mutou and Changping in Dong Guan. *Human Geography.* 2018, 33(04):113–120.

Fernandez *et al.* Frail Elderly as Disaster Victims: Emergency Management Strategies. *Prehospital and Disaster Medicine.* 2002, 17(2): 67–74.

Fan Le, Wang Yanyu, *et al.* Safety Resilience Promotion Strategies for Mountainous Neighbourhoods in Southwestern China Based on Crowd Evacuation Behaviour. *Journal of Tsinghua University (Science and Technology).* 2020, 60(01):32–40.

Guo Sirun, Duan Xiang, Zhang Ying. Research on classified guidance construction strategy of small towns in Wuhan[J]. Journal of Wuhan Polytechnic University, 2022, 41(02):80–85+106.

Manyena S. B., O Brien G., O Keefe P. *et al.* Disaster Resilience: ¿a Bounce Back or Bounce Forward Ability?. *Local Environment.*, 2011, 16(05): 417–424.

Constructional Engineering and Ecological Environment – Chih-Huang Weng (Ed)
© 2024 The Author(s), ISBN 978-1-032-53198-4

Systematic analysis of PMV and SET* in an imagined normal environment

Xiaodan Huang*, Yi Xun, Cheng Zhao & Yixuan Peng
School of Art and Design, Guangdong University of Technology, Guangzhou, China

Wenzhi Guo
Architectural Design & Research Institute of South China University of Technology Co., Ltd, Guangzhou, China

ABSTRACT: The Predicted Mean Vote (PMV) and Standard Effective Temperature (SET*) are widely used models in international standards to quantify thermal comfort. Nonetheless, literature is scarce on the quantification of the interrelationship between these two models, demanding a systematic study. Therefore, in this study, the interrelationship between PMV and SET* in an assumed normal environment is investigated. For this purpose, a large set of data samples was generated by considering various thermal variables at different levels within certain threshold limits and employed in the designated model formulation of PMV and SET*. The mathematical models were established to quantify the interrelationship between PMV and SET* for various clothing insulation (I_{cl}). In addition, frequency distributions of PMV and SET* based on established models were evaluated, which showed that the relevant relationship between the proposed models was substantially based on computation outputs rather than computation/heat variables. The parametric analyses showed that these models were significantly and synergistically influenced by different heat variables, and their selection may play an important role in defining thermal comfort using these two models. The results of both models were influenced by air velocity (V) at the lower operating temperature (T_{op}) and relative humidity (φ) at the higher Top in the sedentary circumstance. Nevertheless, with the elevation of the M, the results of SET* were merely influenced by φ. Furthermore, the difference between PMV and SET* decreased with the increase in I_{cl}. This study provides a theoretical perspective to study the association between PMV and SET* with different thermal variables. It can also serve as a reference for researchers to apply the models precisely.

1 INTRODUCTION

Thermal comfort (TC) not only influences human health, perception and behavioural features but also affects the energy consumption patterns in buildings. Recently, energy consumption in buildings has been elevated to a significant scale; thus, the control, forecast and assessment of the TC are vital for promoting sustainable energy consumption patterns. This implies that valid modelling for the assessment and forecasting of TC is imperative for environmental evaluation and societal sustainable development. In this regard, several TC models have been developed which are used for different purposes (Auliciem & Szokolay 2007). For instance, these models are used to set up exposure limits (i.e., wet bulb globe temperature, predicted 4-hour sweat rate), evaluate post exposures optimum control measures indices (i.e., heat stress indicator, an indicator of thermal stress and subjective temperature), determine climate classification indices (i.e., equatorial comfort index and tropical summer index) and define the

*Corresponding Author: dandyhuang@gdut.edu.cn

DOI: 10.1201/9781003410843-27

comfort indices (i.e., predicted mean vote (PMV) and standard effective temperature (SET*)). Meanwhile, the selection of an effective TC index is critical to designing a successful and optimal new heating, ventilation and air conditioning (HVAC) control system. Recently, the development of these control systems has shifted to comfort indices-based control; thus, they are of utmost importance for sustainable and productive HVAC design.

The PMV proposed by Fanger (1970) and SET* proposed by Gagge et al. (1986) are the most commonly utilized comfort indices owing to the fact that they are accepted by globally recognized standards (American Society of Heating, Refrigerating and Air-Conditioning Engineers 2020, 2021). PMV is computed based on the steady-state model of thermal exchanges between the human body and its surroundings. Meanwhile, SET* is based on a dynamic two-node model of human temperature regulation, which is different and more complicated than the steady-state model. Substantial efforts have been made to explore the applicability of these two modelling methods in different areas and building types. Cheung et al. (2019) employed the ASHRAE criterion to evaluate the accurateness of PMV to assess TC in different kinds of architectural structures, ventilation systems and climatic regions, which were found to be dissimilar. Further, Broday et al. (2019) employed PMV to quantify thermosensation and observed that it yields inaccurate results in this regard.

Further, different researchers have also worked to reinforce the accurateness of PMV by proposing different modelling methods. For instance, Kim et al. (2015) proposed adaptation PMV modelling methods to improve the TC prediction ability. Similarly, Li et al. (2018) explored heat responses to assess the impact of humidity on TC based on the SET* modelling method and proposed a correction factor to reduce the deviation of SET*. Moreover, Zhang *et al.* (2020) discovered that using the adaptation TC theory to build and modify the SET* model could improve the results by demanding a thorough examination of the TC concerns.

Different researchers have also worked on comparing the performance of PMV and SET*. For instance, Gao et al. (2015) displayed the contrasts between adjusted PMV modelling methods and SET* modelling methods on thermosensation forecast in natural ventilation systems. They observed that the latter was more precise for thermosensation forecasts than the former. Zheng et al. (2021) highlighted the inefficiency of the PMV and SET* model in predicting the TC of prefab buildings. They also extended the applicability of the PMV and SET* modelling methods to prefab buildings by modifying the numerical values in these models.

Moreover, PMV and SET* modelling methods have been utilized to assess the TC in various climatic regions, as documented by various studies (Sirhan & Golan 2021; Zhang & Lin 2021; Zhang et al. 2021). On the other hand, both PMV and SET* have advantages and disadvantages of their own. Moreover, limited studies have been carried out to establish a relationship between PMV and SET*. Due to the vast application of PMV and SET*, it is necessary to establish a relationship between them for various conditions theoretically.

Consequently, the present study is aimed provide a comprehensive analysis of the PMV and the SET* modelling methods in the imagined normal environment. For this purpose, the association between these two models and the roles of heat variables are studied. In addition, certain prevailing issues regarding the performance of these two models are discussed. This study theoretically reveals the interrelationship between these two modelling methods and provides clear orientation and guidance for the utilization of these two modelling methods.

2 MATERIALS AND METHODS

2.1 *Thermal comfort model description*

PMV model

The PMV model was established based on Fanger's formula for the thermal exchange of the human body (Fanger 1970, 1967). Wherein the indoor heat adaptability can be realized under specific environments by using four environment-related variables, i.e., air temperature (T_a) in °C, relative humidity (φ), air velocity (V) in m/s, average radiant temperature (T_r) and two individual variables, i.e., metabolism rate (M) in W/m^2 and clothing insulation (I_{cl}) in clo (Fanger 1970; Schellen et al. 2013). The PMV modelling equation can be given as follows:

$$\begin{aligned} PMV = & \left[0.303 \times e^{(-0.036*M)} + 0.028\right] \times \{(M - W) - 3.05 \times [5.73 - 0.007 \times (M - W) - P_a] \\ & \times 0.42 \times [(M - W) - 58.15] - 0.0173 \times M \times (5.87 - P) - 0.0014 \\ & \times M \times (34 - T_a) - 3.96^{-8} \times f_{cl} \times [(T_{cl}+273)4 - (T_r+273)4] - f_{cl} \times h_c \times (T_{cl} - T_a) \end{aligned} \quad (1)$$

Where W denotes the physical work rate in W/m^2; P_a denotes the pressure of water vapour in the environment in kPa; f_{cl} denotes the ratio of the superficial region covered in clothes to the nude superficial region; T_{cl} denotes the average temperature of clothing in °C; h_c denotes the convective thermal transference coefficient in $W/(m^{2}*°C)$.

A 7-point scale based on PMV is used to define thermal sense as per ARSHER criterion 55 (American Society of Heating, Refrigerating and Air-Conditioning Engineers 2020) (Table 1). Overall, the PMV result of 0 denotes a neutral environment, with an appropriate TC in a range of ± 0.5 PMV. In the past study (Doherty & Arens1988; Humphreys 2007; Howell & Stramler 1981; Parsons 2002), the PMV results were discovered to vary remarkably in contrast to the real thermosensation when the PMV was over 2 (warm) because of the elevation of the evaporation-related thermal dissipation of sweat, revealing the restricted accurateness of PMV. Moreover, the PMV modelling method is restricted to the prediction of TC in a stable environment rather than a kinetic environment (Humphreys & Nicol 2002).

Table 1. The scale of the PMV modelling method.

–3	–2	–1	0	+1	+2	+3
Cold	Cool	Slightly cool	Neutral	Slightly warm	Warm	Hot

SET* model

SET* is an indicator of human thermal stress based on a two-node modelling method. Contrary to PMV, SET* can be utilized to forecast TC in a kinetic environment (Du & Yang 2020). The two-node model considers the body to have two concentric thermal compartments, the body core and skin, wherein metabolic heat is generated at the former compartment and released through respiration and the latter to the environment. Consequently, the basic Equation of the two-node modelling framework can be written as follows:

$$M - W = Q_{sk} + C_{res} + E_{res} + S \quad (2)$$

Where Q_{sk} denotes the overall thermal loss from the skin in W/m^2; C_{res} denotes the convection thermal loss from breathing in W/m^2; E_{res} denotes the evaporation-related thermal loss from breathing in W/m^2; S denotes the thermal storage in W/m^2.

By adjusting the temperature of the skin (T_{sk} in °C) and the wetness of skin (w), which are acquired using six variables (i.e., Ta, φ, V, T_r, I_{cl} and M) into the two-node modelling framework, SET* can be computed as follows:

$$h \times \left(T_{sk} - T_{op}\right) + w \times h_e \times (P_{sk} - P_o) = h' \times \left(T'_{sk} - SET*\right) + w \times h_e \times (P_{sk} - 0.5P_{SET*}) \quad (3)$$

Where h denotes the sensible thermal transference coefficient in W/ (m^2*°C); h_e denotes the evaporation-related thermal transference coefficient in W/(m^2*kPa); P_{sk} denotes the saturated vapour pressure at the superficial region of the skin in kPa; P_o denotes the vapour pressure at the real environment in kPa; P_{SET*} denotes the saturated vapour pressure at the environment temperature, which is equal to SET* in kPa.

It is important to note that SET* is the temperature of the air of an imagined environment at 50% φ, wherein the skin temperature and moisture of an imaginary occupant wearing specific clothing and engaging in certain activities are the same as those of a person in the actual environment (American Society of Heating, Refrigerating and Air-Conditioning Engineers 2020, 2021). However, because the heat dissipation of the skin varies greatly, it is difficult to obtain identical skin temperature and skin moisture for humans in imagined and real environments at a high metabolic rate (over 1 met) (Du & Yang 2020). As a result, the SET* model is only adequate for measuring thermal comfort at low metabolic rates and should be changed in the event of high metabolic rates.

2.2 *Thermal comfort model description*

For the analysis of these two models in the imagined environment, the heat variables related to the PMV model, SET* model, and their levels were chosen. Literature manifests that T_a, T_r, φ, V, M and I_{cl}, are the key influencing factors in the computation of these two models. owing to the close relationship between T_a and T_r, they can be substituted by T_{op} (American Society of Heating, Refrigerating and Air-Conditioning Engineers 2021). Table 2 presents the selected factors and levels in this study. In this study, the T_{op} was considered in a range of 10°C to 40°C in 16 levels; φ ranged from 30% to 90%; V ranged from 0.1 m/s to 0.5 m/s; M was divided into 3 levels (1met, 1.5met, and 2met); I_{cl} was separated into 0.5clo for summer and 1clo for winter.

Each level of parameters and the computed values of PMV and SET* are combined as a data sample in this study. Different combinations of parameters at different levels were used to compute PMV and SET*. Consequently, 3360 data samples were generated to study the association and the frequency distributional status of PMV and SET* and the roles of influencing factors in these two models.

Table 2. Data on heat variables and their levels.

Level / Factor	T_{op} (°C)	φ (%)	V (m/s)	M (met)	I_{cl} (clo)
1	10	30	0.1	1	0.5
2	12	40	0.2	1.5	1
3	14	50	0.3	2	
4	16	60	0.4		
5	18	70	0.5		
6	20	80			
7	22	90			
8	24				
9	26				
10	28				
11	30				
12	32				
13	34				
14	36				
15	38				
16	40				

3 MATERIALS AND METHODS

3.1 *The equations between PMV and SET**

PMV equations

Figure 1 presents the relationship between PMV with SET* for various I_{cl}, i.e., 0.5clo, 1clo and total. It can be observed that generally, PMV increases as the SET* increases for all I_{cl}. The rate of increase is higher initially and then tends to reduce, showing a quadratic relationship between PMV and SET*. Consequently, the equations display the mathematical relationship to predict PMV using SET* as the predictor for different I_{cl} (0.5clo, 1clo and total). It is important to note that the PMV scale between -3 and +3 is significant. Therefore, these equations are derived for this threshold range only. Table 3 presents predicted values of PMV at various I_{cl} based on SET* using equations. It can be observed that the linearity fitting of these two models displays a curvature association and a positive relationship based on the above results. The results of PMV rise in tandem with the results of SET*.

Furthermore, the slope of PMV in 0.5clo is the highest, referring to the fact that the rate of increase in PMV is larger in summer clothing than in winter clothing. Furthermore, when the PMV result is greater than 7, the SET* results cannot be computed. Even if a PMV result of more than 3 is not substantial, such a situation is inescapable.

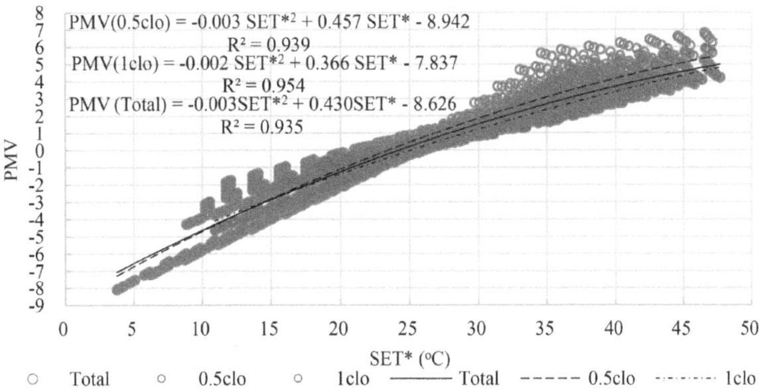

Figure 1. Modelling PMV based on SET* for different I_{cl}.

Table 3. Predicted values of PMV for various I_{cl} using SET* as a predictor.

SET* (°C)	PMV(0.5clo)	PMV(1clo)	PMV(Total)
15	–2.8	–2.8	–2.9
18	–1.7	–1.9	–1.9
21	–0.7	–1.0	–0.9
24	0.3	–0.2	0.0
27	1.2	0.6	0.8
30	2.1	1.3	1.6
33	2.9	2.1	2.3

SET* equations

Figure 2 presents a change in SET* in connection with PMV for different I_{cl} (i.e., 0.5clo, 1clo and total). It can be observed that SET* also increases with the increase in PMV for all I_{cl}; the rate of increase in SET* value increases with the increase in the PMV. Consequently, these equations present the mathematical models of SET* based on PMV for different I_{cl} (0.5clo,

1clo and total). Table 3 presents predicted values of SET* at various I_{cl} based on PMV using these equations. The results show a direct relationship between these two models with a curvature association. The SET* findings are accompanied by an increase in PMV results.

Furthermore, the fastest rate of SET* development is in 1clo, indicating that SET* results increase faster in winter garments than in summer clothes. Furthermore, the outcomes of these equations are more logical than those in the equations in Figure 1, owing to the fact that each PMV value may be used to compute the matching SET* value using given equations if the SET* is larger than 0. As a consequence, models in Figure 2 and corresponding outcomes are used in the subsequent analysis of PMV and SET* for the sake of research precision.

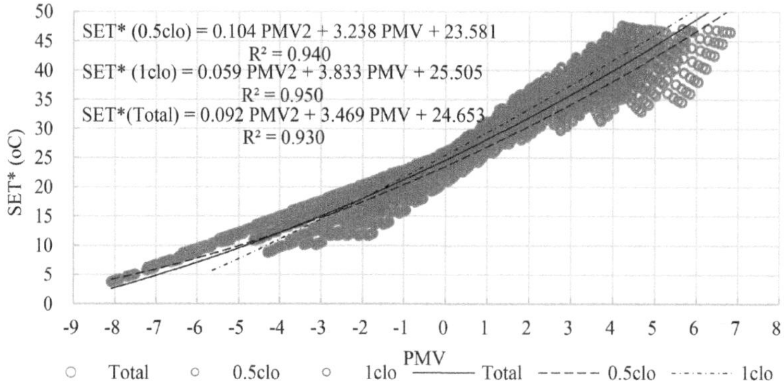

Figure 2. Modelling SET* based on PMV for different I_{cl}.

3.2 *The equations between PMV and SET**

The frequency distributions of PMV and SET* models are shown in Figure 3, based on the corresponding relationship between SET* and PMV, as given in Table 4. The frequency distributional status shows that each scale of these two models occupies a different number of samples, with the scale of x ≥ 3(35.9) occupying the most samples and the scale of (15.1)-3 ≤ x<-2 occupying the least (18.1). Furthermore, on a scale of (32)2 ≤ x<3(35.9), the frequency of these two models is the most similar, with just a 0.1 percent difference. Meanwhile, the least similarity (2.6% of difference) is observed on the scale of (21.3)-1 ≤ x<0(24.7).

The findings show that the relationship between these two models is substantially based on computation outputs rather than computation variables. For instance, for a set of heat variables related to the computation (T_{op}= 32°C, φ= 30%, V= 0.1m/s, I_{cl}= 0.5clo and M=1 met) the computed outcome of PMV was 2 (Equation in Figure 1) and SET* was 30.4°C (Equation in Figure 2), which are not in the identical scale even with the same heat variables. However, it could be matched by the computational outcomes using the equations in Figure 2.

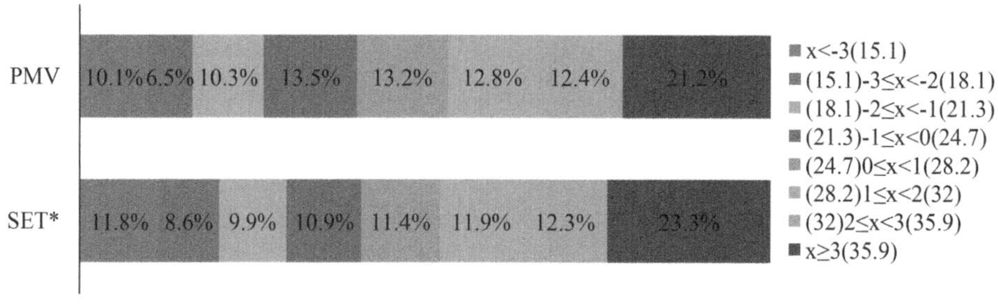

Figure 3. Frequency distributions of PMV and SET*.

3.3 *Roles of heat parameters in these two models*

The impact of different heat variables (i.e., φ, V, I_{cl}, and M) on variations of PMV and SET* results with varying T_{op} are presented in Figure 4. The heat variables were intuitively used as maximal and minimal ones in the current study to reflect the relationship between TC and heat variables more explicitly. Therefore, φ was used as 30% and 90%, V was 0.1 m/s and 0.5 m/s, the M was 1met and 2met; I_{cl} was 0.5clo and 1clo. Further, for uniformity, the PMV values were translated into SET* results using the equations in Figure 2.

Table 4. Predicted values of SET* for various Icl using PMV as a predictor.

PMV	SET*(0.5clo) (°C)	SET*(1clo) (°C)	SET*(Total) (°C)
+3	34.2	37.5	35.9
+2	30.5	33.4	32.0
+1	26.9	29.4	28.2
0	23.6	25.5	24.7
–1	20.4	21.7	21.3
–2	17.5	18.1	18.1
–3	14.8	14.5	15.1

At lower and higher T_{op}, respectively, the values of PMV and SET* increase with the increase in T_{op}, which is mostly influenced by V and φ (Figure 4(a)) when M and I_{cl} are set at 1met and 0.5clo, respectively. Furthermore, starting at a temperature of 24°C, the differentiation between SET* values for 30 and 90 percent φ becomes more obvious. Furthermore, at the high T_{op}, the rate of increase in PMV values is higher than SET* values (above 30°C) (Figure 4(a)). Moreover, it can be observed that PMV and SET* values in Figure 4(b) are higher than those in Figure 4(a) as the M is kept at 1met and I_{cl} is increased to 1clo. Nonetheless, when the T_{op} rises, the rate of increase in PMV and SET* decreases in Figure 4(b). Meanwhile, values of PMV and SET* in the case of I_{cl} equals 1clo at 40°C are comparable to those in the case of 0.5clo at 40°C. Meanwhile, as shown in Figure 4(b), the values of PMV and SET* are mostly influenced by V at the lower T_{op} and φ at the higher T_{op}. Furthermore, from a T_{op} of 24°C, the difference in SET* caused by φ rises. Furthermore, the difference between PMV and SET* at 1met M and 0.5clo I_{cl} is less than the difference between 1met and 0.5clo of M and I_{cl}, respectively.

Moreover, increasing V and φ at lower and higher Top, respectively, impact the PMV values, as the M is increased to 2 met (Figure 4(c)); however, φ is the only influencing element for SET* values. Furthermore, with an increasing Top from 16°C, the difference in SET* caused by φ increases. The value of PMV at the lower T_{op} (i.e., 10°C and 24°C) is greater than that of SET*, but the growth rate of PMV is slower than that of SET*. The growing rate of PMV decreases with increasing T_{op} as the M increases (Figure 4(c)). Moreover, the PMV and SET* values increase as the I_{cl} is increased to 1clo while keeping M at 2met, but the growth rate slows down as the T_{op} rises (Figure 4(d)). Similar to (Figure 4(c)), V at the lower T_{op} and φ at the higher T_{op} has the main influence on PMV values, whereas SET* values are merely affected by φ. Furthermore, with an increase in T_{op} from 10°C, the difference in SET* impacted by φ rises. Moreover, the difference between PMV and SET* at 2met M and 1clo I_{cl} is lower than the difference between 2met and 0.5clo of M and I_{cl}, respectively. Thus, SET* and PMV models are significantly and synergistically influenced by different heat variables, and their selection may play an important role in defining TC using these two models.

4 DISCUSSION

Previously, field surveys were used to investigate the relationship between the comfort vote (CV) and SET*. Gonzalez et al. (1974) investigated the relationship between the anticipated

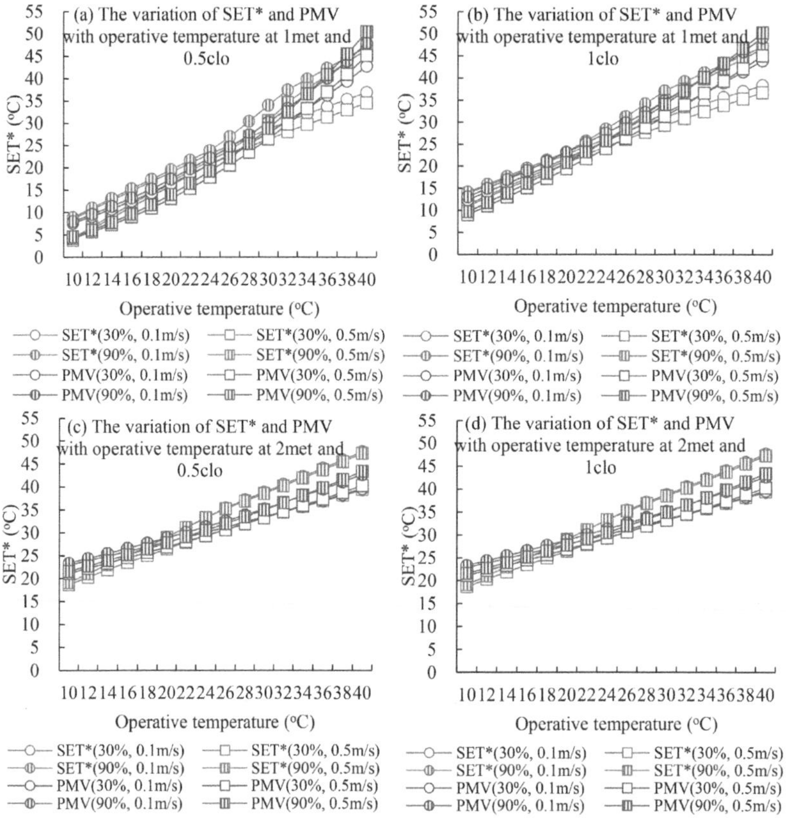

Figure 4. The variation of SET* and PMV with operative temperature.

CV and SET* using physiology and physics-based theory. Gagge (1971) demonstrated the link between the proposed valid temperature scale for inactive people, thermosensation, physiologic responses, and health variables. Furthermore, Gagge et al. (1986) discovered a link between the ambient temperature under saturated conditions, PMV, and the SET* in moderately active adults. Prior to summarizing the findings of previous studies, ASHRAE (Auliciem & Szokolay 2007) postulated a relationship between SET*, CVs, thermosensation, and physiological activity in the sedentary environment. Table 5 displays the SET* for different I_{cl} (i.e., 0.5clo, 1clo and total) obtained in this study and SET* based on ASHRAE in correspondence to the CV to analyze the relationship between the CV and SET* in the imagined normal condition and the real circumstance.

The SET* for ASHRAE is comparable to the SET* results in the current work at the CV=0. Wherein a marginal difference between SET* based on ASHRAE against SET* for 0.5clo, 1clo and total I_{cl} in this study was observed, which was around 0.3°C, 1.6°C, and 0.8°C, respectively. At a CV of 3, the SET* for ASHRAE was greater in comparison to the SET* in the current study except for SET* for 1clo of I_{cl}. Furthermore, at CV<0, SET* for ASHRAE is observed to be much lower than SET* in the current paper, with the highest difference at CV=-3. Furthermore, the findings of SET* for 0.5clo and 1clo of I_{cl} show coherence at the CV of -3.

Further, the increasing rate of SET* for 1clo of I_{cl}. was faster in comparison to that for 0.5clo of I_{cl}. with the increases in CV. The real human thermal perception in SET* for ASHRAE is used to compute the CV, whereas the PMV equation in SET* for different I_{cl} (i.e., 0.5clo, 1clo and total) is used to calculate the CV. Thus, these results demonstrate that the SET* in the imagined and real conditions differed due to the difference in the acquisition methodology of CV. These findings imply that the corresponding relationship between the CV and SET* could be acquired for CV=0, and it should be taken into consideration in other comfort stats as per subjected conditions.

Table 5. The comparison between SET* for various I_{cl} in the present study and SET* for ASHRAE in correspondence to the CV.

CV	SET*(0.5clo) (°C)	SET*(1clo) (°C)	SET*(Total) (°C)	SET* (ASHRAE) (°C)
+3	34.2	37.5	35.9	37.5
+2	30.5	33.4	32.0	34.5
+1	26.9	29.4	28.2	30.0
0	23.6	25.5	24.7	23.9
–1	20.4	21.7	21.3	17.5
–2	17.5	18.1	18.1	14.5
–3	14.8	14.5	15.1	10.0

5 CONCLUSIONS

The following are the main outcomes of the current study.

1) The mathematical models were established for the interrelationship of PMV and SET* under the I_{cl} of 0.5clo for summer, 1clo for winter, and total for the entire year based on regression analyses of 3360 specimens in the assumed normal environment. The models to predict SET* by using PMV as a predictor were regarded as the logically accurate models in this study.
2) Statistical distribution analyses showed that the interrelationship between the PMV and SET* models was substantially based on computation outputs rather than computation/heat variables.
3) SET* and PMV models were significantly and synergistically influenced by different heat variables, and their selection may play an important role in defining TC using these two models. The results of both models were influenced by V at the lower T_{op} and φ at the higher T_{op} in the sedentary circumstance. Nevertheless, with the elevation of the M, the results of SET* were merely influenced by φ. Further, the values of both models increased with the elevation of I_{cl} whereas the increase rate slowed down. Furthermore, the difference between PMV and SET* decreased with the increase in I_{cl}.
4) SET* value based on ASHRAE was found to be comparable to the SET* results in the current study at CV=0. Meanwhile, a difference in SET* is observed at other CVs for the current study and ASHRAE values, which corresponds to the difference in the acquisition methodology of CV at imagined and real conditions.

DECLARATION OF COMPETING INTEREST

The authors declare that they have no known competing financial interests or personal relationships that can have appeared to affect the work reported in this study.

ACKNOWLEDGMENTS

This work was supported by the Guangdong Philosophy and Social Science Planning Project (No. GD21YYS03). The subjects who volunteered for this survey were gratefully acknowledged.

REFERENCES

ASHRAE Standard 55-2020: *Thermal Environmental Conditions for Human Occupancy*. Atlanta GA: American Society of Heating, Refrigerating and Air-Conditioning Engineers, Inc. (2020).

ASHRAE: Chapter 9, Thermal Comfort, ASHRAE Handbook – fundamentals. American Society of Heating, Refrigerating and Air Conditioning Engineers. Inc, Atlanta. (2021).

Auliciem A. and Szokolay S.V.: PLEA Note3: Thermal Comfort. PLEA in Association with the Department of Architecture University of Queensland, Qld, Australia. (2007).

Broday E.E., Moreto J.A., Xavier A. *et al.*: The Approximation Between Thermal Sensation Votes (TSV) and Predicted Mean Vote (PMV): A Comparative Analysis. *International Journal of Industrial Ergonomics*. 69:1–8 (2019).

Cheung T., Schiavon S., Parkinson T. *et al.*: Analysis of the Accuracy on PMV-PPD Model Using the ASHRAE Global Thermal Comfort Database II. *Building and Environment*. 153: 205–217 (2019).

Doherty T.J., Arens E.: Evaluation of the Physiological Bases of Thermal Comfort Models. *ASHRAE Transactions.* 94 (1): 1371–1385 (1988).

Du H. and Yang C.: Re-visitation of the Thermal Environment Evaluation Index Standard Effective Temperature (SET*) Based on the Two-node Model. *Sustainable Cities and Society*. 53: 101899 (2020).

Fanger P.O.: Calculation of Thermal Comfort-introduction of a Basic Comfort Equation. *ASHRAE Transactions.* 73 (2): 1–20 (1967).

Fanger P.O.: *Thermal Comfort–Analysis and Application in Environmental Engineering*. pp. 142–155. Copenhagen: Danish Technology Press, Danish (1970).

Gagge A.P., Fobelets A.P. and Berglund L.G.: A Standard Predictive Index of Human Response to the Thermal *Environment. ASHRAE Transactions.* 92(2B), 709–731 (1986).

Gagge A.P.: An Effective Temperature Scale Based on a Simple Model of Human Physiological Regulatory Response. *ASHRAE Transactions.* 77 (1): 247–262 (1971).

Gao J., Wang Y. and Wargocki P.: Comparative Analysis of Modified PMV Models and SET Models to Predict Human Thermal Sensation in Naturally Ventilated Buildings. *Building and Environment*. 92: 200–208 (2015).

Gonzalez R.R., Nishi Y., Gagge A.P.: Experimental Evaluation of Standard Effective Temperature a New Biometeorological Index of Man's Thermal Discomfort. *International Journal of Biometeorology*. 18(1): 1–15 (1974).

Howell W.C. and Stramler C.S.: The Contribution of Psychological Variables to the Prediction of Thermal Comfort Judgments in Real World Settings. *ASHRAE Transactions.* 87 (5): 609–621 (1981).

Humphreys M.A. and Hancock M.: Do People Like to Feel 'Neutral'?: Exploring the Variation of the Desired Thermal Sensation on the ASHRAE Scale. *Energy and Buildings*. 39 (7): 867–874 (2007).

Humphreys M.A. and Nicol J.F.: The validity of ISO-PMV for predicting comfort votes in everyday thermal environments. *Energy and Buildings*. 34 (6): 667–684 (2002).

Kim J.T., Ji H.L., Sun H.C. *et al.*: Development of the Adaptive PMV Model for Improving Prediction Performances. *Energy and Buildings*. 98: 100–105 (2015).

Li B., Du C., Tan M. *et al.*: A Modified Method of Evaluating the Impact of Air Humidity on Human Acceptable Air Temperatures in Hot-humid Environments. *Energy and Buildings*. 158: 393–405 (2018).

Parsons K.C.: The Effects of Gender, Acclimation State, the Opportunity to Adjust Clothing and Physical Disability on Requirements for Thermal Comfort. *Energy and Buildings*. 34 (6): 593–599 (2002).

Schellen L., Loomans M., Kingma B. *et al.*: The Use of a Thermos-physiological Model in the Built Environment to Predict Thermal Sensation: Coupling With the Indoor Environment and Thermal Sensation. *Building and Environment* 59: 10–22 (2013).

Sirhan N., Golan S.: Efficient PMV Computation for Public Environments With Transient Populations. *Energy and Buildings*. 231: 110523 (2021).

Zhang D., Qu B., Liu P. *et al.*: Comprehensive Evaluation and Optimization of Rural Space Heating Modes in Cold Areas Based on PMV-PPD. *Energy and Buildings*. 246: 111120 (2021).

Zhang S. and Lin Z.: Predicted Mean Vote With Skin Wittedness From Standard Effective Temperature Model. *Building and Environment*. 187: 107412 (2021).

Zhang S. and Lin Z.: Standard Effective Temperature Based Adaptive-rational Thermal Comfort Model. *Applied Energy*. 264: 114723 (2020).

Zheng Z., Zhang Y., Mao Y. *et al.*: Analysis of SET* and PMV to Evaluate Thermal Comfort in Prefab Construction Site Offices: A Case Study in South China. *Case Studies in Thermal Engineering*. 26: 101137 (2021).

Constructional Engineering and Ecological Environment – Chih-Huang Weng (Ed)

Habitat improvement of waterbirds in LianHua Beach in Xixi Wetland

Lin Chen
Hangzhou Xixi National Wetland Park Ecology & Culture Research Center, Hangzhou, China

Ziming Wang*
Zhejiang Institute of Hydraulics & Estuary (Zhejiang Surveying Institute of Estuary and Coast), Hangzhou, China

Kekan Yao
Hangzhou Xixi National Wetland Park Ecology & Culture Research Center, Hangzhou, China

Lei Hua
Zhejiang Institute of Hydraulics & Estuary (Zhejiang Surveying Institute of Estuary and Coast), Hangzhou, China

Congcong Zhang
Hangzhou Xixi National Wetland Park Ecology & Culture Research Center, Hangzhou, China

Aiju You
Zhejiang Institute of Hydraulics & Estuary (Zhejiang Surveying Institute of Estuary and Coast), Hangzhou, China

ABSTRACT: Wetlands are important habitats for birds. Based on the current status of bird habitats in Xixi Wetland, this paper reconstructs the bird habitats in Lianhua Beach in Xixi Wetland from the perspective of hydrodynamics and builds a two-dimensional hydrodynamic model. The measured data validates the model, on which basis, through micro-terrain reconstruction, a foraging area and a safety island suitable for bird survival and avoidance were constructed, and the water exchange time in the Lianhua beach area before and after habitat reconstruction was analyzed. The results show that the micro-terrain reconstruction project does not change. The hydrological process of Lianhua Beach has relatively little impact on its ecology. It is of positive significance for the reconstruction and construction of the existing habitat of Xixi Wetland to create a habitat that can increase the species and number of waterbirds in the wetland.

1 INTRODUCTION

Wetland is an important ecosystem on earth, known as the "kidney of the earth" and "biological supermarket" (Li *et al.* 2016). Xixi National Wetland Park is located in Xihu District, Hangzhou City, Zhejiang Province (30°14′57″~.30°16′57″N, 120°02′20″~120°05′32″E), with a total area of 11.5 square kilometres and wetlands. The park is rich in ecological resources, simple in the natural landscape and profound in cultural heritage. It is currently the only national wetland park in China that integrates urban wetlands, agricultural wetlands and

*Corresponding Author: wangziming1301@126.com

DOI: 10.1201/9781003410843-28

cultural wetlands (Chen 2010; Liu *et al.* 2015; Yu *et al.* 2007). Xixi Wetland is located in a subtropical monsoon climate, with an annual average temperature of 15.17°C, an annual average relative humidity of 78%, and an annual average precipitation of 1399 mm. The climate is mild, the rainfall is abundant, the sunshine is long, and the four seasons are distinct (Shi *et al.* 2014). A typical representative of an artificial complex wetland ecosystem. As an internationally important wetland, there are many studies on the current status of Xixi wetland resources (Sun 2019), fish resources (Wang *et al.* 2018), wetland water quality (Ni & Wang 2012; Wang & Yang 2009), and phytoplankton community characteristics (Jia *et al.* 2010). There has been no report on the upgrading and transformation.

2 REGIONAL OVERVIEW

Lianhua beach Wetland is located in the southeast corner of Hangzhou Xixi Wetland, the main habitat and viewing area for birds in Xixi Wetland. Lianhua beach is connected to the backbone channel of Xixi Wetland along Shanhe River and Jiangcun Port. There is no gate control in the middle, and the water level of Lianhua beach fluctuates with the surrounding rivers. According to the monitoring data of Xixi Wetland Hydrological Station, the normal water level of mountain rivers outside Xixi Wetland is basically about 1.7 m.

According to biodiversity monitoring data (Zhejiang Natural Museum 2021), the species and number of birds on Lianhua beach in Xixi Wetland have been increasing year by year in

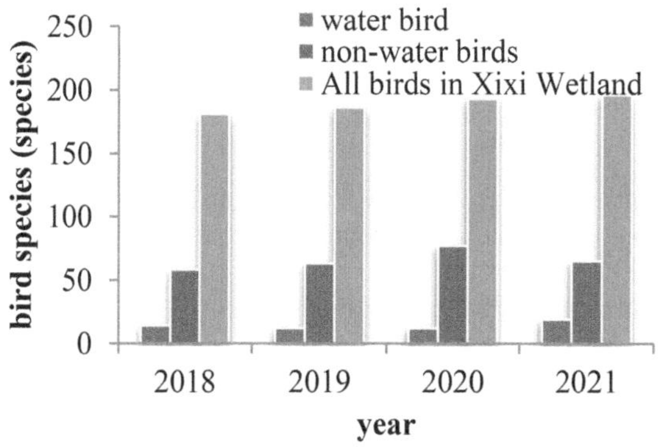

Figure 1. The number of species of birds.

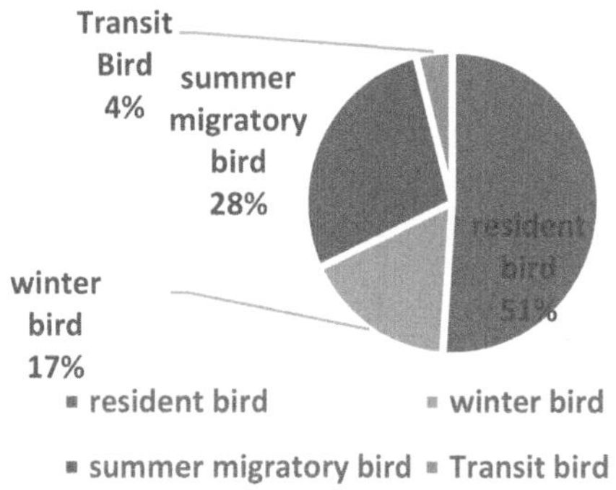

Figure 2. The proportion of birds.

recent years, and most of them are non-water birds; in 2021, 65 species and 806 non-water birds were recorded in field observations, 19 species of waterfowl with a number of 406. Among the types of birds in the Lianhua beach area, resident birds are the majority, with 41 species, accounting for more than half (51%) of the statistical bird species; 22 species of summer migratory birds, accounting for 28%; 14 species of winter migratory birds, accounting for 17%; 3 migratory transit birds' species, accounting for 4%. The ten dominant bird species in Lianhua Beach are: Spot-billed Duck, Little Phoebe, Night Heron, Blackwater Chicken, White-headed Bulbul, Strong-footed Tree Warbler, Black Thrush, Magpie, Brown-headed Ravenfinch, and Brown-backed Shrike.

3 RESEARCH PROPOSAL

In this paper, the bird habitats of Lianhua Beach in Xixi Wetland are reconstructed from the perspective of hydrodynamics. Therefore, based on the two-dimensional hydrodynamic numerical model of Lianhua Beach in Xixi Wetland, the model is used to classify the different water depths of the bird feeding areas. Statistics and evaluate the hydrodynamic impact after the project through the model so as to create a habitat suitable for the survival of different birds.

3.1 *Model establishment*

Construct a large-scale model of Xixi Wetland involving the first, second and third stages of Xixi wetland connecting channels and ponds. Based on the verification of the large-scale model, a small-scale model is constructed by intercepting the local area of Lianhua Beach. The model adopts a mixed quadrilateral and triangular mesh; among them, the large-scale model has 82,911 mesh elements, 92,571 nodes, and the minimum mesh size is about 1 m; the Lianhua Beach model is within the area of the red line, with a total of 18,887 mesh elements.

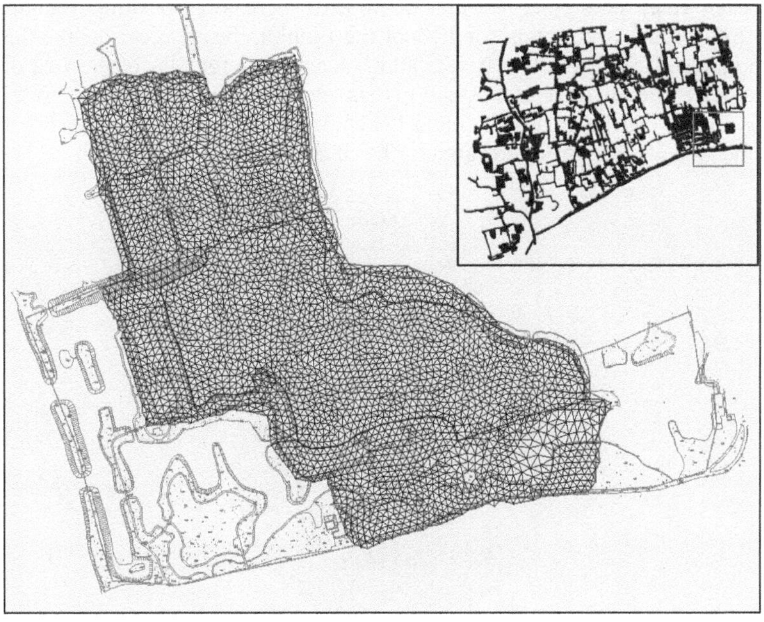

Figure 3. Scope of the Lianhua beach model.

3.2 *Model validation*

On November 11, 2021, the water level and flow at the entrance and exit of Xixi Wetland were monitored. The monitoring period was from 10:00 to 17:00. The inflow flow of Dongmuwu Stream was monitored at 11.2m^3/s on that day, and the regulation of Xixi Wetland sluice gates: five of the six sluice gates in the first phase of the encirclement were closed, of which Qianjinyang was open from 9:00 to 9:51, and closed at the rest of the time; The sluice in the garden enclosure was closed, and the sluice in the third phase of the enclosure was opened. This time, the measured data in 2021 was used to verify the large model. The calculated value of the outlet flow of Yanshan River and Jiangcun Port is basically the same as the measured value, and the error is within 10%. After verification, the model roughness ratio of 0.03.

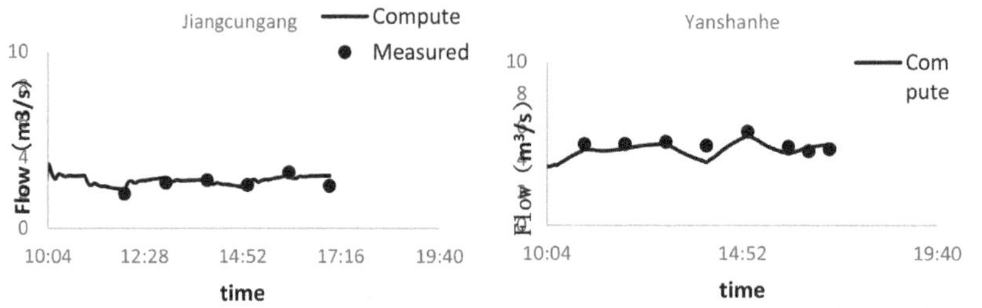

Figure 4. Hydrodynamic verification results.

4 HABITAT SCHEME RESEARCH

4.1 *Foraging area*

Under the constant water level after reconstruction, the water depth area of 0-0.1 m increased from the original 0.22 million m^2 to 0.42 million m^2, accounting for 8% of the Lianhua beach area; the 0.1-0.2 m water depth area increased from the original 4,500 m^2 increased to 6,100 m^2, accounting for 12% of the Lianhua beach area; the 0.2-0.3 m water depth area increased from 1,700 m^2 to 5,500 m^2, accounting for 11% of the area of Lianhua Beach; the water depth area of 0.3 ~1 m has increased from^{2}3,200 m^2 to 27,200 m^2, accounting for 54% of the area of Lianhua Beach; the water depth area above 1 m remains basically unchanged, accounting for about 10% of the Lianhua beach area.

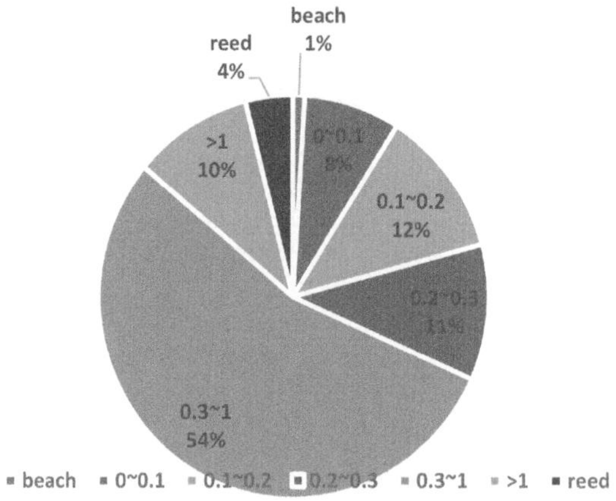

Figure 5. The proportion of different water depths under the constant water level.

Table 1. Areas of different water depths at 1.7 m water level after micro-topographic reconstruction.

Water depth (m)	Transformation of the former		After transformation	
	Area (m^2)	Proportion	Area (m^2)	Proportion
open beach	0	0%	708	1%
0~0.1	2155*	4%	4240*	8%
0.1~0.2	4508	9%	6128	12%
0.2~0.3	1731	3%	5533	11%
0.3 ~ 1 _	23272	46%	27211	54%_
> 1	5034	10%	5049	10%
Reed beach	1 4406	26%	2237	4%

4.2 *Safety island*

Three independent small islands with a total area of 210 m^2 are constructed using the small island on the north side of Lianhua Beach, far away from the crowd activity area. At the same time, in order to facilitate waterfowl foraging and other activities, the islands need to artificially maintain local beach features to create a minimal intervention for birds. A safe island can create an ideal bird habitat. The location of the safety island is shown in the figure below.

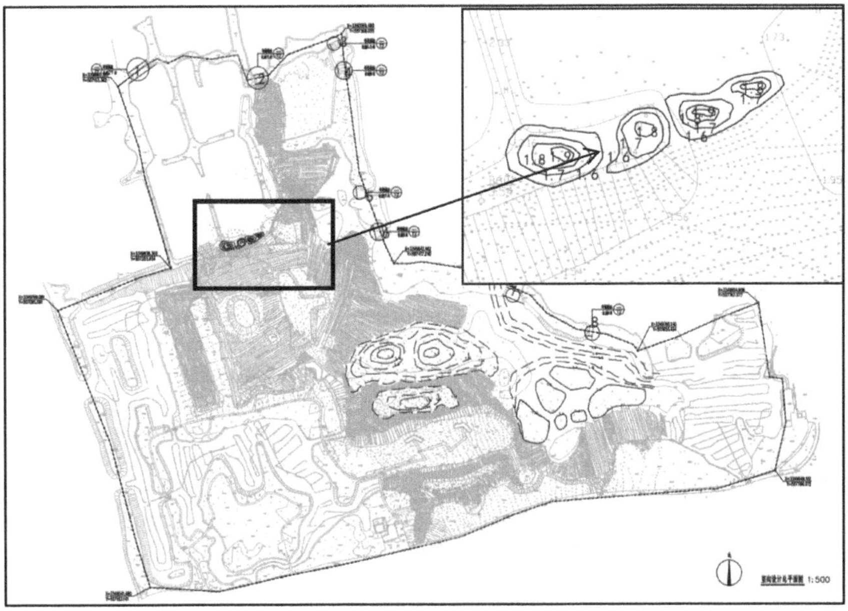

Figure 6. Schematic diagram of the location of the safety island.

4.3 *Water exchange time*

Lianhua Beach is affected by the surrounding waters, and the water level fluctuates greatly, which leads to the instability of the habitat environment of waterbirds. In order to lay a foundation for the convenience of further regulation of the water level in Lianhua Beach, it is planned to build a polder to stabilize the water level of Lianhua Beach. At present, the water flow direction of Lianhua Beach is from south to north and west to east. Under the current conditions, the bottom elevation of the river channel on the northeast side is about 1.4~1.6m, the water flow velocity is almost 0 due to the influence of aquatic plants, and the positions 11~14# are basically blocked. The mainstream mainly enters Lianhua beach from the south and flows out from the west, so it is planned to build a polder on the northeast side. Numerical calculations show that the water exchange time in the Lianhua beach area under the current normal water diversion conditions is more than 25 days, and the construction of the polder on the northeast side has little effect on the water exchange time in this area.

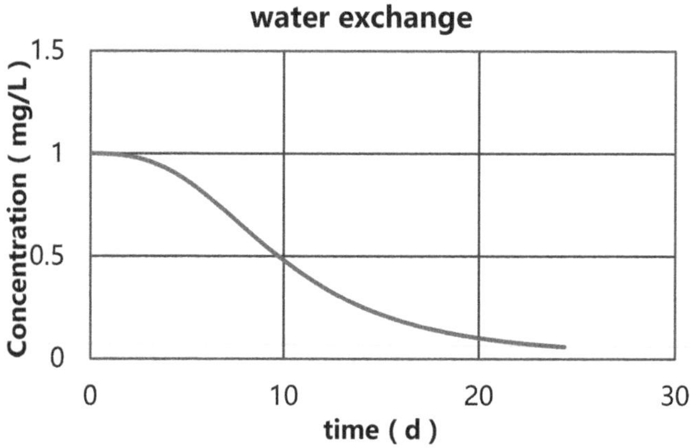

Figure 7. Water exchange time in the core area of Lianhua beach.

5 CONCLUSION

Based on the results and discussions presented above, the conclusions are obtained as below:

(1) On the basis of the current situation, the project will carry out a micro-topographic reconstruction of Lianhua Beach. When the normal water level of Xixi Wetland is 1.70m, part of the reed beach will be cleaned, focusing on the construction of a 0.1~0.3 m water depth area. After the reconstruction, the terrain distribution is more uniform in the water depth range of 0~0.6m under the normal water level; the area within the water depth range of 0.1 m every interval within the water depth range of 0~0.6 m accounts for about 10% of the core area of Lianhua beach, which is basically suitable for satisfying It meets the water depth requirements of different birds when the water level changes.

(2) We build a safe island with a partial beach style to create an ideal bird habitat. Three independent small islands with a total area of $210m^2$ are constructed using the small island on the north side of Lianhua Beach, far away from the crowd activity area.

(3) Numerical calculations show that the water exchange time in the Lianhua beach area under the current normal water diversion conditions is more than 25 days, and the construction of the polder on the northeast side has little effect on the water exchange time in this area.

ACKNOWLEDGMENTS

Lin Chen: Data Curation, Writing-Original Draft. ZimingWang: Model&Validation, Resources, Data Curation, Writing-Review & Editing.Kekan Yao: Investigation, Formal analysis.Lei Hua: Conceptualization, Supervision. Congcong Zhang: Validation. Aiju You: Writing-Review & Editing.

This work was financially supported by the Hangzhou West Lake Scenic Area Science and Technology Project (2021-014) and special project supported by Zhejiang provincial scientific research institutes (ZIHEYW22001).

REFERENCES

Chen Bojun. Xixi Wetland, a Green Kidney in Paradise on Earth. *Wetland Science and Management*. 2010,6 (04):30–32.

Jia Xinghuan, Wu Ming, Shao Xuexin, and Jiang Keyi. Characteristics and Influencing Factors of Phytoplankton Community in Closed Ponds of Xixi Wetland. *Journal of Ecology*. 2010, 29(09): 1743–1748. DOI: 10.13292/ j.1000-4890.2010.0299.

Li Bo, Wang Yunyun, Shi Lili, Ying Jiawei, Lu Ning, and Ye Xuhong. Water Quality Evaluation and Change Trend Analysis of Xixi Wetland in Hangzhou. *Zhejiang Chemical Industry*. 2016, 47(08): 51–54.

Liu Jiahong, Li Wenwen, Lai Zhouxiang, Qu Qianxi, Huang Junlin, Hu Tangao, Zhang Dengrong. Analysis of the Driving Force of the Change Mechanism of Xixi Wetland in the Past Ten Years. *Journal of Hangzhou Normal University (Natural Science Edition)*. 2015, 14(03): 269–275.

Ni Hanbin, and Wang Le. A Preliminary Study on the Impact of Coral Sand Diversion on the Water Quality of Xixi Wetland. *Science and Technology Bulletin*. 2012, 28(07): 160–164. DOI: 10.13774/j.cnki.kjtb.2012.07.029.

Shi Jian, Liao Xinfeng, Fang Xiaobo, Zhou Luyan, Yu Haixia, Yao Lin, and Hong Yifan. Analysis of Temporal and Spatial Changes of Water Quality in Xixi Wetland in Four Seasons and its Influencing Factors. *Environmental Pollution and Prevention*. 2014,36(06): 39–46.DOI: 10.15985/j.cnki.1001-3865.2014.06.012.

Sun Yongtao. The Current Situation and Protection Countermeasures of Hangzhou Xixi Wetland Resources. *Wetland Science and Management*. 2019,15(03):38–41.

Wang Yingying, Chu Yujiang, and Jin Zhi. Fish Resources and Conservation Strategies in Xixi Wetland. *Wetland Science and Management*. 2018, 14(01): 22–25.

Wang Jian, and Yang Yanyan. Discussion on Improving the Water Environment of Hangzhou Xixi Wetland by Water Diversion and Distribution. *Zhejiang Water Conservancy Science and Technology*. 2009(04):15–16+19.

Yu Minjie, Wu Jianjun, Xu Jianming and Shi Chun. Research on the Change of Landscape Pattern of Hangzhou Xixi Wetland in the Past 15 Years. *Science and Technology Bulletin*. 2007(03):320–325.

Zhejiang Natural Museum. 2021 Annual Monitoring Report on Biodiversity of Hangzhou Xixi National Wetland Park [R]. Zhejiang Natural Museum. 2021.

Constructional Engineering and Ecological Environment – Chih-Huang Weng (Ed)
© 2024 The Author(s), ISBN 978-1-032-53198-4

A "Spongification" plan for Guangzhou's historic district based on the SWMM model: The Chuanghua street example

Xuan Xie* & Guanyi Luo
School of Architecture and Allied Art, Guangzhou Academy of Fine Arts, Guangzhou, China
Guangzhou Municipal Engineering Design & Research Institute Co., LTD, Guangzhou, China

ABSTRACT: In recent years, urban flooding has frequently occurred in China, necessitating the urgent optimization and renovation of urban drainage systems. This study promotes the construction of pilot sponge cities and explores a LID strategy for the historic district of Chuanghua Street in Guangzhou with the help of an SWMM model. The study aims to explore a design approach for sponge construction measures in the urban drainage system and a feasible path to saponification micro-renovations. The waterways heritage and water safety approach of the region was built considering the problems of rainfall and flooding that the historic centre is currently facing. Through design simulation and visualization analysis, the regional water environment can be comprehensively optimized to achieve the goal of a resilient neighbourhood. This study sorts out the changes in the water system pattern and historical culture of Changhua Street in Guangzhou. Then it proposes a sponge transformation strategy suitable for the local environment and cultural resources under the historical, cultural and hydrological security pattern. And through the SWMM model, two generalized models of the original base and the pattern after comprehensive sponge planning are constructed, and the effect of ecological maintenance is verified by visual simulation.

1 INTRODUCTION

The rising frequency of natural disasters because of climate change and rapid urbanization is a serious concern for human residential environments (Wei 2014). In this context, the concept of sponge cities represents a measure against urban flooding caused by heavy rainfall (Chan 2018). A sponge city resembles a sponge adapting to environmental changes and responding to natural disasters with good "resilience." The most significant part of a sponge city is the realization of its "sponges" such as rivers, lakes, ponds, and other water systems, but also city parks and small green areas, permeable pavements, and other "green" infrastructures (Fu et al. 2022). The main function of the sponges is to allow rainwater to touch the soil first so that it can go through a process of purification, percolation, retention, storage, and reuse through the sponge. Therefore, the realization of sponge cities can enhance urban areas' natural capacity for rainwater storage and alleviate the problem of urban flooding by substantially reducing the city's ground-level hard paving (Qiao et al. 2020). The preliminary measures of rainwater purification of the sponge city can reduce the number of stormwater pollutants in the city and improve the urban stormwater drainage systems. The measures of rainwater storage and reuse promoted in a sponge city can also mitigate the problem of urban water shortage.

In the study of "urban saponification," the stormwater management model (SWMM) can anticipate the effectiveness of the transformation (Yuan et al. 2022). It is, therefore, a

*Corresponding Author: kyc@gzarts.edu.cn

 DOI: 10.1201/9781003410843-29

significant method for quantitative analysis in sponge city construction. The rainfall input in the software can be based on the actual rainfall measurements in the region or on the rainfall data provided by the local government (Randall et al. 2019). The software can systematically simulate the complete urban rainfall occurrence in terms of regional water storage, water discharge, and the dynamic loads in the piping system (Pochwat et al. 2017). The U.S. Environmental Protection Agency (EPA) developed the SWMM digital model as early as the 1970s for water quality monitoring and rainfall management. Ever since, the unique advantages of SWMM have been widely used by many scholars all over the world for simulations, for planning the layout of drainage systems in sponge city construction, and for LID design. Chinese scholars first conducted SWMM model-related research after 1990, starting with areas in China that were subject to stormwater hazards or serious urban diseases (Bai et al. 2018). Tianjin scholar Liu Jun applied the SWMM model to the sponge city planning and construction in Tianjin (2001), constructing a city-level model of Tianjin's runoff yield and collection, providing a rigorous pattern of the confluence of rainwater for the urban renovation and restoration of the city. Other scholars used the SWMM model for the construction of residential areas, such as the Wenzhou residential area (2012), to form a visual digital model of the urban drainage system, which facilitated the update and arrangement of the underground pipe network. As the LID module in the SWMM model adapts well to the construction of sponge cities, many scholars, such as He Shuang and Jia Haifeng, have evaluated and analyzed the effectiveness of stormwater control provided by single or combined LID facilities through the SWMM model, and proposed a series of optimal combinations (2013); Che Wu et al. used the SWMM model and summarized a situation of regional stormwater control after a LID remodulation; Li Hanyu and Wu Xu used the SWMM model to make a simulation and a comparison of an urban park before and after "saponification" (2018).

However, in reality, waterlogging occurs frequently in many cities in China, and the urban drainage system urgently needs to be optimized and improved. Especially in the historical and cultural heritage areas of many cities, these areas are often densely populated, have outdated facilities, and have low flood control capabilities. Flood control projects in these areas often have to consider the protection and safety of historical heritage. This study takes the sponge city construction pilot—Guangzhou City as an empirical research case, and uses the SWMM model to explore and study the layout of low-impact development measures in Changhua Street, a historic district in Guangzhou. The purpose of this study is to explore a design method of sponge measures in urban drainage system, which can also be understood as a feasible path of sponge micro-retrofit. The main contributions of this study are: combining the current rain and flood problems faced by historical urban areas, constructing the regional water system heritage and water security pattern; through simulation design and visual analysis, comprehensively optimize the regional water environment, and achieve the goal of a resilient community.

2 SPATIAL SUBSTRATE ANALYSIS OF CHUANGHUA STREET FOR SPONGE CITY PLANNING

2.1 *The waterways system of the old city of Guangzhou*

Guangzhou city is located in the centre of China's southern coast, and the confluence of the three river arteries of Beijiang, Xijiang and Dongjiang at the river mouth has created Guangzhou's advantageous geographical location and unique water culture characteristics of Lingnan. Prior to the era of the Ming and Qing dynasties, Guangzhou was a city that could be traversed by boats on the rivers (Dąbrowski et al. 2021). During Guangzhou's ancient urbanization, a network of waterways used to run inside and outside the old city. In times of flooding, it provided a network of diversion channels and storage for surface runoff that would be rapidly and completely drained; when the city was short of water, natural water bodies such as wet ponds and lakes inside and outside the city provided backup water for working and living in the city (Ma 2004). This kind of natural storage system that combines diversion and infiltration of rainwater is the insight of the early sponge city created in the old city of Guangzhou in response to disasters. However, the natural terrain, which is prone to flooding, can also pose a serious threat to the old city of Guangzhou during the particularly heavy and continuous rainfall that may occur during the summer season. The continuous development of the city has changed the original water environment, and the drainage function within the city has gradually weakened, leading to an increased frequency of natural disasters (Ruan et al. 2021). The densely populated waterway system of the inner city was the first to be affected. This gradually widened to the suburban water system west of the city walls, which was prone to extensive flooding because the suburbs owing to the lack of a developed flood control system.

The research object of this study is Chuanghua Street in Liwan District, in the historic centre of Guangzhou. Chuanghua Street originated from what is commonly known as "Xiguan" by the old Guangzhou people—the suburban area located outside the ancient city walls to the west. Historically, it was a suitable location for the detached villas and parks of the imperial palace, where rivers and streams crossed through the expanse of the major palaces and gardens. The district has an important waterways heritage, Lychee Bay, named after the beautiful and elegant scenery of lychee forests that once stood along the river. The administrative division of Chuanghua Street covers an area of 1.62 square kilometres, with a population of 31,871 (Zhou et al. 1998). It is located in the central north of the Liwan district in the west of Guangzhou, next to the Fengyuan district to the east. It is linked to Duobao to the south and overlooks the Zhujiang river and the Qiaozhong bridge on its west. The Caihong district is to its north, and the Nanyuan district is the border. There are seven neighbourhoods in the district: Liwan Lake, Ruyi Fang, Changhuayuan, Saikwan Mansion, and Pun Tong (Figure 1).

In addition to the Lingnan traditional gardens, the Renwei Temple, and other historical attractions, the district retains a rich cultural heritage of water landscapes along Xiguan Creek. The waterways heritage of Chuanghua Street comprises the historic waterways heritage and the waterways heritage of the historic urban landscape. The waterways heritage had Lychee Bay as its maximum extension and probably included the Xiguan Creek and other branches flowing through the area. The artificial Liwan lake was dug by the Guangzhou Municipal Government in 1958; it is of immense significance given the rainy environment of Chuanghua Street and is the main water system heritage in the area (Figure 2). The urban waterways heritage mainly presents a landscape of the Lingnan architecture, developed from the ancient

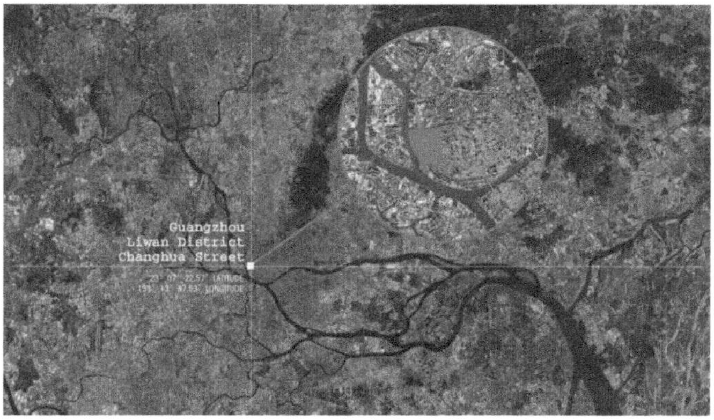

Figure 1. Schematic diagram of the location of Changhua Street (made by the author).

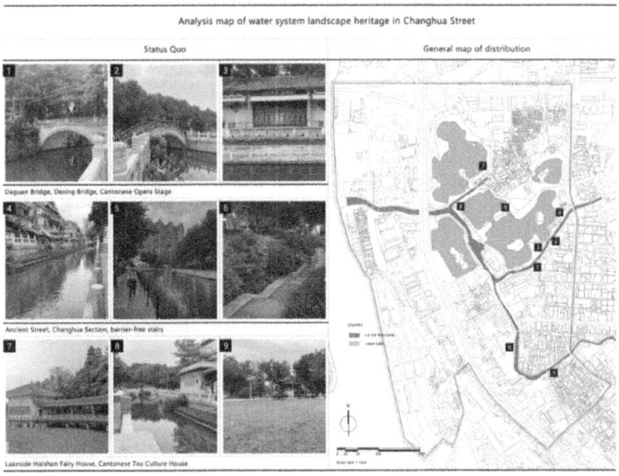

Figure 2. Analysis map of water system landscape heritage in Changhua Street (made by the author).

river culture. These include eight old bridges that connect the two artisanal districts on either side of the river, such as the Longjin bridge, Huiyuan bridge, and Dexing bridge. Additionally, the arcades of Longjin West Street, the remnants of the Lingnan gardens architecture, such as the Haisan Xianguan and the Xiaohuafang Study, and the old Renwei Temple are also unique urban heritage characteristics of Lingnan.

2.2 *Rapid urbanization brings flooding risks to Chuanghua street*

With the rapid modernization of the city, Chuanghua Street, which originally comprised fields and swamps, has undergone major changes. In the initial years of the city, Guangzhou had already formed a preliminary grid road, but to build roads, many canals and creeks within the Chuanghua Street district were filled. The construction of residential and office buildings contributed to the gradual reduction of ecological land in the area, such as agricultural land and water systems. In the early 21st century, it experienced a real-estate boom (Tian & Ma 2020). As the pace of urban construction continued, intensive and inordinate construction caused major changes to the Lychee Bay area and the architecture of Lingnan. Rapid construction also led to ground imbalance, which has indirectly aggravated the regional stormwater problem. Chuanghua Street has been continuously facing flooding issues because of the rainy summer climate and its geography: low-lying wetlands in the western suburbs of the city, with a dense network of rivers (Jordan 2020). In this study, a water safety approach to the site was created using the ArcGIS software, combined with the SWMM water collecting simulation, supported by the daily stormwater data provided by the Liwan hydrological bureau, and combined with the DEM topographic data of the site. The analysis revealed a major risk of flooding on Longjin West Road, Duobao Road, Pun Tong, and along the river (Figure 3).

3 THE SPONGIFICATION MICRO-RENOVATIONS STRATEGY OF THE CHUANGHUA STREET BASED ON THE SWMM MODEL

3.1 *Building a spongification plan for the Chuanghua street*

Before building an extensive pattern, it is necessary to determine the low-lying areas within Chuanghua Street that are most likely to create potential water hazards based on the current

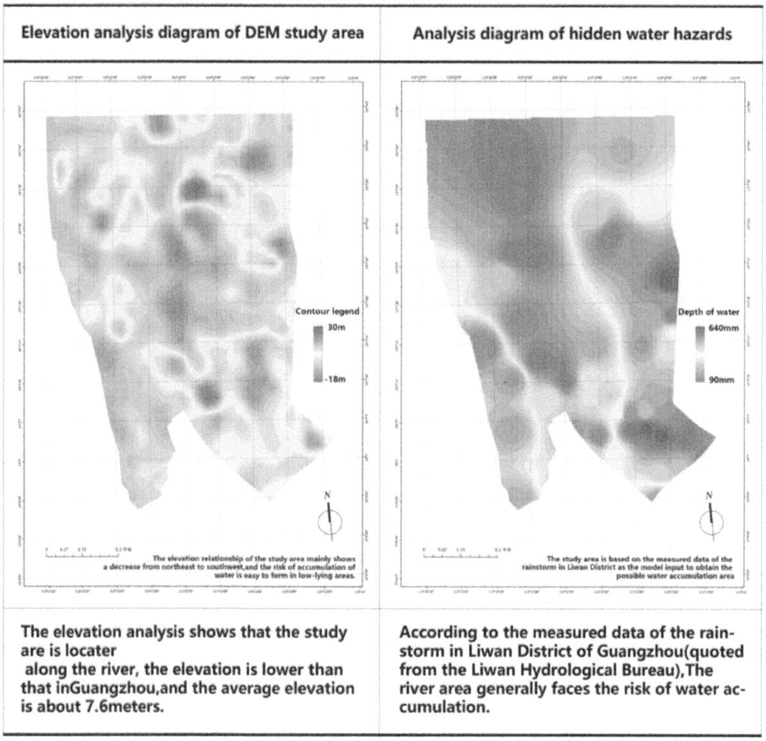
Elevation analysis diagram of DEM study area
Analysis diagram of hidden water hazards
Contour legend
30m
-18m
N
The elevation relationship of the study area mainly shows a decrease from northeast to southwest,and the risk of accumulation of water is easy to form in low-lying areas.
Depth of water
640mm
90mm
N
The study area is based on the measured data of the rainstorm in Liwan District as the model input to obtain the possible water accumulation area
The elevation analysis shows that the study are is locater
along the river, the elevation is lower than that inGuangzhou,and the average elevation is about 7.6meters.
According to the measured data of the rain-storm in Liwan District of Guangzhou(quoted from the Liwan Hydrological Bureau),The river area generally faces the risk of water ac-cumulation.

Figure 3. Analysis of the current elevation and hidden water hazards in Changhua Street (made by the author).

runoff situation and an elevation difference of the site and then conduct a detailed analysis of the overall runoff direction and runoff type. First, the runoff analysis was conducted by taking the Digital Elevation Model (DEM) of the area as the main data; this was combined with ArGIS hydrological analysis auxiliary functions. The advantage of this method is that it can produce a visual model of the surface runoff convergence at the site by superimposing the single DEM data of flow direction, flow rate, and runoff yield and collection.

Second, software simulation can be used for examining the overall direction of rainwater runoff on the site, the hydrology at the source for a certain rainfall time, the secondary flush convergence and diversion of rainwater that occurs across the site and study the hydrology of the end of the site after a certain time interval.

Finally, based on the runoff analysis results (in a rainstorm process), in the initial rainfall process, the Inner Ring Road functions as a watershed for rainwater. Part of the flow goes west of the Inner Ring Road, along the Xinfeng harbour, and along the southwest sloping terrain toward the Zhujiang River. Most of the stormwater runoff, after flowing into the neighbourhood of Ruyi Fang in the form of multi-level drainage, is remitted to the municipal pipe network. A small part of it converges and stays in the neighbourhood green spaces and eventually goes through percolation and dissipation. In contrast, the remaining rainwater flows southward toward the next drainage process. The overflowing rainwater in the neighbourhood of Pun Tong that is not dissipated through the drainage pipes and green spaces flows southwest into Liwan lake; the rainwater in the neighbourhood of Chuanghua Avenue flows southwest into Lychee Bay creeks, and the rainwater in the neighbourhood of the Saikwan Mansion flows northwest into the river. However, as the buildings in the

Saikwan Mansion and Changhua Avenue occupy a dense area, and there are very few green spaces, the overflow of rainwater can accumulate and flood the neighborhood. Rainwater from the Chongbian cross-street neighbourhood sinks into the river to the northeast, where the main neighbourhoods form small drainage units, each responsible for the diversion of rainwater. Conversely, most of the primary road drainage is discharged through the green ditches and gutters on the sides of the road because of the different forms of road separation and drainage (Figure 4).

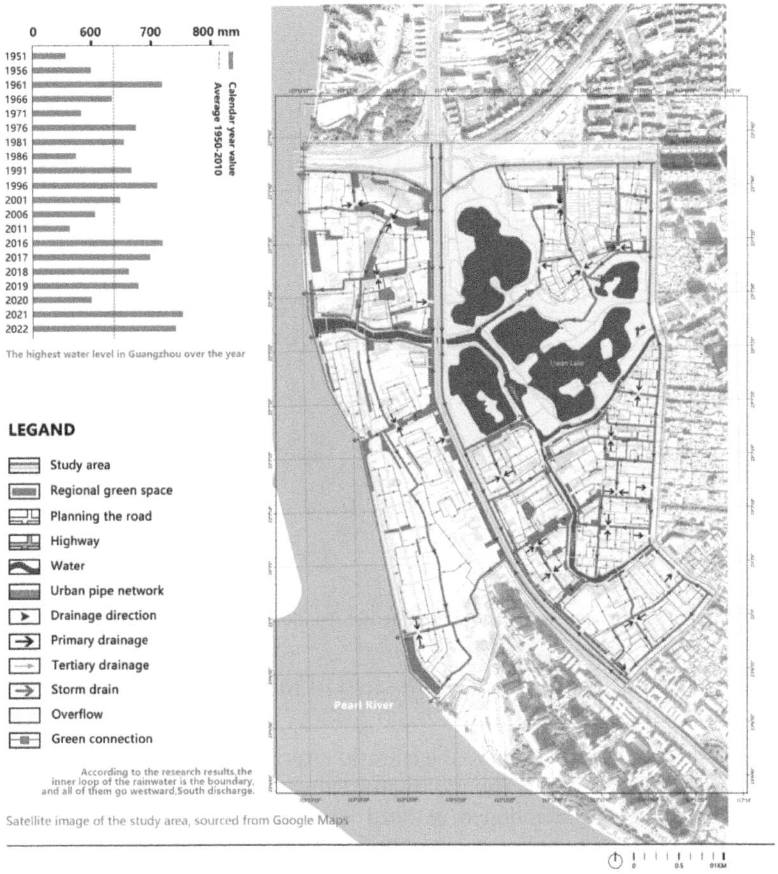

Figure 4. Analysis of the overall runoff from Changhua Street (made by the author).

The analysis delivered an integrated pattern blueprint, which is significant for the construction of the saponification plan for the area of Chuanghua Street. This is the combination of the overall landscape heritage, the water ecological safety approach, and the overall runoff flow analysis results (Wang et al. 2022). The heritage pattern includes the heritage of waterways landscapes and urban landscapes. In the overall saponification reform, the best recreation paths and radiation areas of the heritage layout can be well connected with the water ecology. Furthermore, the social and cultural values of the historic heritage can be reflected to the greatest extent through the saponification micro-renovations of the district (Figure 5).

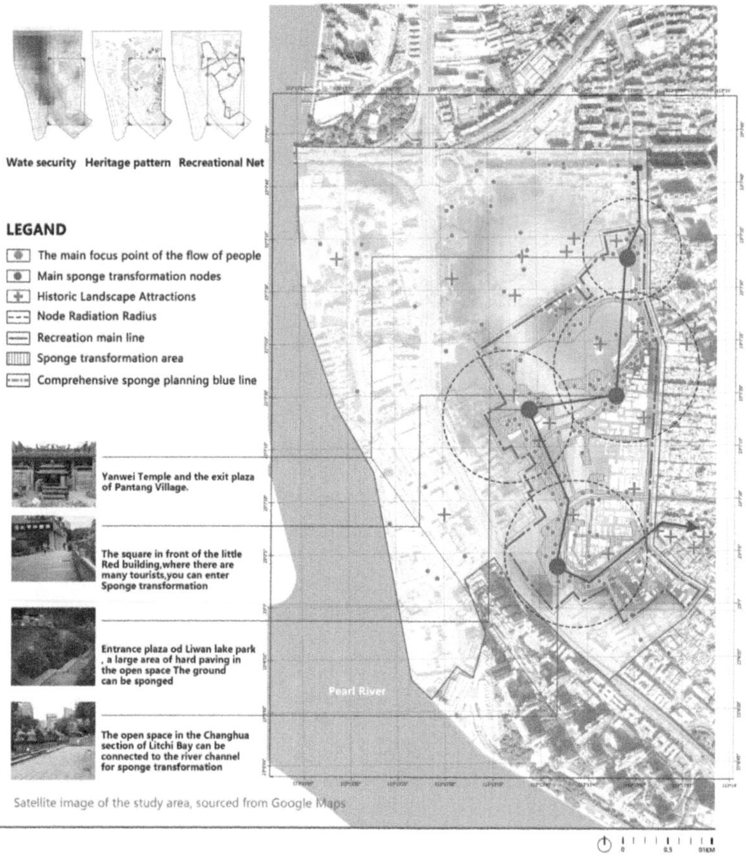

Figure 5. A saponification plan for Chuanghua Street (made by the author).

3.2 *Simulation analysis of the overall effectiveness of comprehensive spongification of the Chuanghua street*

The objective of this plan is to preserve the cultural imprint of the regional water system and activate its social value through sponge micro-renovations along the route. The specific approach is to destroy the abandoned or hard landscape space along the historic Liwan creek, integrating sponge facilities in communion with the river. The existing hard riverbanks can be softened in the controllable scope of the plan so that the space along them and the surrounding cultural and landscape architectures can participate in the management of rainwater infiltration and purification, creating an eco-friendly historic district with water culture characteristics and a perfect sponge system.

The sponge construction plan of Chuanghua Street is based on the saponification plan of the area, which demarcates restricted areas on the premise of safeguarding the ecology of the typical historic district within Chuanghua Street. The transformation is intended to address the current water environment problems. The Lychee Bay river and Liwan Lake are among the sponge facilities that are significant objects in this planned renovation. The sponge construction design intends to break the hard landscapes on both sides of the river, excavate various forms of small wetland parks for infiltration and storage of rainwater before it is discharged into the river, and build soft revetments on the river banks (Figures 6 and 7).

The plan revolves around the development of the traditional water interconnection between Lychee Bay creeks and Liwan Lake, as its atmosphere carries the characteristics of

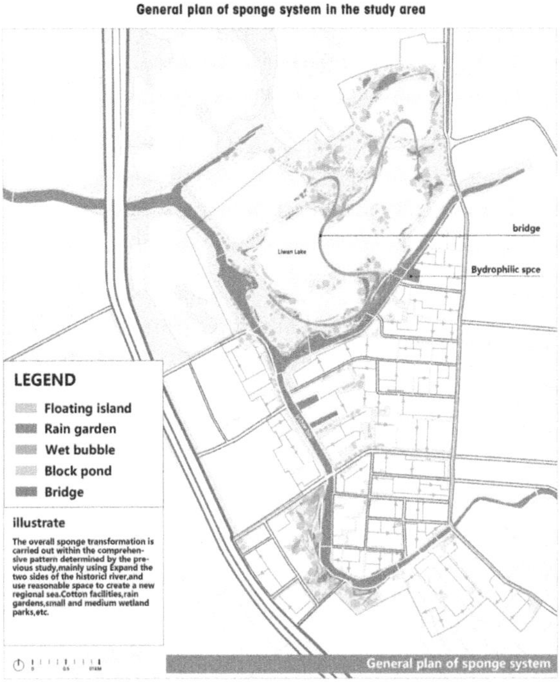

Figure 6. Map of sponge city planning for Changhua Street (made by the author).

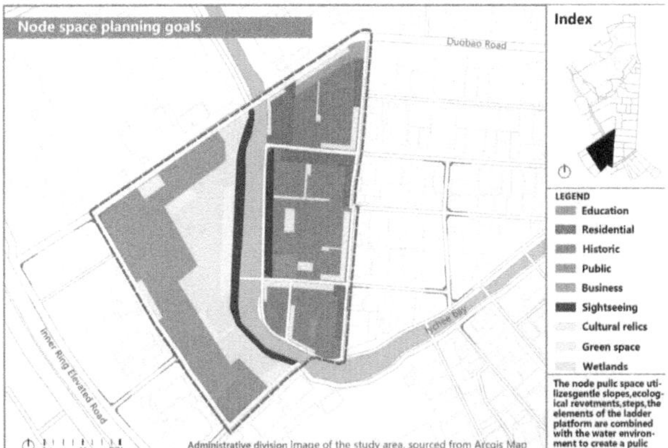

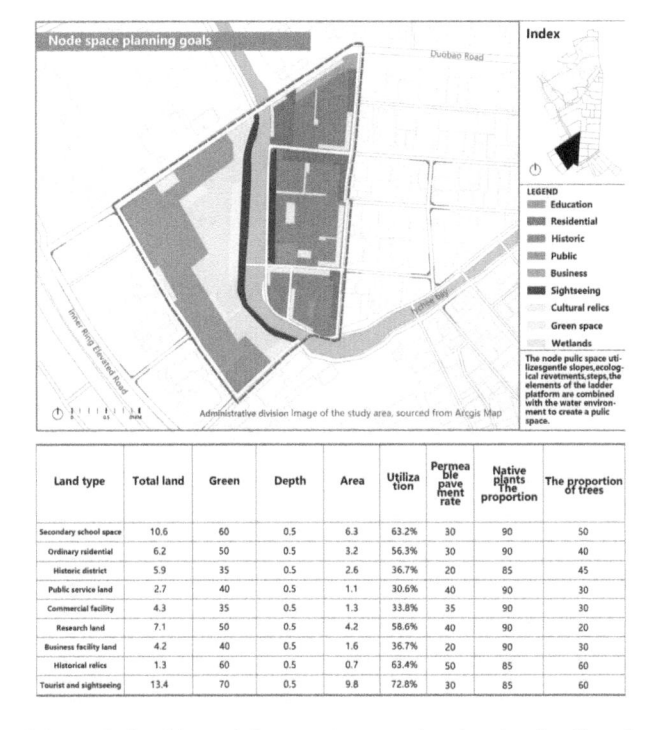

Land type	Total land	Green	Depth	Area	Utilization	Permeable pavement rate	Native plants The proportion	The proportion of trees
Secondary school space	10.6	60	0.5	6.3	63.2%	30	90	50
Ordinary rsidential	6.2	50	0.5	3.2	56.3%	30	90	40
Historic district	5.9	35	0.5	2.6	36.7%	20	85	45
Public service land	2.7	40	0.5	1.1	30.6%	40	90	30
Commercial facility	4.3	35	0.5	1.3	33.8%	35	90	30
Research land	7.1	50	0.5	4.2	58.6%	40	90	20
Business facility land	4.2	40	0.5	1.6	36.7%	20	90	30
Historical relics	1.3	60	0.5	0.7	63.4%	50	85	60
Tourist and sightseeing	13.4	70	0.5	9.8	72.8%	30	85	60

Figure 7. Map of the analysis of the nodal areas of sponge city planning for Changhua street (made by the author).

the Chuanghua culture. The saponification of the clearings along the river can increase the overall soft landscaping for the protection of the traditional river. Towpaths, walks, and boat moorings can be added between the rivers. Appropriate Liwan cultural symbols along the shoreline can provide people walking by the river with a feel of the bygone days of fishing and singing in Liwan.

For the interoperability of the ecological sponge facilities with the original sponge facilities, the old river in Chuanghua Street needs to be dredged, widened, and connected to the internal water system. This will form the basis of Liwan park and a fundamental framework of regional saponification. Increasing public space in wetlands and construction areas can also enhance the recreation experience. In terms of overall ecology, the sponge facilities can support the control of total runoff in Changhua Street and play a role in flood peak attenuation and rainwater purification; they can be divided into three classes (Figure 8). The primary sponge facilities include green ditches, rain gardens, green roofs, revetments with vegetated buffers, permeable paving, etc., which can be arranged on roads, small spaces, residential areas, and riverbanks. Intermediate sponge facilities include wet ponds, wetlands, etc., which can be excavated at corresponding locations along rivers and lakes, and create fully functional wetland parks. Advanced sponge facilities include large wetlands such as rivers and lakes. The Lychee Bay river and the Liwan Lake already have the qualities of advanced sponge facilities and only need to be better optimized in accordance with the original design.

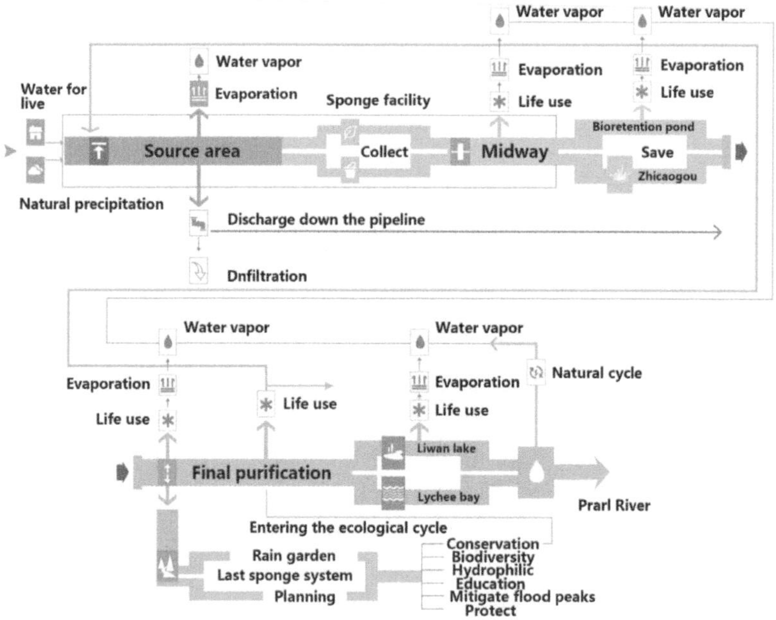

Figure 8. Post-planning operation path of the sponge system (made by the author).

4 THE SWMM MODEL DATA SIMULATION AND VISUALIZATION

4.1 *Construction of the SWMM model for the study area*

To construct the SWMM model of the area, first, each sub-catchment area is delineated using the surface drainage units of neighbourhoods and municipal roads as a fundamental element, relying on the contour areas and neighbourhood demarcation lines as the basis and combining them with the overall runoff analysis map of the Chuanghua Street drawn in the previous section. The Manning values for the land cover of the whole area in the model before construction were set to default values, and the infiltration model was dominated by the Horton equation.

In the process of building the SWMM model, each drain outlet in the region corresponds to a sub-catchment area. Data generalization is a significant part. Rainwater collection pipes and manholes are generalized as nodes, drainage pipes as pipes, drain outlets along the river and municipal manholes as discharge outlets. It mainly consists of various graded sub-catchment areas, such as the Saikwan Mansion, Chuanghua Street, Pun Tong, and City No. 1 Middle School, which are combined with drainage pipes and catchment areas to form the frame of the model. Among them, the S10-S11-S9 catchment area of the Saikwan Mansion neighbourhood mainly relies on J3-J13-J14 drainage pipes for stormwater discharge, and the S18-S17 catchment area of the Changhuayuan neighbourhood mainly relies on J6-J5 drainage pipes for stormwater discharge. The final outlet of these two typical neighbourhood drainage units is the PFK2 node along Longjin West Road to the south. The S16-S13 catchment area of the Affiliated Hospital of Medical University mainly relies on J6-J7-J21-J22-J15 drainage pipes for stormwater discharge, and the outlet node is PFK4. The S20-S23 catchment area of the Puiying Middle School and City No. 1 Middle School relies on J16-J17-J18 drainage pipes, and the final outlet is PFK3. All of the above are significant catchment units and sponge micro-renovation design areas in this SWMM model, which have a certain effect on the data of the entire simulation study.

In the SWMM generalized model of Chuanghua Street, the blue lines of sponge city planning were divided into a total of 29 sub-catchment areas and 18 basic drainage pipes (Figure 9). The generalized model is divided into two scenarios: the current situation of the comprehensive sponge city planning within the blue lines and the scenario after saponification micro-renovations. The focus of the micro-renovations is on the softening of the river channels in the Lychee Bay river, reasonable river widening, and the addition of different levels of sponge facilities such as wetlands and green areas that can reduce local runoff. Therefore, the main difference between the two models is the area of soft landscape on the river and the borders of the catchments near the river.

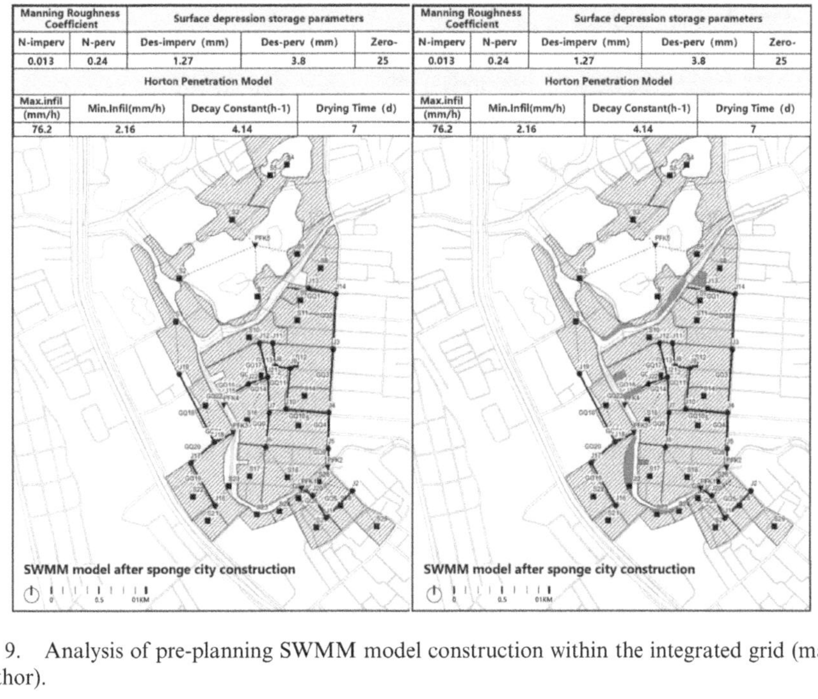
Manning Roughness Coefficient
Surface depression storage parameters
N-imperv
N-perv
Des-imperv (mm)
Des-perv (mm)
Zero-
0.013
0.24
1.27
3.8
25
Horton Penetration Model
Max.infil (mm/h)
Min.Infil(mm/h)
Decay Constant(h-1)
Drying Time (d)
76.2
2.16
4.14
7
SWMM model after sponge city construction
Manning Roughness Coefficient
Surface depression storage parameters
N-imperv
N-perv
Des-imperv (mm)
Des-perv (mm)
Zero-
0.013
0.24
1.27
3.8
25
Horton Penetration Model
Max.infil (mm/h)
Min.Infil(mm/h)
Decay Constant(h-1)
Drying Time (d)
76.2
2.16
4.14
7
SWMM model after sponge city construction

Figure 9. Analysis of pre-planning SWMM model construction within the integrated grid (made by the author).

The SWMM model for the two scenarios was simulated by setting the rainfall coefficient to the September 2021 stormwater data published by the Liwan hydrological bureau as the background and comparing the one-year rainfall intensity with the five-year rainfall intensity. The SWMM model of the existing scenario can begin running the simulation directly on this basis, whereas the saponification micro-renovations scenario requires setting up the LID module. Different LID modules have corresponding infiltration parameters within the SWMM model, and their quantitative values are based on the floor area ratio of the module. The following were added: 5% of falling rainwater collection pipes; 20% of rainwater collection devices such as rainwater tanks; 30% of rain gardens, which can be scattered and set up again in the district with an interconnected and multiplying infiltration effect; 10% of tree grates, mainly because currently the trees (including old trees) are surrounded by large hard paving areas; 30% of water storing wetlands, with main focus on the newly widened river space; and 20% of permeable pavement—currently the site presents a hard pavement with extremely high surface runoff coefficient; high requirements for the construction of permeable pavement are needed.

4.2 *Simulation and comparison of the effect of spongification in different models*

With regard to the LID facilities within the Blue Line grid, a simulated comparative analysis of the two models before and after the saponification micro-renovations of the Chuanghua Street, with a total rainfall of 2067mm, output series data (Figure 10), shows that the total water storage of the site before the transformation is 1055.637mm, the total external discharge is 1011.763. The annual runoff total control rate is 51.06%. It indicates that the large area of construction seriously reduces the rainwater infiltration effect on the site, half of the total rainfall is basically discharged outward, and it is not possible to make effective use of the rainwater. The data simulation analysis of the current situation reveals the need for saponification micro-renovations of the site.

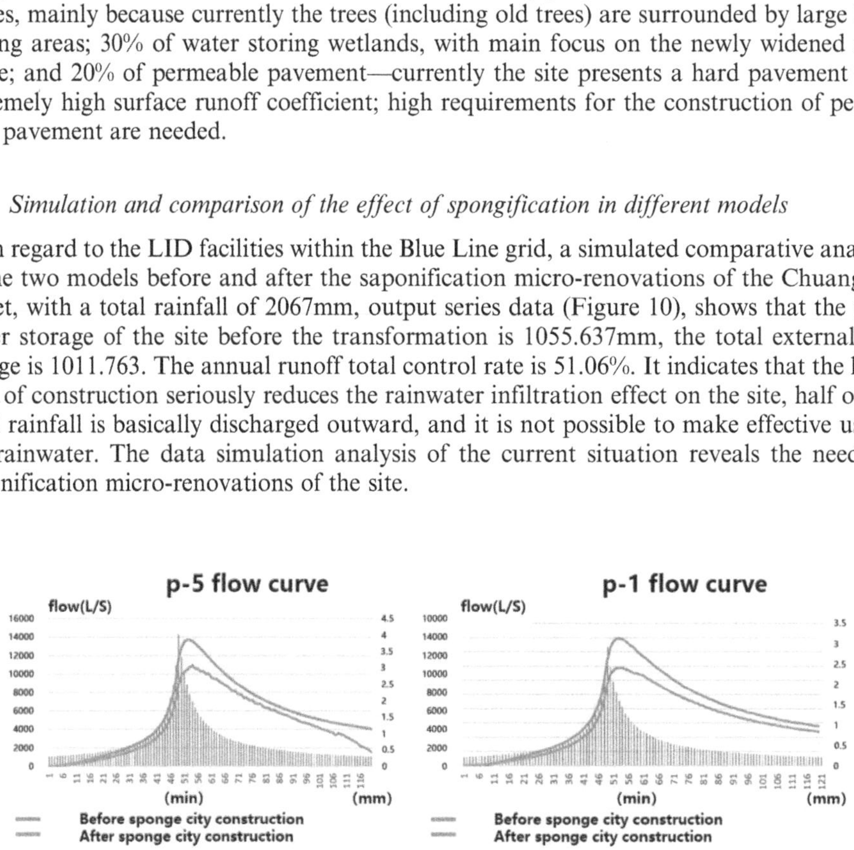

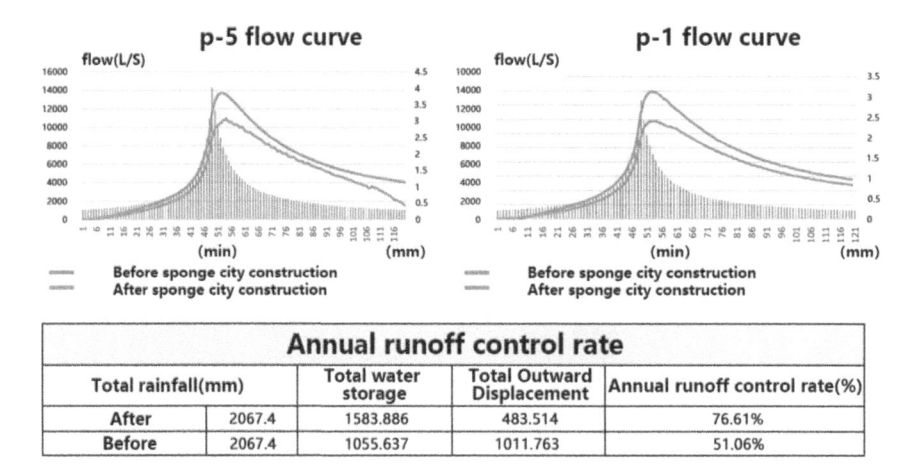

Annual runoff control rate					
Total rainfall(mm)		Total water storage	Total Outward Displacement	Annual runoff control rate(%)
After	2067.4	1583.886	483.514	76.61%
Before	2067.4	1055.637	1011.763	51.06%

Figure 10. Analysis of simulation results of the overall SWMM model for the post-planning scenario (self-drawn by the authors).

Simulation of the scenario of the site after comprehensive saponification yielded a total water storage volume approximately 50% higher than before renovations, reaching 1583.886mm. Rainfall was infiltrated and stored in situ, the total external discharge was only 483.514mm, the annual runoff total control rate reached 76.61%, and the final simulated runoff control rate result is in line with the 78% runoff total control rate that the Liwan District needs in preparation for the sponge city of Guangzhou 2035. Combining the various comprehensive data derived from the simulation, the average annual runoff volume and average runoff days after the renovation are lower than before. The maximum rainwater retention was twice as high as the value before the renovation. The study design results fundamentally meet the requirements of saponification micro-renovations in the SWMM model simulation.

Additionally, regarding the two kinds of rainfall intensity, the comparative analysis of rainfall flow and flooding before and after site saponification reveals that in once-a-year heavy rainfall intensity, the scenario before sponge construction will reach a peak flow of 15000 (L/s) in 48 minutes of rainfall. In the scenario after sponge construction, a peak flow of only 10,000 (L/s) was gradually reached only after 55 minutes of rainfall. The rainfall curve is slower, indicating that the process of runoff reduction is better compared to before the construction. In the case of a once-in-five-year storm, the curves before and after construction are relatively more bonded, indicating that the saponification effect can better play the role of rainwater retention and drainage in response to a once-a-year storm. In case of a once-in-five-year rainstorm, the sponge facilities in the area do not function optimally.

In general, the hydrological process of Chuanghua Street under the control of traditional drainage systems is reflected in the traditional model data. It mainly relies on the existing drainage network, the narrow ancient river, and the Liwan Lake for control and utilization of rainwater. In contrast, under the saponification micro-renovations model, LID facilities with optimal integration connectivity can be implemented on the site on the basis of the existing drainage systems. These facilities include bioretention facilities, ecological revetments, as well as green ditches and wetland bubbles, which complete the control and utilization of rainwater with the original drainage system. Thus, the advantage of the integrated saponification model is the overall hydrological process that is formed after the integrated sponge city planning has been implemented and a series of interconnected sponge facilities have been added.

5 CONCLUSIONS

In today's modern city constructions, fast, engineered, standardized models favour the construction of urban drainage systems, as well as the massive use of gray building materials, such as cement. This has led to the weakening of urban ecological functions and the subsequent increase in the potential risk of urban flooding and inundation. In the context of historical culture and hydrological safety, we propose a saponification strategy suitable for the local environment and cultural resources. The saponification strategy will be implemented through a modification of the waterways system and the historical and cultural reorganization of Chuanghua Street, a historical district in Guangzhou. Through the SWMM model, we built two generalized models: one of the original ground and one after the integrated sponge city plan, with a visual simulation to verify the ecological preservation effect. This study maintains that the renovation of Chuanghua Street through comprehensive sponge city planning not only has ecological value but also inherits the cultural value of its unique Lingnan traditional water landscape. Additionally, in today's historical urban areas, the problem of backward drainage infrastructure standards still exists. This study combines the historical heritage pattern with exploration the of sponge city construction. It will be a useful reference for the current stormwater management problems caused by rapid urbanization and for the protection of the cultural value of the water system.

REFERENCES

Bai, Yiran, *et al.*: Storm Water Management of Low Impact Development in Urban Areas Based on SWMM. *Water* 11.1, 33 (2018).

Chan, Faith Ka Shun, *et al.*: "Sponge City" in China—a Breakthrough of Planning and Flood Risk Management in the Urban Context. *Land Use Policy*. 76, 772–778 (2018).

Dąbrowski, Marcin, *et al.*: Adaptive Capacity of the Pearl River Delta Cities in the Face of the Growing Flood Risk: Institutions, Ideas and Interests. *Urban Studies*. 58.13, 2683–2702 (2021).

Fu, Guangtao, *et al.*: Are Sponge Cities the Solution to China's Growing Urban Flooding Problems?. *Wiley Interdisciplinary Reviews: Water*. e1613 (2022).

Jordan, Pamela.: The International Conference on Climate Services 6: Advancing the Knowledge and Practice of Climate Services for Climate Resilience. (2020).

Ma N. Lizhiwan shihua (Stories of Lychee Bay). *Lingnan Wenshi*. 56 (04), 52–53, (2004).

Pochwat, Kamil, Daniel Słyś, and Sabina Kordana.: The Temporal Variability of a Rainfall Synthetic Hyetograph for Dimensioning Stormwater Retention Tanks in Small Urban Catchments. *Journal of Hydrology*. 549, 501–511 (2017).

Qiao, Xiu-Juan, Kuei-Hsien Liao, and Thomas B. Randrup.: Sustainable Stormwater Management: A Qualitative Case Study of the Sponge Cities Initiative in China. *Sustainable Cities and Society*. 53, 101963 (2020).

Randall, Mark, *et al.*: Evaluating Sponge City Volume Capture Ratio at the Catchment Scale Using SWMM. *Journal of Environmental Management*. 246, 745–757 (2019).

Ruan, Jieer, Yingbiao Chen, and Zhiwei Yang. "Assessment of Temporal and Spatial Progress of Urban Resilience in Guangzhou Under Rainstorm Scenarios." *International Journal of Disaster Risk Reduction*. 66, 102578 (2021).

Tian, Li, and Wenjun Ma.: Government Intervention in City Development of China: A Tool of Land Supply. *Land Use Policy*. 26.3, 599–609 (2009).

Wang, Xiaoyan, *et al.*: Dynamic Simulation and Comprehensive Evaluation of the Water Resources Carrying Capacity in Guangzhou, *China. Ecological Indicators*. 135, 108528 (2022).

Wei, Yehua Dennis, and Xinyue Ye.: Urbanization, Urban Land Expansion and Environmental Change in China. *Stochastic Environmental Research and Risk Assessment*. 28.4, 757–765 (2014).

Yuan, Yingwei, *et al.*: Evaluation of Comprehensive Benefits of Sponge Cities Using Meta-Analysis in Different Geographical Environments in China. *Science of The Total Environment*. 836, 155755 (2022).

Zhou, Chunshan, Xueqiang Xu, and Szeto Sylvia. "Population Distribution and its Change in Guangzhou City." *Chinese Geographical Science*. 8.3, 193–203 (1998).

Constructional Engineering and Ecological Environment – Chih-Huang Weng (Ed)
© 2024 The Author(s), ISBN 978-1-032-53198-4

A summary and prospect of the research on Guilin traditional villages in recent ten years

Wanjing Li*
Guilin University of Technology, Yanshan District, Guilin City, Guangxi Province, China

ABSTRACT: As the main carrier of the traditional culture of the Chinese nation, traditional villages have become the focus of attention of all sectors of society in recent years. Guilin, the most important gathering place of traditional villages in Guangxi, has formed a large number of traditional villages with local characteristics because of its long history and rich cultural relics. Through the research and systematic sorting of various research achievements of traditional villages in Guilin in the past ten years, this paper discusses their historical evolution, spatial form, cultural landscape, protection methods and development strategies. Then it points out the deficiencies in the current research process, research methods and research system. It puts forward prospects, hoping to have some inspiration for the future research and development of traditional villages in Guilin.

1 INTRODUCTION

As the largest legacy left by the agricultural civilization of our country, the traditional villages have accumulated rich and colourful historical information and cultural landscapes, which are the spiritual matrix and cultural origin of all ethnic groups, and also the extremely valuable cultural and tourism resources that cannot be copied and regenerated at present (Duan 2020). This paper takes Guilin as a city area. Because of its unique natural resources and well-known beautiful environment, Guilin provides a place for people of different nationalities to live in harmony. Because of the arrival of different nationalities, the development of traditional villages in Guilin has been greatly promoted, and a large number of villages with a long history and profound culture have been built.

2 RESEARCH STATUS OF TRADITIONAL VILLAGES IN GUILIN

2.1 *Analysis of research data*

This paper searched CNKI China Knowledge Network with “Guilin Traditional Villages” as the subject term respectively. The time range of the literature search was set from January 1, 2011, to December 31, 2020. A total of 132 pieces of literature were retrieved, including 65 academic journals, 47 master’s degree papers, 8 doctoral degree papers, 3 international conference papers, 1 newspaper paper, and the rest 7 papers were featured journal papers.

*Corresponding Author: 282651855@qq.com

DOI: 10.1201/9781003410843-30

2.2 *Analysis of the research stage*

Judging from the publication time of the literature, the research on Guilin traditional villages started in 2011 and was in a relatively slow stage from 2011 to 2015. Since the 2016 Speech at the Sixth Session of the Fourth Guilin Conference of the CPPCC put forward that the protection of traditional villages in Guilin should be urgent, traditional villages in Guilin have attracted public attention, and relevant research has begun to grow rapidly. According to the analysis of the above data and a large number of documents, the study of Guilin traditional villages can be divided into three periods.

The report of the 18th National Congress of the Communist Party of China in 2012 put forward the slogan of building a beautiful China in order to implement the spirit of the 18th National Congress of the Communist Party of China on building a beautiful China and further promoting the construction of a new countryside (Wei 2012). Guilin actively carries out the activity of "beautiful Guangxi and clean countryside", aiming at beautifying the countryside, cultivating new styles and benefiting the masses, and strengthening the construction of traditional villages in Guilin, which is an irresistible trend and situation (Xia 2016). During this period, the research on Guilin's traditional villages was at the initial stage. The overall number of research documents was small, and the research direction was also relatively limited. Most of them focused on the construction and protection model.

Research and development period (2016-2018): In 2016, Guilin had 49 traditional villages in China, so normative documents such as the Measures for the Protection and Utilization of Traditional Villages were formulated to make the protection and utilization of traditional villages and ancient buildings step into the track of legalization, standardization and routinization (Chu 2018), and traditional villages in Guilin are increasingly attracting the attention of the public and experts and scholars. At this stage, the number of research results has increased significantly, and the depth of research has gradually deepened from the surface. From the perspective of research content, from the level of protection of traditional villages to the level of culture, the proportion of research on the cultural connotation of traditional villages has been greatly increased. At the same time, a large number of in-depth studies have been carried out from the perspective of ethnic minorities.

Research promotion period (2019-2020): As the urbanization process continues to accelerate, the protection and development of traditional villages are also facing the shackles of ecological degradation and restoration, loss and protection of architectural style, cultural fault and continuity (Li 2018). At the same time, the report of the 19th National Congress of the Communist Party of China put forward the strategy of rural revitalization, which stimulated the vitality of traditional villages. In this stage, a large number of documents were studied based on the strategy of rural revitalization. At the same time, the protection and construction of traditional villages were once again noted for the academic community. The study of traditional villages was no longer limited to a small direction but was broader in scope and more comprehensive in content.

3 ANALYSIS OF THE RESEARCH CONTENT OF GUILIN TRADITIONAL VILLAGES

3.1 *Study on the spatial form of Guilin traditional villages*

Guilin's traditional villages have been evolving with the passage of time. The study of their spatial form starts from a single village space to Guilin as a geographical unit. Zheng Jingwen (2005). (Zheng 2005).Deng Chunfeng et al. (2007) believed that the formation of urban spatial form in northern Guangxi was the result of the joint action of various constituent elements under the specific background of the natural environment, cultural connotation, social economy and transportation mode. (Deng *et al.* 2007) Duan Sijia *et al.* (2020) This paper classifies the villages according to their natural environment, history, humanities, spatial organization, and other characteristics. It explores the spatial characteristics of Guilin's traditional villages (Duan & Wang 2020).

3.2 *Research on the cultural landscape of Guilin traditional villages*

Guilin traditional villages have a long history, culture and rich natural resources, but due to the rapid development of modernization, the landscape of traditional villages has been destroyed in a large area. Liu Hui et al. (2017) proposed that the reason for the loss of cultural landscape in traditional villages in Guilin is that people have different levels of needs under different cultural backgrounds and regional environments. Traditional villages cannot meet the inner needs of villagers in a short period of time, causing villagers to leave the village. Going to a different place, the landscape of a traditional village without the humanities will cease to exist (Liu *et al.* 2017). Wang Yiqi (2019) established the relationship between the aesthetics of landscape painting and traditional rural landscape by extracting the aesthetic characteristics of traditional landscape painting and using the method of comparative analysis. The construction of the traditional village landscape provides a positive reference (Wang 2019). In the research on the landscape aesthetics of traditional villages in Guilin, Tang Tiantian (2019) expounded on the development and changes of the villages, the layout and construction of the village space through the analysis of its formation reasons and spatial characteristics and obtained the traditional village landscape in northern Guilin. Space aesthetics is divided into theoretical aesthetics and practical aesthetics of traditional Feng Shui type and natural ecological type (Tang 2019).

3.3 *Research on the protection and development of Guilin traditional villages*

With the prosperity of Guilin's tourism industry driving its economic development, the unique natural scenery and rich and long cultural history of Guilin's traditional villages have attracted numerous Chinese and foreign tourists. However, at the same time as the development of tourism and the gain of economic benefits, the ecological culture of traditional villages has been strongly impacted, and the landscape environment has suffered a lot of damage. In order to cater to tourists, the original architectural appearance has been changed. There is a strong commercial atmosphere, so the protection of traditional villages is imminent. Zhou Kaibao (2014) believes that the protection of traditional villages and ancient buildings in Guilin is mainly composed of two forces: one is the official government agency; The strength of private capital is maintained, and the remediation and utilization of natural environment resources around ancient villages and towns are strengthened (Zhou 2014). Xie Jin (2016) believed that the protection of ancient buildings in traditional villages should not only improve the protection awareness of residents but also need a comprehensive census, in-depth research, strengthening protection planning, and timely repairs. The urban and rural style reconstruction is organically combined (Xie 2016). Qiu Meirong (2019) discussed some problems existing in the protection and development of traditional villages in Guilin. She believed that only by finding the key to the problem according to her actual situation could the problem be solved fundamentally, and she made recommendations for the protection and development of traditional villages in Guilin as some theoretical guidance (Qiu 2019).

4 CONCLUSION AND OUTLOOK

4.1 *Contribution and deficiency of Guilin traditional village research*

According to the collation and analysis of existing relevant literature, Guilin's traditional villages have begun to enter the focus areas of experts and scholars from all walks of life.

Their related research has also made certain achievements. In terms of the overall research process, the number of research results has increased year by year, the depth and breadth of research content have been improved, and research methods and ideas have been expanded. However, there are still some deficiencies in the study of Guilin traditional villages:

From the existing literature, the research on traditional villages in Guilin mostly selects typical villages with ethnic minority characteristics or around scenic spots, and the content tends to be templated. The lack of research on atypical villages in different regions of Guilin leads to an imbalance in the research area. From the perspective of research content, the research on traditional villages in Guilin is mainly limited to individual research on a certain village. Based on different realistic backgrounds, there is a lack of macro research on the Guilin area. From the perspective of the research system, the research on traditional villages in Guilin is basically based on the mode of "expounding the concept, research background and principles of traditional villages in Guilin → describing the current situation or existing problems of a traditional village → proposing strategies for protection and development", lacking systematic research on traditional villages. From the perspective of the research field, many disciplines such as architectural science, tourism, agricultural economics and so on have been involved in the research of traditional villages in Guilin, but most of them are studied and analyzed only by a single discipline, and the cross-disciplinary research literature is particularly insufficient.

4.2 *Prospect of Guilin traditional villages*

In view of the problems existing in the study of Guilin traditional villages, in order to make further research and better development of Guilin traditional villages in the future, the following points need to be strengthened:

Balance the study area and consider village differences. At present, it is far from enough to focus on the ethnic or well-known villages in Guilin. We should expand the scope of research and include the traditional villages in the marginal areas of Guilin. In addition, in the study of protection and development, villages in different regions have different development conditions, and their differences should be fully considered. Improve research and attach importance to cultural development. Guilin minority culture is an important part of Guilin's history and culture, and the clan is an indispensable organization in Chinese society. Experts and scholars should recognize the importance of minority cultures and the need to protect and develop them. They should further explore and discuss these weaker topics to promote the inheritance and development of these traditional minority cultures. Strengthen the research system and consolidate the research framework. Guilin should improve the systematization and diversification of the study of traditional villages, carry out detailed and detailed research from different angles, and the level of research should also be deepened. As a flood-prone area near the Lijiang River, the research on disaster prevention and reduction system in Guilin is very important. Therefore, the research and protection measures on disaster prevention systems should be further strengthened and consolidated. With the rapid development of modern technology and digital technology and the continuous improvement of laws and regulations, the digital research of traditional villages will gain more technical reserves and scientific support. At the same time, it will also open up the ideas and dimensions of theoretical research.

REFERENCES

Chu Jinlong. Study on the Spatial Distribution and Evolution of Traditional Villages in Ancient Huizhou. *Journal of Anhui Jianzhu University*. 2018, 26 (03): 26–34

Deng Chunfeng, Feng Bing, Gong Ke, and Liu Shengwei. Spatial Form and Landscape of Urban Settlements in Northern Guangxi. *Urban Issues*. 2007 (09): 62–68

Duan Sijia and Wang Jingwen. Analysis of Spatial Morphological Characteristics of Traditional Villages in Guilin. *China Urban Forestry*. 2020,18 (03): 84–89

Duan Sijia, Research on Spatial Form of Traditional Villages in Guilin [D]. *Beijing Forestry University*. 2020

Li Jiulin. Spatial Evolution Characteristics and Driving Mechanism of Traditional Villages in ancient Huizhou. *Economic Geography*. 2018, 38 (12): 153–165

Liu Hui, Guo Huilin, and Fan Yaming. Protection and Development of Traditional Village Landscape in Northern Guangxi Under the Background of Beautiful Rural Construction. *Industry and Technology Forum*. 2017, 16 (10): 30–32

Qiu Meirong. Reflections on the Protection and Development of Traditional Villages in Guilin. *Times Economy and Trade*. 2019 (07): 56–57

Tang Tiantian is a sweet Comparative Study on Landscape Space Aesthetics of Traditional Villages in Northern Guangxi [D]. Guangxi University, 2019. Wang Pan Shuikou Garden in Huizhou Village Anhui Architecture. 2011 (03): 55–56

Wang Yiqi. *Research on Rural Landscape Construction Strategy Based on Traditional Landscape Painting Aesthetics* [D]. Guilin University of Technology. 2019

Wei Xueyuan. Clan Control and the Village Pattern of Huizhou in the Ming and Qing Dynasties *Journal of Suihua University*. 2012 (04): 68–70

Xia Shujuan. An Analysis of the Internal Mechanism of the Construction of Traditional Villages in Huizhou *Journal of Social Sciences of Jiamusi University*. 2016, 34 (02): 155–157

Xie Jin. Construction and Protection of Traditional Villages – Taking Majiafang Village in Guilin as an Example. *Tourism Overview (Second Half of the Month)*. 2016 (02): 139

Zheng Jingwen. Analysis of the Space of Ethnic Minority Settlements in Northern Guangxi [D]. *Huazhong University of Science and Technology*. 2005

Zhou Kaibao. Discussion on the Mode of Protection and Development of Traditional Villages in Guilin. *Guangxi Urban Construction*. 2014 (11): 34–45

Constructional Engineering and Ecological Environment – Chih-Huang Weng (Ed)
© 2024 The Author(s), ISBN 978-1-032-53198-4

Analysis of the factors influencing the flexural strength of steel slag eco-type ultra-high performance concrete

Jieling Ma*, Xianyuan Tang, Jie Luo & Binbing He
School of Architecture and Transportation Engineering, Guilin University of Electronic Technology, Guilin, China

ABSTRACT: In order to study the flexural performance of ecological ultra-high performance concrete (UHPC) prepared by replacing quartz powder with steel slag powder, the orthogonal test method was used to optimize the design of the fitting ratio of steel slag UHPC, and the flexural strength tests were carried out under different fitting ratios in order to investigate the effect of four factors of silica fume, steel slag power, river sand and steel fibres on the flexural strength of steel slag UHPC. The test results show that the volume dosing of steel fibres has the most significant effect on the flexural strength of the steel slag UHPC. In contrast, the dosing of river sand and steel slag powder has a greater effect, and the dosing of silica fume has a smaller effect. The flexural strength of UHPC increases and then decreases with the amount of silica fume, river sand and steel fibres, decreases and then increases as the amount of steel slag powder mixed increases.

1 INTRODUCTION

Ultra-High Performance Concrete (UHPC) is a new type of cementitious composite material with excellent properties such as ultra-high strength, ultra-high toughness and excellent durability, which is widely used in marine engineering, large span bridges and other structures with special requirements for concrete (Chen 2019: Shao 2021). UHPC is mostly made of high-strength cement + mineral dopants + fine aggregates + steel fibres + high-efficiency water-reducing agents, together with special mixing technology and maintenance methods, due to its high cost of raw materials, formulation difficulties and other problems, resulting in its high project cost, to a certain extent, limiting its promotion and application (Xu 2021). China is a major steel-producing country, with crude steel production exceeding 1 billion tons in 2020, but the current utilization rate of steel slag is only about 40% (Wu 2021). Steel slag will be processed and applied to replace quartz powder and part of the cement to develop ultra-high performance concrete made of green environmental protection building materials, which will help to improve the effective use of industrial waste slag, turning waste into treasure.

To reduce the cost of preparing ultra-high-performance concrete, some researchers have proposed that industrial and agricultural waste materials be blended into UHPC after reasonable processing (Yang 2020). In the study of UHPC with steel slag powder, Wang H. (2016) investigated the effect of steel slag powder on the mechanical properties of UHPC, showing that steel slag powder can be used as a cementitious material for the preparation of ultra-high-performance concrete. Tang X.Y(2021) demonstrated the feasibility of replacing quartz powder with steel slag powder in the formulation of UHPC by conducting tests on the performance effects of steel slag powder in replacing quartz powder in the formulation of UHPC. Zu Q.H (2019) investigated the effect of coarse particle size interval of steel slag on the performance of UHPC by formulating UHPC with coarse particle size interval of steel slag powder as admixture.

*Corresponding Author: majielingzc1128@163.com

 DOI: 10.1201/9781003410843-31

Most of the studies have considered the effect of a single factor on steel slag UHPC. They have not focused on the joint effect of changes in the combination of lots of different raw material mixes on its flexural strength. In order to optimize the fitting ratio of steel slag UHPC, this paper adopts the design method of orthogonal test based on the modified Andreasen & Andersen model to prepare UHPC by varying the dosing of silica fume, steel slag powder, river sand and steel fibre, and to test its flexural strength in order to analyze the significant effect law of varying the dosing of various raw materials on the flexural strength of steel slag UHPC.

2 RAW MATERIAL TESTS

2.1 *Source of raw materials*

Cementitious materials: P-O 42.5 grade ordinary silicate cement in bags; silica fume using grey powder silica fume with SiO_2 content greater than 93%; steel slag with 18% grade II fly ash, obtained by different grinding time and ball milling medium, whose free calcium oxide (f-CaO) does not exceed 1%.

Other materials: natural river sand with a fineness modulus of 2.70; steel fibre with a diameter of 0.22mm, length of 13mm and tensile strength of 2500MPa; water reducing agent is poly-carboxylic acid high-efficiency water reducing agent mother liquor with a water reduction rate of more than 30%; mixing water is tap water.

The basic properties and chemical composition of the main materials are shown in Table 1.

Table 1. Main material properties and chemical composition.

Materials	Bulk density (g/cm^3)	Specific surface area (m^2/kg)	Chemical compositions (%)						
			SiO_2	Al_2O_3	Fe_2O_3	CaO	MgO	Burning loss	Mud content
Cement	1.60	300	21.9	4.22	2.68	63.7	2.12	1.71	–
Silica fume	1.52	2200	93.6	0.5	0.59	1.95	0.27	1.30	–
Steel slag powder	1.78	620	17.1	5.64	22.69	43.38	5.98	1.56	–
River sand	1.55	–	98.7	–	0.07	–	–	0.9	0.08

2.2 *Raw material grading and SEM analysis*

Natural river sand was sieved according to construction sand, and cementitious materials such as silica fume, steel slag powder and cement were sieved using negative pressure. The grading sieving curve is shown in Figure 1. The silica fume and steel slag powder were subjected to microscopic analysis using SEM, the pattern of which is shown in Figure 2.

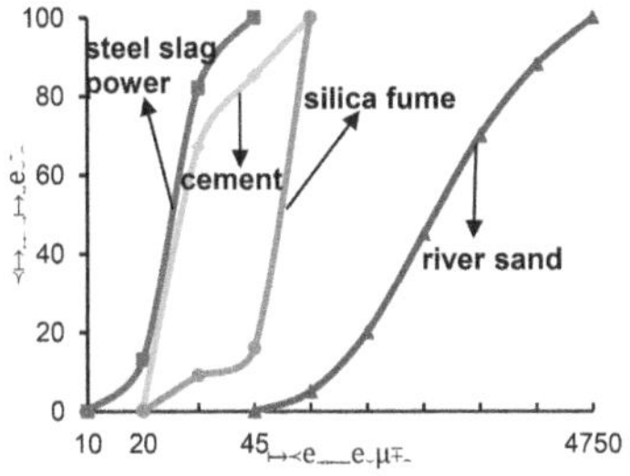

Figure 1. Cumulative raw material sieving curves.

(a) Silica fume (b) Steel slag power

Figure 2. SEM microscopic analysis chart.

As can be seen from Figure 2: under the scanning electron microscope, the silica fume crystal particles are small, and the addition of silica fume in the preparation process can fill the pores between the large particles of material and improve the pore size distribution, which has a positive effect on the improvement of concrete strength and also reduces the permeability of the hydrated slurry. The distribution of steel slag powder particles is irregular, with no specific morphology. From the figure can see the mineral crystalline crystals, crystal surface is relatively smooth and flat, and its RO phase (steel slag powder fused with FeO, MgO and other divalent metal oxides such as MnO to form a wide range of solid solution) crystallization degree is better, indicating that the steel slag powder is less active, not easy to react with water.

3 ULTRA HIGH-PERFORMANCE CONCRETE TESTS

3.1 *Base ratio design*

In this test, the base mix ratio design is based on the closest packing theory, and the Modified Andreasen & Andersen model (MAAM)(YU 2020) is used to calculate and analyze the base mix ratio of steel slag UHPC with the following equation.

$$P(D) = \frac{D^q - D^q_{\min}}{D^q_{\max} - D^q_{\min}} \times 100\% \tag{1}$$

Where: P(D) is the cumulative percentage of sieve residue of particles with a particle size less than D, %. D is the particle size, μm; $D_{\max}$ and $D_{\min}$ are the maximum, and minimum particle sizes in the system, μm; q is the particle size distribution modulus, which is taken 0.24 in this paper.

Using python to model and analyze Equation (1), the least squares method (LSM) was used to calculate the raw material ratios according to the particle distribution curves of different raw materials and the MAAM target curve in Figure 1. The stacking curve was made as close to the target curve as possible by adjusting the ratio of each material. The calculated steel slag UHPC base mix ratio is detailed in Table 2, with a fixed water-to-glue ratio of 0.18.

Table 2. Basic mix ratio.

	Raw material usage /(kg/m^3)							
Group	Cement	Silica fume	Steel slag power	River sand	Steel fibre	Water	Water reducing agent	Water-to-glue ratio
L	720	380	140	760	0	225	38	0.18

3.2 *Orthogonal experimental design*

An experiment to investigate the effect of four main factors - (A) silica fume, (B) steel slag powder, (C) river sand and (D) steel fibres - on the flexural strength of steel slag powder ultra-high performance concrete. Three levels were set for each factor, and an orthogonal experimental design was used to reduce the number of trials, as the full-scale test required $3^4 = 81$ trials. Group L was used as the base mix ratio on which the factor levels were increased or decreased. A total of nine mix ratios were tested according to the L_9 (3^4) orthogonal test design table. The factors and levels were set as shown in Table 3, and the orthogonal test protocol is shown in Table 4.

Table 3. Factor level table.

Level	A Silica fume Dosing (kg/m^3)	B Steel slag Dosing (kg/m^3)	C River sand Dosing (kg/m^3)	D Steel fibres by volume/ (%)
1	342(-10%)	126(-10%)	684(-10%)	0
2	418(+10%)	154(+10%)	836(+10%)	+1.5
3	494(+30%)	182(+30%)	988(+30%)	+2

Table 4. Orthogonal test protocols.

Group	A/(kg/m^3)	B/(kg/m^3)	C/(kg/m^3)	D/(%)
L	380	140	760	0
T1	342	126	684	0
T2	342	154	836	1.5
T3	342	182	988	2
T4	418	126	836	2
T5	418	154	988	0
T6	418	182	684	1.5
T7	494	126	988	1.5
T8	494	154	684	2
T9	494	182	836	0

3.3 *Preparation, maintenance and performance testing*

During the test, the components are calculated and weighed according to the mix ratio. In the mixing process, first put all the dry powder into the concrete mixer and mix for about 3 min, then add the weighed water and water-reducing agent evenly. After 8 min mixing add the steel fibres. The UHPC mix is then mixed for 2 min to complete the preparation.

The specimens were left to stand for 1d in a room at a temperature of 20°C±5°C and a relative humidity of >50% before being demolded and numbered, followed immediately by high-temperature maintenance at 90°C±5°C for 2d, then standard maintenance and removed for testing when 28d was reached. The cube compressive strength test was carried out using 100×100 ×100 mm size test blocks, and the flexural strength test was carried out using 100× 100×400 mm prisms. The tests were carried out in strict accordance with the requirements of the mechanical properties tests.

4 EXPERIMENTAL ANALYSIS

4.1 *Flexural strength analysis*

To investigate the flexural properties of steel slag UHPC, the orthogonal test steel slag UHPC 28d flexural strength test was plotted as a comparative analysis, shown in Figure 3.

Analysis of Figure 3 shows:

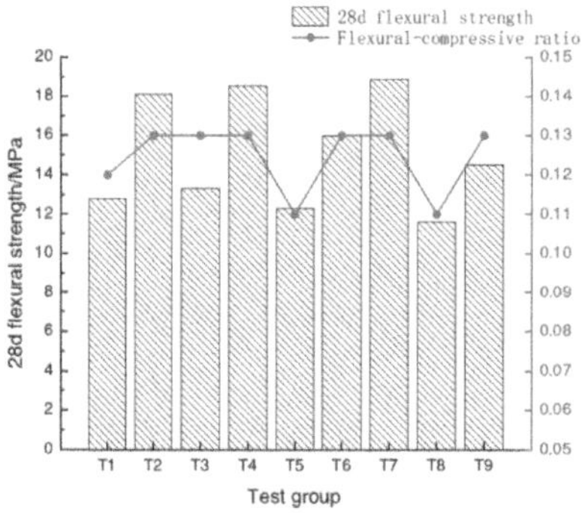

Figure 3. Comparative analysis of flexural strength.

(1) The 28d flexural strength of steel slag UHPC with different fitting ratios in the test group reached above 11.0 MPa, with an average flexural strength of 15.1 MPa, indicating that steel slag UHPC has a strong flexural resistance. The ratio of 28d flexural strength to the cubic compressive strength of the steel slag UHPC specimens with different fitting ratios were all stable between 0.11 and 0.13, with an average value of 0.124.
(2) The flexural strength of the steel slag UHPC varies greatly among the different mix ratios. Among them, group T7 has the highest strength, with 28d flexural strength up to 18.9 MPa, while group T8 has the lowest strength, with 28d flexural strength of only 11.6 MPa, with a difference of 7.3 MPa between them. This means that the raw material mix ratio has a greater influence on the steel slag UHPC. When the mix ratio is not reasonable, the raw materials with different particle sizes cannot reach the maximum dense accumulation, resulting in a larger spacing, which will be more likely to produce cracks when under stress, making the flexural strength smaller and thus leading to a larger gap in flexural strength.

4.2 *Analysis of the range*

The range calculations for the effect of the four factors of silica fume, steel slag powder, river sand and steel fibre on flexural strength were collated to Table 5.

Analysis of Table 5 gives the following:

(1) The magnitude of the R-value of the 28d flexural strength of steel slag UHPC is in the following order: $R_D>R_C>R_B>R_A$, indicating that each factor influences the flexural strength in the following order: D (steel fibre), C (river sand), B (steel slag powder) and A (silica fume).
(2) The range value (R_D) of steel fibre is the largest, reaching 13.4MPa, which indicates that the dosing of steel fibre has a significant effect on the flexural strength of steel slag UHPC. River sand and steel slag fines had the next largest effect, with extreme difference

Table 5. Analysis of extreme differences in flexural strength.

Range	28d Flexural strength (MPa)			
	A	B	C	D
K_1	44.2	50.2	40.4	39.6
K_2	46.8	42.0	51.1	53.0
K_3	45.0	43.8	44.5	43.4
k_1	14.7	16.7	13.4	13.2
k_2	15.6	14.0	17.0	17.6
k_3	15.0	14.6	14.8	14.4
R	2.6	8.2	10.7	13.4

where: K_i is the sum of the test results at the level of each factor i; k_i is the mean of the test results at the level of each factor i; R is the range.

values of 10.7 MPa and 8.2 MPa, respectively, while silica fume had little effect on the flexural strength of UHPC, with a range value (R_A) of 2.6 MPa.

(3) In this test, the index is the 28d flexural strength of steel slag UHPC. The larger the index value, the better, so the level corresponding to the maximum value of each factor, K_i is selected to form the optimal solution. Factor A: $K_2>K_3>K_1$, Factor B: $K_1>K_3>K_2$, Factor C: $K_2>K_3>K_1$, Factor D: $K_2>K_3>K_1$.Therefore the theoretically preferred solution for steel slag UHPC at 28d flexural strength is $A_2B_1C_2D_2$.

4.3 *Analysis of influencing factors*

The above superior solutions were obtained by theoretical analysis. They were derived under the conditions of given factors and levels, and it is possible to obtain better test solutions if the given levels are not limited. In order to analyze the effect of the variation of the dosing of the four raw materials on the flexural strength of steel slag UHPC, and then to determine the optimal flexural performance of the optimal solution. Using the factor admixture as the horizontal coordinate and the mean 28d flexural strength k_i as the vertical coordinate, the variation pattern of each factor admixture versus flexural strength is plotted on a trend diagram, as shown in Figure 4.

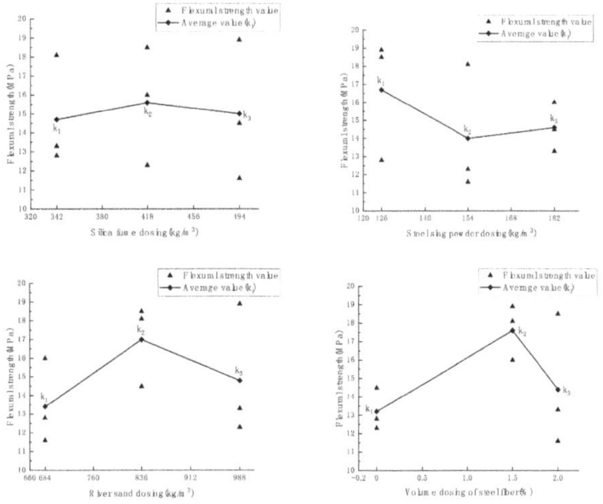

Figure 4. The trend of variation in flexural strength by dosing factor.

As can be seen in Figure 4:

(1) The flexural strength of steel slag UHPC increases and then decreases with the amount of silica fume, river sand and steel fibres mixed increases. It decreases and then increases as the amount of steel slag powder mixed increases. This indicates that the flexural strength of steel slag UHPC can be improved by increasing the amount of silica fume, river sand and steel fibre and decreasing the amount of steel slag fibre on the basis of the base ratio.
(2) The amount of silica fume dosing had little effect on the flexural strength of steel slag UHPC. The flexural strength only increased by 0.9 MPa when the dose was increased from 342 g/m^3 to 418 kg/m^3 and decreased slightly by 0.6 MPa when the dose was further increased to 494 kg/m^3.
(3) The dosing amount of steel slag powder has a small effect on the flexural strength of UHPC. The flexural strength decreased by 16.1% after increasing the dose from 126 kg/m^3 to 154 kg/m^3. The flexural strength of UHPC increased slightly by 4.3% after increasing the dose to 182 kg/m^3 compared to 154 kg/m^3, and the change in strength was not obvious.
(4) The amount of river sand has a greater influence on the flexural strength of steel slag UHPC. The flexural strength increased by 26.8% after the dosage was increased from 684 kg/m^3 to 836 kg/m^3 and decreased when the dosage was increased to 988 kg/m^3, the flexural strength decreased by 12.9%, and the strength changed significantly.
(5) The influence of steel fibre doping on the flexural strength of steel slag UHPC is obvious. The flexural strength of steel fibre was the highest at 1.5% by volume, which was 33.3% and 22.2% higher than that without steel fibre and at 2% by volume, respectively. This indicates that the steel fibres play a good bridging role in the steel slag UHPC, which can effectively prevent the sudden fracture of the specimens so that the specimens still have a certain flexural strength after destruction, which to a certain extent compensates for the shortcomings of the brittle fracture of UHPC.
(6) According to the trend of each factor, when the dose of silica fume is 418 kg/m^3 (level 2), the dose of steel slag powder is 126 kg/m^3 (level 1), the dose of river sand is 988 kg/m^3 (level 2) and the volume dose of steel fibre is 1.5% (level 2), the flexural strength of steel slag UHPC is the highest, the best solution is $A_2B_1C_2D_2$.

In summary, the preferred solution for the steel slag UHPC flexural strength ratio was selected as $A_2B_1C_2D_2$.

5 BEST MIX RATIO PERFORMANCE TESTS

In order to verify the reasonableness of the optimized steel slag UHPC mix ratio, steel slag UHPC was prepared according to the best mix ratio scheme $A_2B_1C_2D_2$ and tested for slump and expansion, and the corresponding cubic compressive strength and flexural strength were tested when the test period age was reached. The test results are shown in Table 6.

Table 6. Superior solution performance test results.

Group	Slump (mm)	Expansion (mm)	Cubic compressive strength (MPa)			28d flexural strength (MPa)
			3d	7d	28d	
TS	200	320	126.2	141.9	152.0	21.8

As can be seen from Table 6, the slump of steel slag UHPC prepared by the best ratio solution is 200 mm, the expansion degree is 320 mm, the 28d compressive strength reaches 152.0 MPa, and the flexural strength is up to 21.8 MPa, which shows that the best ratio is reasonable and has good working performance and mechanical properties to meet the requirements of engineering design.

6 CONCLUSIONS

(1) The 28d flexural strength of steel slag UHPC all reached above 11.0 MPa, with an average flexural strength of 15.1 MPa, indicating that steel slag UHPC has a strong flexural resistance.
(2) The flexural strength of steel slag UHPC under different mix ratios varied greatly. The change in the mixing amount of each factor had a significant effect on the flexural strength of steel slag UHPC. In this test, $R_D>R_C>R_B>R_A$, the degree of influence of each factor on flexural strength in the order of predominance is: steel fibre>river sand>steel slag powder>silica fume.
(3) The flexural strength of steel slag UHPC under the combined effect of silica fume, steel slag powder, river sand and steel fibre increase with the amount of silica fume, river sand and steel fibre. Then it decreases with the amount of steel slag powder.
(4) The optimum ratio for the flexural strength of steel slag UHPC is 418 kg/m^3 of silica fume, 126 kg/m^3 of steel slag, 836 kg/m^3 of river sand and 1.5% of steel fibre by volume. The 28d flexural strength of steel slag UHPC prepared using the preferred mix ratio was up to 21.8MPa.

ACKNOWLEDGEMENTS

This work was supported by the National Natural Science Foundation of China [42067044] and the Innovation Project of Guang Xi Graduate Education [YSCW2021171].

REFERENCES

Chen B.C., Wei J.G., Su J.Z. *et al.* (2019). State-of-the-art Progress on the Application of Ultra-high Performance Concrete. *Journal of Architecture and Civil Engineering.* 36(02):10–20.

Shao X.D., Fan W. and Huang Z.Y. (2021). Application of Ultra-High-Performance Concrete in Engineering Structures. *China Civil Engineering Journal.* 54(01):1–13.

Tang X.Y., Guo B., MA J.L. *et al.* (2021). Effect of Steel Slag Powder on the Properties of Ultra-high Performance Concrete (UHPC). *J. Chi Con Cem pro.* (11):82–89.

Wang H., Guo L.P., Sun W., *et al.* (2016). Steel Slag Powder on Ultra-high Performance Cementitious Composites Performance. *J. Chi Con Cem Pro.* (12):13–19.

Wu Y. D., Peng B., Wu L., *et al.* (2021). Review on the Global Development of Treatment and Utilization of Steel Slag. *J. Environmental Engineering.* 39(01):161–165.

Xu Y.B., Yu Y., Jin Z.Q., *et al.* (2021). Review on Effects of Microstructure and Mechanical Properties of Ultra-High Performance Concrete By Curing Regimes. *J. Bulletin of the Chinese Ceramic Society.* 40 (09):2856–2870.

Yang R., Yu R., Shui Z.H. *et al.* (2020). Feasibility Analysis of Treating Recycled Rock Dust as an Environmentally Friendly Alternative Material in Ultra-High Performance Concrete (UHPC). *Journal of Cleaner Production.* 258(C):120673–120673.

Yu R., Fan D.Q., Shui Z.H., *et al.* (2020). Mix Design of Ultra-High Performance Concrete Based on Particle Densely Packing Theory. *J Chin Ceram Soc.* 48(08):1145–1154.

Zu Q.H., Zang J. and Shen X.D. (2019). Study on the Application of Steel Slag Powder in UHPC in the Coarse Particle Size Interval. *J. Chi Con Cem Pro.* (08):1–4.

Constructional Engineering and Ecological Environment – Chih-Huang Weng (Ed)
© 2024 The Author(s), ISBN 978-1-032-53198-4

Research on wind avoidance and attitude adjustment of photovoltaic tracking bracket under wind load

Zhiwen Guo
Jiangxi University of Science and Technology, Ganzhou, China
Zhejiang Wanli University, Ningbo, China

Lirong Huang
Jiangxi University of Science and Technology, Ganzhou, China

Wen Liu
Zhejiang Wanli University, Ningbo, China

Sen Liu
Ningbo Powernice Intelligent Technology Co., LTD, Ningbo, China

Ning Zhang & KunHao Yan
Zhejiang Wanli University, Ningbo, China

ABSTRACT: To address the problem of low reliability of PV tracking brackets under extreme wind loads, ANSYS fluid-structure coupling is applied to analyze the PV tracking system under different operating angles in terms of wind pressure distribution, structural stress, modal vibration and dynamic response, to establish a reliability performance model, to determine the attitude adjustment operating angle of PV tracking brackets under extreme wind loads, and to provide a reference for the design of PV tracking system reliability and wind avoidance The results of the study show that the higher the working angle, the higher the wind protection angle. The results show that the larger the working angle is, the smaller the wind pressure on the surface of the PV module, the smaller the maximum stress value of the PV tracking bracket, the smaller the inherent frequency of the PV tracking bracket, and the smaller the maximum amplitude of wind-induced vibration. The larger the working angle of the PV tracking bracket, the higher the reliability of the structure. Through the reliability performance model established in this paper, the working condition angle in the wind protection state can be determined according to the demand, balancing the power generation and reliability of the PV system.

1 INTRODUCTION

In the 14th Five-Year Plan, by 2025, incremental renewable energy generation will account for more than 50% of the incremental electricity consumption of society as a whole, and wind and solar power generation will double, making green and clean energy such as photovoltaic gradually become the key development direction of the future energy industry (National Energy Bureau 2021). In solar power generation systems, the use of tracking PV mounts can increase the power generation efficiency by 10-15% compared to fixed PV mounts (Li et al. 2021; Li & Kong 2019), but PV tracking systems are usually applied in Xinjiang, Inner Mongolia and other outdoor geographical expanses, without the shading of buildings and trees leading to significant effects of wind loads on PV tracking structures. It is of significance to study the wind protection status of PV tracking brackets during extreme wind loads for the reliability and stable operation of PV power generation (Kilikevičius et al. 2016; Mireille et al. 2019).

 DOI: 10.1201/9781003410843-32

Several scholars have studied the performance of wind load on the structure of PV supports. Yu (2012) simulated the force of photovoltaic modules under strong wind conditions based on the fluid analysis method to study different turbulence models and verified the feasibility and validity of the computer simulation by comparing the measured results. Warsido (Warsido et al. 2014) investigated the effect of PV array spacing on wind pressure on PV structures. The wind load coefficient increased with increasing panel column spacing, and the wind load coefficient of roof arrays decreased with increasing building edge perimeter. All the above scholars can provide good suggestions for the design of wind resistance of PV tracking bracket structure.

In this paper, we consider the effect of wind load on the performance of PV tracking bracket structure under different working angles and use the adjustability of the PV tracking bracket to reduce the surface pressure difference of the PV module, the extreme value of the PV bracket stress and the amplitude of wind-induced vibration as the target to determine the adjusting working angle when the PV tracking bracket is protected against wind under extreme wind load, so as to improve the reliability of PV tracking system and provide the basis for real-time intelligent control algorithm.

2 WIND LOAD THEORY AND MODELS

When the wind load acts on the PV tracking bracket, the wind load is actually a dynamic load (Qiao et al. 2016). Firstly, the wind load is analyzed theoretically to determine the boundary conditions of the fluid-solid coupling model, so that the simulation analysis is closer to the actual situation.

2.1 *Wind pressure coefficient*

Define the wind pressure coefficient of the PV module as:

$$C_p = \frac{\sum_{i=1}^{u} C_{pi}(t) A_i}{7XY} \tag{1}$$

Where: $C_{pi}(t)$ is the wind pressure coefficient at point i; $P_{mi}(t)$and $P_{ni}(t)$are the upper surface wind pressure and lower surface wind pressure at point i, respectively; C_p is the wind pressure coefficient of the PV module; A_i is the area represented by point i; X and Y are the length and width of the PV module, respectively.

2.2 *Simulation model and parameters*

In this paper, SolidWorks is used to model the PV tracking bracket in 3D, and the model is imported into ANSYS to simulate the flow-solid coupling of the PV tracking bracket. The computational model is shown in Figure 1.

As shown in Figure 2, the boundary condition at the entrance is set to a uniform wind speed of 20 m/s, the exit boundary condition is set to a fully developed boundary condition

Figure 1. Photovoltaic tracking bracket experiment bench.

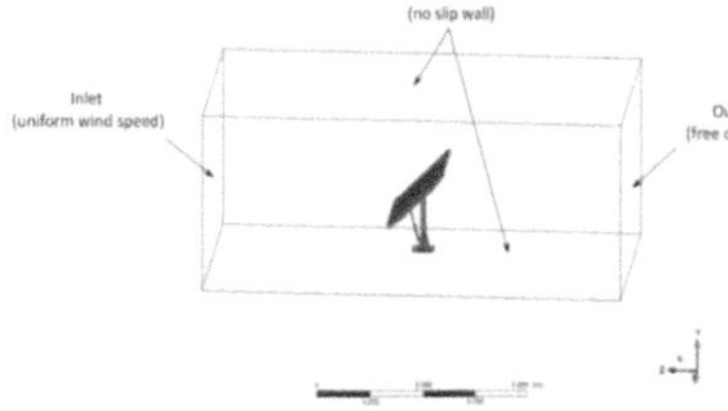

Figure 2. Simulation schematic.

outflow, and the four sides are set to a no-slip wall condition wall, which is kept relatively stationary with the fluid.

3 WIND LOAD SIMULATION ANALYSIS AND DISCUSSION

3.1 *Wind pressure distribution analysis*

As the working condition of the PV tracking bracket changes, it will change the wind area of the PV module, thus changing the wind pressure on the module surface. As shown in Table 1, the wind pressure on the PV module panel is the highest near the front end, and gradually decreases at the back end, showing a gradient distribution along the downwind direction. The wind pressure at the edge of the PV module is smaller compared to the center because the incoming wind is obstructed and bypassed through the edge of the PV module, as shown in Figure 3, the wind flow produces a three-dimensional bypass phenomenon when passing through the edge of the PV module, forming an angular vortex at the back of the PV module to generate back pressure.

Table 1. Wind pressure distribution clouds on the surface of PV modules under different working conditions.

Work angle	Wind pressure distribution	Work angle	Wind pressure distribution
30°		60°	
35°		65°	
40°		70°	
45°		75°	
50°		80°	
55°		8°	

Considering the pressure distribution as trapezoidal distribution, this paper divides the PV module surface into five areas, as shown in Figure 4, with five rows and one column. The trapezoidal distribution can easily describe the wind load distribution law of PV modules and can consider the influence of bending moment around the X-axis on solar mounts.

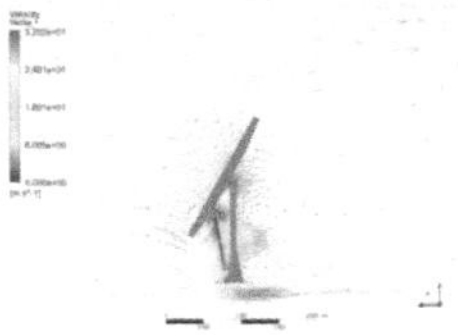

Figure 3. 30° working angle flow field profile.

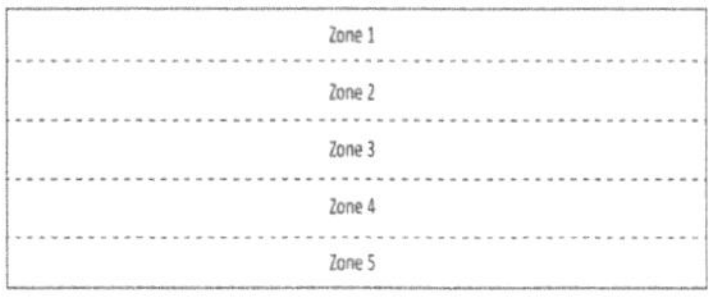

Figure 4. Schematic diagram of trapezoidal distribution.

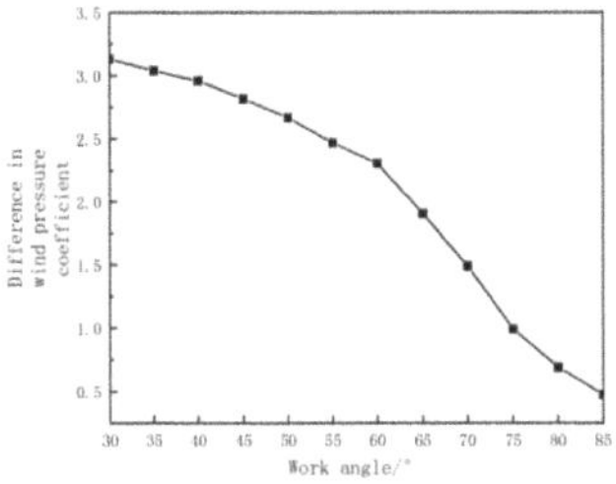

Figure 5. Difference in wind pressure coefficient at different working conditions.

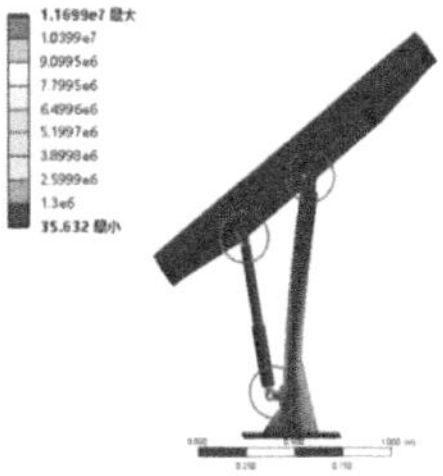

Figure 6. Wind load stress cloud of PV tracking bracket.

In this paper, the difference between the wind pressure coefficient in zone 5 and zone 1 defines the magnitude of the overturning force of the PV module. It can be seen from Figure 5 that the wind pressure coefficient difference of the PV module decreases gradually with the increase of the working condition angle. The maximum wind pressure coefficient difference is 3.132 when the working condition angle reaches 30°.

3.2 *Wind load force analysis of photovoltaic tracking bracket*

The stress distribution of the PV bracket structure in different working conditions is calculated by the fluid-solid coupling finite element analysis, as shown in Figure 7. Three locations show the stress concentration phenomenon. These three positions are the most easily damaged, and it is necessary to increase the size of the structure or improve the stiffness of these three positions to ensure the normal and stable operation of the PV tracking bracket.

Under the action of the same wind load, the maximum stress of the PV tracking bracket structure under different working conditions is different. As shown in Figure 8, the

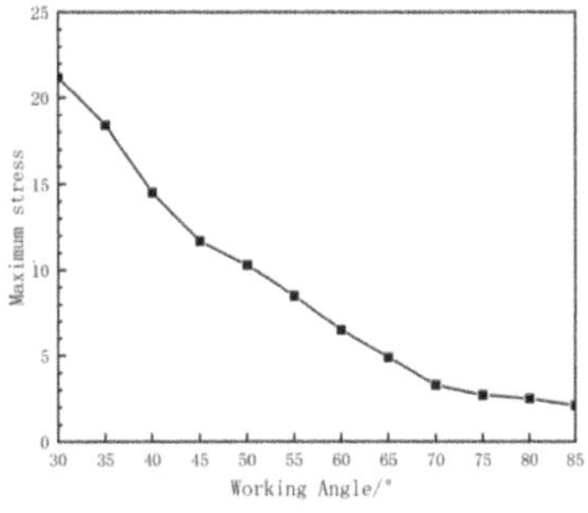

Figure 7. Maximum stress under different working conditions.

maximum stress value of the PV tracking bracket structure gradually decreases with the increase of the working angle. In order to avoid damage to the PV tracking bracket, you should try when adjusting to a large working angle attitude.

3.3 *Modal analysis*

Firstly, the first six orders of inherent frequencies of the PV tracking bracket under different working conditions are analyzed. As can be seen from Table 2, among the first six orders of modes, the first three orders of vibration inherent frequencies have the same trend and are more densely distributed, concentrated between 2 and 7 Hz, and can produce a variety of vibration superposition states under the same excitation frequency. The growth rate of inherent frequencies gradually increases in the 4th to 6th orders of vibration, showing the opposite trend. When the inherent frequency of the structure is far from the vibration frequency generated by wind load on the structure, it can effectively avoid the resonance phenomenon and resonance damage, and the relevant frequency can provide the basis for verifying the safety of the structure.

Table 2. The first six orders of inherent frequency of PV tracking bracket at different working conditions /HZ.

Order angle	30	35	40	45	50	55	60	65	70	75	80	85
1	2.6	2.5	2.4	2.3	2.2	2.2	2.2	2.2	2.1	2.1	2.0	1.9
2	3.9	3.9	3.9	4.0	4.1	4.1	4.1	4.2	4.2	4.2	4.3	4.3
3	7.0	7.0	6.9	6.8	6.7	6.7	6.6	6.6	6.6	6.5	6.5	6.5
4	14.4	14.4	14.4	14.4	14.3	14.3	14.3	14.3	14.3	14.3	14.3	14.3
5	21.0	22.3	24.7	25.8	25.8	25.8	25.8	25.6	24.9	24.1	23.9	23.8
6	25.8	25.9	26.1	26.2	26.2	26.1	26.1	26.0	25.9	25.8	25.7	25.7

3.4 *Dynamic response analysis*

The inherent frequency and modal vibration pattern of the PV tracking bracket under different working conditions were determined by modal analysis. In order to further determine whether the wind load on the structure of the bracket resonates under different working conditions of the PV tracking bracket, the vibration response characteristics of the wind load on the structure of the PV tracking bracket were studied. The displacement, velocity, acceleration and dynamic stress of the PV tracking bracket under different working conditions at the minimum requirement of 20 meters per second for triggering the wind protection state were calculated. The horizontal displacement, velocity and acceleration response curves of the top 64816 nodes of the PV tracking bracket (position shown in Figure 1) under 30°, 45°, 60° and 75° operating conditions are shown in Figures 10–12.

From the displacement response curves in Figure 10, it can be seen that the generated vibration amplitude, velocity and acceleration all decrease gradually with time, which is due to the suppression effect of the system damping on the vibration, and the PV tracker cannot maintain the equal amplitude vibration. The vibration amplitude gradually decreases, and no resonance phenomenon occurs. As the working angle increases, the amplitude gradually decreases.

4 DETERMINATION OF WIND PROTECTION STATE WORKING CONDITIONS

In this paper, the wind pressure distribution, structural stress, modal vibration pattern and dynamic response of PV tracking bracket with different working conditions at a wind speed of 20 m/s are analyzed, as shown in Figure 12. The larger the working condition angle is, the more stable the PV tracking bracket structure is.

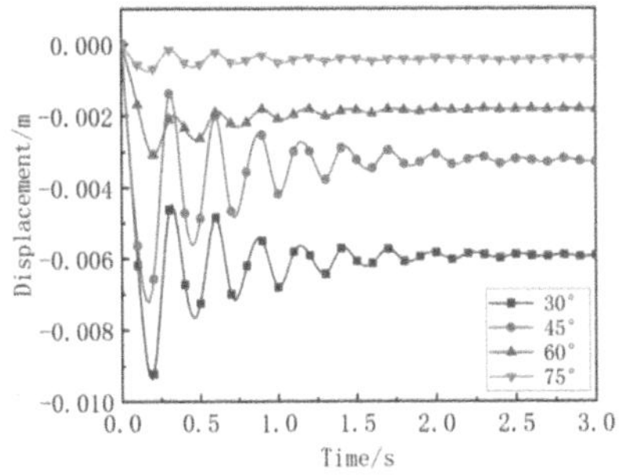

Figure 8. 64816 node horizontal displacement response curve.

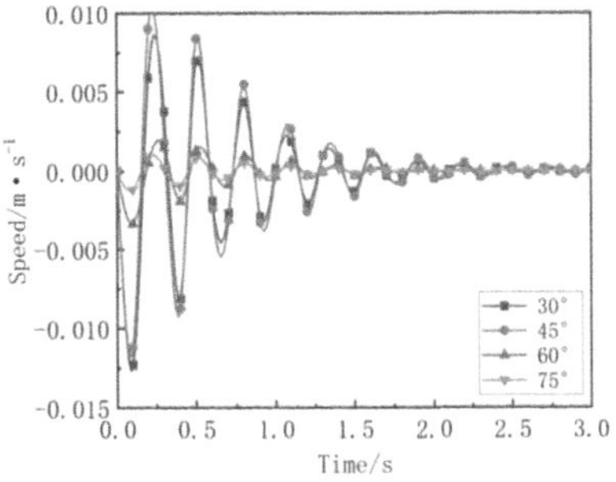

Figure 9. 64816 node horizontal direction velocity response curve.

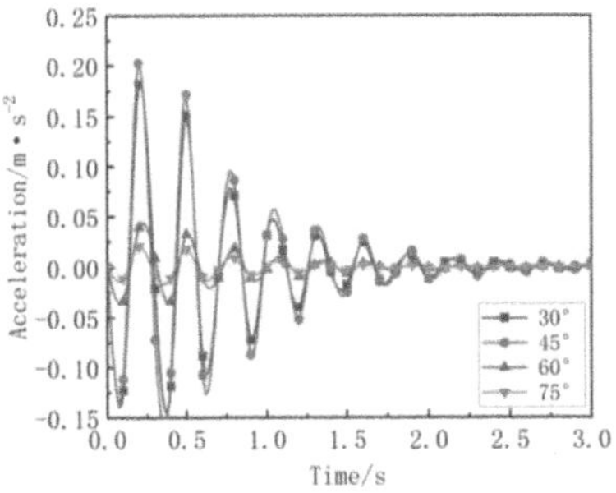

Figure 10. The horizontal acceleration response curve of 64816 node.

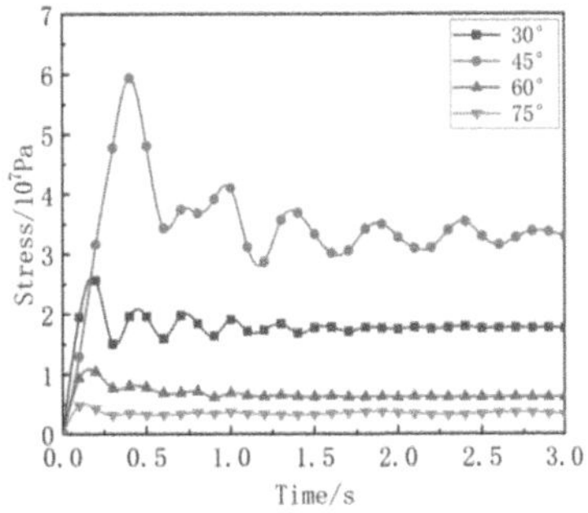

Figure 11. The stress response curve of node 84271.

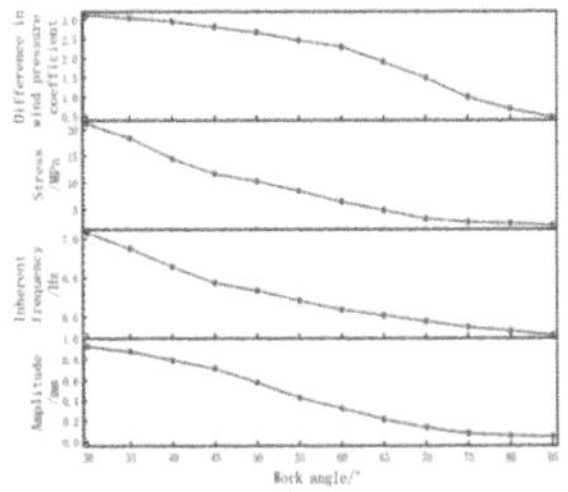

Figure 12. The state of the PV tracking bracket under different working angles.

Through the above analysis, the wind pressure coefficient difference, stress, inherent frequency and amplitude of PV tracking bracket structures under different working conditions are normalized to comprehensively evaluate the reliability of PV tracking brackets at different working angles and establish the normalized performance index Q:

$$Q = \sqrt{\frac{\Omega \cdot S \cdot U}{\omega}} \tag{2}$$

Where: Ω is the wind pressure coefficient difference, S is the stress coefficient, ω is the intrinsic frequency coefficient, and U is the amplitude coefficient.

When Q = 1, the power generation is not considered, and the PV system reliability is completely dominated. For example, if Q meets 0.8, the working angle of the PV tracking system should be adjusted to at least 75°when it enters the wind protection state so that the wind pressure difference between the front and rear of the PV module, the maximum stress of PV tracking bracket structure, the inherent frequency and the wind vibration amplitude can reach the safe range to ensure the reliability and life of PV tracking system while meeting the power generation of a PV system as much as possible.

5 CONCLUSION

The change in working angle has a great influence on the wind pressure distribution on the surface of the PV module, the maximum stress value and the wind response value of the PV bracket. The difference of wind pressure coefficient between the front and rear ends decreases with the increase of working angle; the smaller the working angle, the higher the maximum stress of the bracket structure and the easier it is to damage; the larger the working angle, the smaller the displacement, velocity and acceleration generated by the wind load and the less likely it is to produce structural vibration. Comprehensive wind pressure distribution, structural stress, modal vibration pattern and dynamic response are analyzed in four aspects: wind load on the PV tracking bracket structure under different working conditions, and the reliability performance model is obtained through normalization, which can determine the working condition angle when entering the wind protection state according to the demand and balance the power generation and reliability of PV system. Taking Q satisfying 0.8 as an example, the wind protection state needs to be adjusted to at least 75°or more.

ACKNOWLEDGMENT

Fund Projects:

1. Ningbo Science and Technology Innovation 2025 Major Special Project (2021Z075)
2. Zhejiang Basic Public Welfare Research Program (LGG22E050042)
3. 2022 Science and Technology Innovation Activity Plan for College Students in Zhejiang Province (2022R420A008)
4. 2022 Science and Technology Innovation Activity Plan for College Students in Zhejiang Province (2022R420A001).

REFERENCES

Kilikevičius A., Čereška A. and Kilikevičienė K. Analysis of External Dynamic Loads Influence to Photovoltaic Module Structural Performance. *Engineering Failure Analysis*. 2016, 66: 445–454.

Li H., Liu D. and Yao D. A Study on the Development of China's Power System With the Goal of Carbon Neutrality. *Chinese Journal of Electrical Engineering*. 2021, 41(18): 6245–6259.

Li Y. & Kong L. Development of Solar and Wind Power Generation Technologies to Accelerate China's Energy Transition. *Journal of the Chinese Academy of Sciences*. 2019, 34(04): 426–433.

Mireille B., Tadie F., Arman H., *et al.* Performance of Turbulence Models in Simulating Wind Loads on Photovoltaics Modules. *Energies*. 2019, 12(17): 3290.

National Energy Bureau. Renewable Energy Development Plan for the 14th Five-Year Plan. http://zfxxgk.nea.gov.cn. 2021.10.21

Qiao S., Wang X., Wang Z., *et al.* Numerical Simulation of Equivalent Static Wind Load on Large-span Tensioned Truss Canopy Structure. *Journal of Shanghai Jiaotong University*. 2016, 50(01): 59–64.

Warsido W.P., Bitsuamlak G.T., Barata J. *et al.* Influence of Spacing Parameters on the Wind Loading of Solar Array. *Journal of Fluids and Structures*. 2014, 48(6): 295–315.

Yu Y. Numerical Simulation of Wind Load on Roof Mounted Solar Panels[D]. Windsor, Canada: *University of Windsor*. 2012.

Constructional Engineering and Ecological Environment – Chih-Huang Weng (Ed)

Application of modern raw soil materials in rural reconstruction: Taking building 5 of Doushui lake as an example

Junxin Song*
Faculty of Humanities and Arts, Macau University of Science and Technology. Avenida Wai Long, Taipa, Macau, China

ABSTRACT: This paper takes the utilization of modern raw soil materials as the core and takes the reconstruction of Doushuihu Building 5 as an example to discuss. The study found that retaining the raw soil materials of the original buildings in the process of village reconstruction has the best effect of integrating into the village environment. First, the article analyzes the characteristics of raw soil materials and then summarizes the application of raw soil materials in the design of architectural appearance transformation, in order to provide some reference for the transformation of traditional buildings.

1 INSTRUCTIONS

1.1 *Use of raw soil materials in China*

Raw soil is one of the traditional construction materials with the longest history and the most extensive application in China and even the world. (Guo 2022) Especially in China, its application can be traced back to the Neolithic Age at least 8000 thousand years ago. (Yan 2010) In traditional Chinese society, "the work of civil engineering" is the general name of all construction projects. (Gao 2021) It can be seen that raw soil, like wood, plays an important role in China's traditional construction technology and architectural and cultural heritage. From the plaster of straw mud used to insulate and decorate the cave walls in the Dadiwan culture period, the straw-reinforced mud waterproof roof and load-bearing wooden skeleton mud wall in the transition from semi-cave dwellings to above-ground buildings. (Gao 2021) In the Yangshao cultural period, the rammed earth city wall and rammed earth platform in large numbers. In the Longshan cultural period, the ancient Great Wall and large high platform palace buildings were built with rammed earth. In the Spring and Autumn Period and the Warring States Period, through the comprehensive development of the Qin and Han Dynasties to the early Tang Dynasty, based on straw reinforced mud, wooden skeleton mud wall. The civil mixed structure in the form of adobe (or clay), rammed earth, straw mud mound, and other raw soil materials has become systematic and normative. (Guo 2021) They have gradually become the mainstream construction technology. It is widely used in the construction of urban and rural houses, palace offices, temples, and altars in the north of the Yangtze River, as well as the construction of the Great Wall, city walls, tombs, dams, and other construction facilities. With the migration of a large number of people across regions and the exchange and integration of different nationalities in history, the raw soil construction technology has gradually expanded from the Yellow River basin to the south of the Yangtze River. (Hu 2019) Under the influence and role of different natural conditions,

*Corresponding Author: 1173009896@qq.com

DOI: 10.1201/9781003410843-33

economic technologies, and social cultures in various regions, it has presented a wide range of application forms and technological characteristics. In this process, a variety of traditional native residential buildings have gradually formed, which are widely distributed throughout the country. (Liu 2016) According to the 2010-2011 national rural housing census data of the Ministry of Housing and Urban-Rural Development, the application of traditional raw soil construction technology in rural housing construction has spread across all provinces. (Wang 2015) At least 60 million people still live in various forms of raw soil dwellings in China. Most of them are concentrated in relatively poor rural areas in the Loess Plateau, Southwest, East China, Qinghai Tibet Plateau, Xinjiang, and other places.

Due to the differences in climate, resources, customs, and other factors in different regions, traditional raw soil materials have a variety of different application forms, mainly including rammed earth, soil clay, adobe, straw mud plastering, wood/bamboo skeleton mud wall, and earth covering in China. Among them, straw mud as an auxiliary material is widely used in adobe or clay bonding, wall plastering, wood skeleton mud wall, and other raw soil processes; Earth covering refers to the well-known traditional cave dwellings, such as mountain leaning kilns, underground pit kiln, and Gu kiln, which are mainly used in the Loess Plateau; The application of wood bone mud wall or bamboo bone mud wall is mostly seen in the traditional houses in the south, mainly used as the light partition wall of the wood structure house or even the external wall of the building. Both adobe and clay belong to adobe formed by mold processing, but the former refers to adobe formed by mold using mud or straw mud as raw material, which is very common in regions with relatively abundant rainfall or water resources in China; The latter refers to the soil bricks formed by ramming damp soil into molds with rammers, which are mainly used in areas where water resources are relatively scarce, such as the Loess Plateau. Tamping refers to the construction process of using tools such as a tamping hammer to impact, compress and compact the moist soil under the constraint of formwork and then tamp it layer by layer to form a wall. Relatively speaking, due to its relatively good mechanical integrity, local materials, one-time forming, and other advantages, rammed soil can be said to be the most common traditional raw soil construction technology nationwide.

1.2 *Differences between rammed earth used in modern buildings and traditional rammed earth construction*

In China's traditional rural society, the frequency of house renewal is generally in intergenerational units. (Wang 2012) Compared with the requirements of "durability" and "heritage," rural people generally pay more attention to the convenience and cost performance of housing construction. The reason why the traditional ramming technology is widely used is that it has a series of advantages that just meet the needs of this building construction: ramming soil can use local materials and adjust measures to local conditions; Simple and fast construction and low cost; Rammed earth wall has outstanding heat storage performance, which can make the house warm in winter and cool in summer. Its moisture absorption capacity is more than 30 times that of conventional industrial building materials, which can effectively balance indoor humidity. (Wei 2011) The effect is more obvious in the humid and hot south and dry and cold northern climate conditions. Especially from the perspective of today's ecologically sustainable development, rammed earth materials are renewable, and the earth materials after house demolition can be reused or even returned to farmland as fertilizer; The ramming construction process is low in energy consumption and pollution-free. It is estimated that its processing energy consumption and carbon emissions are 3% and 9% of clay bricks and concrete, respectively. However, it is undeniable that there is still a big gap between the material mechanics and durability of the traditional raw rammed soil and conventional wall materials such as sintered brick and concrete.

Just like today's pure cotton products, compared with the traditional hand-spun "earth cloth" in the past, the so-called "modern rammed earth" is relative to the traditional rammed earth.(Yang 2020) The biggest difference between it and the traditional rammed earth is that the modern rammed earth can effectively overcome its inherent defects while maintaining a series of ecological performance advantages of the traditional rammed earth through the introduction of the grading of soil, sand, and stone for ramming raw materials and the introduction of modern machinery based on mechanical ramming.(Zhou 2022) Due to the great difference in soil composition in different regions and even in different locations of the same borrow point, the mechanical, water resistance, dry shrinkage, and other performance defects of traditional raw soil materials are mostly due to the problems of soil particle size distribution in soil materials.(Zhang 2018) The basic principle of modern ramming earth is that, according to the particle size composition characteristics of the soil materials taken, sand and stone with corresponding proportions are mixed into it, and its particle size composition is adjusted to the state of approaching continuous grading. Under the condition of 8%~12% water content of the material, pneumatic or electric rammers are used to apply high-strength ramming to make the aggregates of each particle size closely aggregated. (Zhang 2017) According to the cube compression test results, without any additives, the average compressive strength of modern rammed earth materials can reach 3 MPa, which is 2~4 times the traditional rammed earth strength and can fully meet the strength parameter requirements of load-bearing walls of low-rise buildings and infilled walls of frame structures. (Zhou 2013) At the same time, due to the high polymerization characteristics of soil, sand, and stone, the rammed earth wall can effectively resist the erosion of wind and rain even without any wall protection treatment under the condition of avoiding the concentrated rainwater flowing down the wall.

The optimization mechanism of the above modern rammed earth materials seems simple. However, achieving the quality and effect of high-quality rammed earth walls still requires the coordination of multiple links, such as technical material analysis and allocation, architectural and structural design, construction equipment system, and ramming construction technology. Among them, reducing the use of chemical additives as much as possible is the key to ensuring its ecological performance advantages, such as heat storage and moisture absorption. (Zhang 2022)

1.3 *Research methods*

This research is mainly aimed at a village reconstruction project in Doushui Town, Ganzhou City, Jiangxi Province. First of all, through on-site investigation and analysis, the direction of the building reconstruction design is determined. Secondly, the investigation of the village development and the study of the overall village style finally determined to use modern rammed earth as the final design material.

2 SITE CONDITIONS

2.1 *Location*

Located in Doushui Town, Ganzhou City, Jiangxi Province, China, the building is located at a higher place in the village, with a large viewing area and landmarks. It can interact with the opposite mountain, sea, and lake.

2.2 *Building status*

The village is located the downstream of Yangming Lake Scenic Area. Inside, there are complete buildings such as the Soviet expert building, enterprise culture training center, canteen, guest house, and machine repair workshop in the architectural style of the power plant. The original function of Building 2 was a student dormitory, which was built in 1974.

Figure 1. Current situation in the town.

Figure 2. Status quo of building 5 in Doushui Town.

It is a civil structure with 2 floors and 27 rooms. The building area is 542 m^2, and the use area is 344.6 m^2. It is currently idle.

2.3 *Concept of scheme design*

In the upper planning, the village is positioned as a homestay town, so this building is designated as a homestay. The project adopts the attitude of making the best use of the original buildings, with the main body of the building reserved and some parts divided and added to meet the functional requirements of the homestay.

The rammed earth materials in southern Jiangxi are mainly used for rammed earth, wood, stones, and tiles, while black bricks are only used in a few places. The raw materials for ramming soil are mainly collected from fields or mountains. When used, the rammed soil can be directly transported to the destination without any treatment. The used rammed soil can be transported back to its original place for normal use when not in use, and it is also a good fertilizer. The area of southern Jiangxi is wide, and there are differences due to different soil

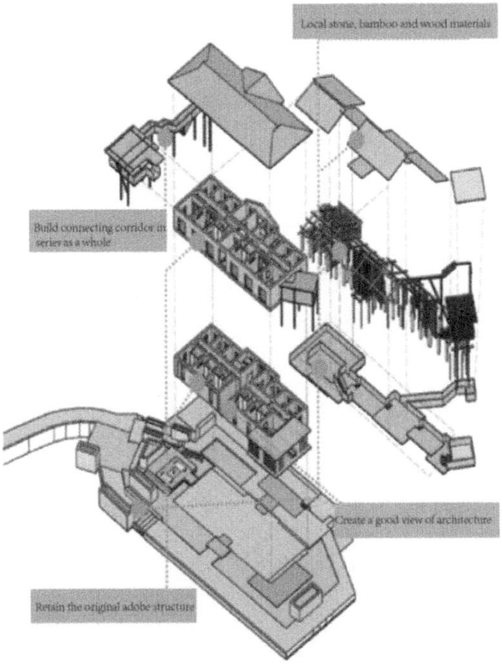

Figure 3. The design concept of building 5 in Doushui Town.

quality, mainly reflected in color. Some areas may be yellow, and some areas may be red. When building rammed earth walls, craftsmen often add rice straws, small stones, pebbles, and bamboo chips to the rammed earth to increase the firmness of the rammed earth walls. The adobe brick is composed of clay and grass, mixed with some additional materials such as fly ash, mixed evenly by manpower or animal power, and then placed in a wood mold for drying. When building the house, the original layered masonry technology shall be adopted to ensure that the bricks are jointed layer by layer.

3 APPLICATION OF MODERN RAW SOIL MATERIALS

Before the transformation, the adobe wall as a load-bearing component was covered with modern plastering materials, and the wooden structure of the pitched roof was out of repair for a long time and could not be reused. Taking advantage of the opportunity of replacing and updating the roof structure, the corresponding strategies were adopted, thereby strengthening and optimizing the original building structure system, fully combining the design of the new structure system, optimizing the indoor space layout, improving the quality and functional flexibility of indoor space, and meeting the space utilization needs of various forms of activities. At the same time, the external facade of the building was stripped of late plastering materials, exposing the original materials of building adobe and strengthening and preserving the cultural characteristics of the times. Among them, the concrete frame placed in the indoor space is the key. The placement of this square table-shaped frame can be said to do more at one stroke: 1. Take over the bearing role of the local original rammed earth exterior wall, bearing the entire roof system. To reduce the load pressure on the adobe exterior wall, the original interior non-load-bearing partition wall is replaced by a concrete frame, which transforms the original exterior wall load-bearing structure into a

Figure 4. Elevation after design.

central frame and exterior wall joint load-bearing structure system; 2. The repaired secondary roof purlin and the wooden ring beam at the top of the wall are tied together between the concrete frame and the rammed earth exterior wall to ensure the overall structural safety of the house; 3. The concrete frame beam and the adobe exterior wall together become the space-defining elements that redefine the interior space and meet the needs of multi-functional elastic use. 4. A roaming corridor is set outside the house to form an alternate space experience of width, height, light, and shade.

4 CONCLUSION

Over the years, in the process of rural practice, builders have been thinking about and exploring architectural design and construction technology paths that are suitable for the current situation of rural development. Especially for the relatively extensive rural construction conditions and organization mode, how to adopt simple, simple, and easy to operate technical measures to solve complex problems and create a rich space experience while meeting functional requirements. The project of Building 2 in Doushui Town is a successful attempt at present.

REFERENCES

Gao Hahui, Li Wei, Gao Zuo and Xu Shuang. Overview of Modification and Development of Raw Soil *Materials. China Science and Technology Information*. 2021 (16): 47–48

Gao Zuo, Li Wei, Gao Haihui and Xu Shuang. Prospects for the Development of Raw Earth Building Materials. *Building Safety*. 2021,36 (07): 55–57

Guo Bingqi. Inheritance and Innovation of Raw Earth Construction. *Urban Architectural Space*. 2022,29 (06): 68–70

Guo Yuting, *Environmental Design Practice of Yuanyang Azeke Villager Activity Center* [D]. Yunnan University. 2021. DOI: 10.27456/d.cnki.gyndu.2021.000523

Hu Chengzheng, *Exploration of Modern Minnan Raw Earth Architecture Design* [D]. Xiamen University. 2019

Liu Chong, Li Yu and Yue Bangrui. Research on Composite Application and Modern Expression of Contemporary Foreign Raw Earth Materials. *Architecture and Culture*. 2016 (08): 218–219

Wang Lijun, Zhang Xi and Yang Kai. Natural Remolding: Research on the Construction Logic of Raw Soil Materials in Contemporary Architectural Design. *Journal of Architecture*. 2012 (S1): 114–117

Wang Shuai, *Research on the Application of Modern Rammed Earth Construction Technology in Architectural Design* [D]. Xi'an University of Architecture and Technology. 2015

Wei Hongyang, Zhang Bo, Li Yuanhao and Wang Long. Summary of Research on Raw Earth Buildings at Home and Abroad. *Value Engineering*. 2011, 30 (14): 116–117. DOI: 10.14018/j.cnki.cn13-1085/n.2011.14.075

Yan Fuqing, *Research on Construction Technology of Raw Earth Buildings in Southern Shaanxi* [D]. Xi'an University of Architecture and Technology, 2010

Yang Xinyu. A New Form of Old Materials - Application Analysis of Raw Soil Materials in Architectural Design. *Western Leather*. 2020, 42 (20): 3–4

Zhang Fupeng, Shi Lei, Liu Simian, Shi Jiaqi, Zhang Ying and Zhang Jinyue. Research on the Progress of Modification of Raw Earth Building Materials in China. *Journal of Railway Science and Engineering*: 1–9 [2022-10-23]. DOI: 10.19713/j.cnki.43-1423/u.T20211180

Zhang Xue, *Research on Sustainable Development of Raw Earth Buildings* [D]. Northeast Normal University, 2017

Zhang Yu, Li Nan, Lin Zhichao. and Zhou Zhiwei. Overview of Raw Soil Material Modification and Its Design Application. *Architectural Skills*. 2018 (01): 112–113

Zhou Liang. Research on Classification and Contemporary Development of Raw Earth Building Construction. *Brick and Tile* 2022 (02): 44–45+48. DOI: 10.16001/j.cnki.1001-6945.2022.02.065

Zhou Xi, *Research on Soil Texture and Creation of Local Cultural Artistic Conception [D]*. Hunan University, 2013

Constructional Engineering and Ecological Environment – Chih-Huang Weng (Ed)
© 2024 The Author(s), ISBN 978-1-032-53198-4

State-of-the-art report on the photocatalytic degradation of automobile exhaust on asphalt road

Zhongxin Ma
Lishui Traffic Construction Development Co., Ltd., Lishui, China

Li Chen
Lishui Traffic Engineering Management Center, Lishui, China

Yidong Xu* & Jinting Wu
NingboTech University, Ningbo, China

ABSTRACT: In view of the serious mobility and dispersion of running automobiles, relatively fixed road materials can be combined with photocatalysts to purify automobile exhaust and improve the road environment. This paper describes the research status of road photocatalytic materials and related road performance tests in asphalt pavement at home and abroad according to three aspects: the modification of road photocatalytic exhaust gas degradation materials, the distribution and degradation test methods of road pollutants, and the comprehensive road performance of photocatalytic composites. Based on the characteristics of road engineering, research on the preparation and modification methods of composite photocatalyst materials for road use will provide a useful reference for the application of photocatalyst materials in road engineering.

1 INTRODUCTION

The development of road traffic not only improves production efficiency, ensures economic development, and facilitates residents' lives but also brings about a substantial increase in automobile exhaust emissions, including more than 90% of NO_x (nitrogen oxides) and PM (particulate matter) and more than 70% of CO (carbon monoxide) and HC (hydrocarbons). These exhaust emissions have created challenging environmental pollution and induced considerable human health problems. Although the natural ecosystem and new energy vehicles have played a certain role in curbing the growth of automobile exhaust emissions, the effect is limited, and it is not enough to completely eliminate automobile exhaust and its harm.

At present, with the significant increase in car ownership, its exhaust emissions pose a threat to environmental pollution and human health, but the natural ecosystem and new energy vehicles alone are not enough to completely eliminate automobile exhaust and the harm caused by it. As the nearest contact source after automobile exhaust emissions, the road takes the road surface as the entry point. It uses photocatalytic technology to load catalytic materials in the pavement materials, becoming a new type of environmentally friendly pavement-photocatalytic asphalt concrete pavement, which can open up a new way

*Corresponding Author: xyd@nit.zju.edu.cn

 DOI: 10.1201/9781003410843-34

to solve automobile exhaust pollution and respond to the national development strategy of "carbon peaking and carbon neutrality."

2 MODIFICATION AND APPLICATION OF ROAD PHOTOCATALYTIC EXHAUST DEGRADATION MATERIALS

With the increasing emission of air pollutants in recent years, the harmless treatment of polluting gases has received unprecedented attention, which has also promoted the research of TiO_2 photocatalysts for the purification of polluting gases. A large number of experimental studies at home and abroad show that micron-sized or nanosized TiO_2 can remove polluting gases such as NO_x and SO_2, as well as volatile organic compounds (VOCs) and so on, in the laboratory by the photocatalytic reaction.

Chen used N-doped TiO_2 as the photocatalyst for asphalt pavement reconstruction, and the results showed that it has a high degradation rate (Chen et al. 2007). Wei Peng measured the maximum degradation efficiency values of nano TiO_2 photocatalyst materials with different concentrations and found that the maximum degradation efficiency was 40%, 60%, 75%, and 75%, respectively, when TiO_2 concentrations were 20%, 25%, 30%, and 35%. The maximum degradation efficiency of photocatalyst materials with different concentrations of nano-TiO_2 showed an increasing trend with the increase of nano-TiO_2 concentration until a certain concentration. The degradation efficiency would not increase (Wei 2008). Cao prepared a Ce-doped TiO_2 photocatalyst with cerium nitrate as the doping source. It was found that the doping of a single rare earth element can improve the photocatalytic performance of TiO_2 for road use, but there is an optimal doping amount, and the sample prepared by surface spraying of a Ce TiO_2 suspension aqueous solution has the highest NO_x degradation rate (Cao et al. 2017). Leng developed a method of coating TiO_2 particles on asphalt pavement surfaces based on suction and lotion effects, which not only increases the NO_x degradation ratio but also helps to improve material durability and maintain degradation efficiency after wear (Leng & Yu 2016). Hassan evaluated the effect of preparing a hot mix asphalt mixture by using a thin coating of TiO_2 suspension with water as a dispersion medium or directly mixing TiO_2 into an asphalt binder. The experimental results show that it is difficult to determine the degradation effect of nano TiO_2 in a hot mix asphalt mixture system by the dynamic method. After the suspension is sprayed, the NO_x degradation effect of TiO_2 on the road surface can be measured by the dynamic method, and the efficiency is between 31% and 55% (Marwa et al. 2014).

Singh et al. studied the feasibility of TiO_2 waterborne epoxy resin as a fog sealing and tail gas degradation material for asphalt pavement. They believed that the relevant photocatalytic reaction functional layer could decompose pollutants, maintain the skid resistance of the pavement (Singh et al. 2016), and improve the impermeability of the pavement (Hu et al. 2016). Wang used a composite material mixed with epoxy resin and fly ash TiO_2 cement mortar to bond to the asphalt pavement surface to obtain a lasting and efficient degradation of nitrogen oxide (Wanget al. 2016).

Relevant tests show that the degradation efficiency of exhaust gas is not related to the void fraction of the asphalt mixture. Ballari et al. studied the change in the NO degradation efficiency of nano TiO_2 coatings at different heights from the road surface. They found that the degradation efficiency at 5 cm from the road surface was 1/3 higher than that at 150 cm (Ballari & Brouwers 2013). Liu et al. found that nano TiO_2 photocatalysts and rubber powder can improve the degradation efficiency through composite technology (Liu et al. 2015). Qian Guoping's research results show that when nano TiO_2 and silane coupling agents are mixed in a certain proportion, the prepared coating material exhibits the best NO degradation efficiency. When the coating is used for ordinary asphalt pavement and a seal coat, the degradation rate of NO in one hour can reach 40% and 30%, respectively (Qian et al. 2016). Hassan et al. studied the influence of the nano-TiO_2 addition mode on the NO degradation effect. The results showed that the degradation efficiency of TiO_2 directly added into asphalt mixtures is far less than the value of nano-TiO_2 coated on asphalt mixture surfaces (Hassan & Asadi 2012).

The product of the TiO_2 photocatalytic oxidation of NO_x in air pollutants is nitrate (NO_3^-), which flows into the surrounding soil after being washed from surfaces by rain, increasing nitrogen deposition and eutrophication, causing secondary pollution to the surrounding land and water ecological environment. This hinders the application of TiO_2 as a photocatalytic material used in the reduction of pollution resulting from exhaust emissions.

Layered bimetallic hydroxides (LDHs) are widely used in catalysis, adsorption, ion exchange, and other fields due to their exchangeable interlayer anions. Paredes S mixed MgAl-LDHs with TiO_2 at different mass fractions and tested its phenol degradation rate in an aqueous solution. The results showed that when the mass ratio of MgAl-LDHs to TiO_2 was 1:1, a higher phenol degradation rate (93%) could be obtained. In contrast, when the mass fraction of MgAl-LDHs further increased, the phenol degradation rate exhibited a downward trend (Paredes et al. 2011). BaiS successfully prepared a Fe_2O_3/$BiVO_4$/NiFe-LDH photocatalyst material with an n-n type heterojunction by electrodeposition, which has excellent visible light response characteristics (Bai et al. 2018). Todorova et al. studied the removal of NO_x (NO and NO_2) by a composite photocatalyst material with a layered double hydroxide and TiO_2 weight ratio of 1:1 under ultraviolet and visible light irradiation. The results showed that the composite exhibited better photocatalytic activity than TiO_2 under two kinds of radiation (Todorova et al. 2014). TiO_2@LDHs nanocomposites prepared by Wang et al. exhibit very high photocatalytic activity for the degradation of toluene under simulated and real light conditions, with removal rates of 85.9% and 91.7%, respectively (Wang et al. 2019). Suh's experimental results show that the combination of delaminated LDHs with TiO_2 can be used as a solid adsorbent to maximize the exposure of the positively charged surfaces of the composite materials to pollutants. The adsorption capacity of these composite materials to the pollutant's methyl orange and 2,4-dichloro phenoxy acetic acid increased by 16.0 times and 76.7 times, respectively (Suh et al. 2019). Jo et al. believed that TiO_2/LDH core-shell hybrids have a strong light collecting capacity, large surface area, porous structure, and excellent CO_2 adsorption capacity, and TiO_2/LDHs based on unique core-shell geometry have a large interface contact area, providing a broader platform for effective charge transfer (Jo et al. 2020). Therefore, it is worthwhile to study how to reduce the recombination of the "electron-hole pair" and improve its photocatalytic activity by combining TiO_2 and LDHs to construct heterojunctions. At the same time, we make NO_3^- generated in the process of photocatalysis be adsorbed by LDHs and use it in the construction of photocatalytic pavement.

3 EXPERIMENTAL METHOD FOR THE DISTRIBUTION AND DEGRADATION OF ROAD POLLUTION GASES

The reaction process of road-contaminated gas involves gas-phase reactions and gas-solid phase photocatalytic degradation, and the reactant phase is considerably different from the soluble pollutants. The photocatalytic performance test of road photocatalytic materials and loaded structural materials is limited by the size of the road construction materials, and the sample plate is usually flat. Accordingly, the test equipment for measuring the photocatalytic rate or the parameters of the rate equation is mostly flat plate DC structure photocatalytic reactors.

Asadi et al. also increased the change in environmental conditions, such as humidity, while increasing the reaction chamber to study the degradation performance and reaction mechanism of road photocatalytic materials under more complex environmental changes (Asadi et al. 2012). Han Xiangchun developed a set of automobile exhaust reaction test equipment with a cylindrical reaction box and light source in the cylinder core, which is composed of a gas chamber, sensor, computer-aided test system, and automobile exhaust introduction system (Han et al. 2005). Li Jianfei independently developed and designed a set of static testing equipment, which is composed of a quartz glass container, an FGA-4100 (5G) automobile exhaust analyzer, a UV-A ultraviolet radiometer, an ultraviolet lamp light source box, a gas collection device, an air pump, and a collection bag. The system has a large reaction chamber volume, controllable light source energy density, and high compatibility with road engineering material specimen size (Li et al. 2010). Relevant personnel have also developed an indoor simulation test device, which is composed of an automobile exhaust gas supply part, an automobile exhaust gas reaction part, an automobile exhaust gas detection part, and a connection part. Among them, the former has the defect that it cannot accurately reproduce natural conditions by using a solar-simulating ultraviolet radiation system as the light source. The small gasoline engine used in the latter may do harm to the health of testers.

To verify and study the degradation effect of materials on the road site, the research team led by Dylla and Osborn also proposed that the degradation effect of photocatalytic road exhaust on the site can be determined by measuring the residual amount of NO_3 on the road surface after reaction (Dylla et al. 2012; Osborn et al. 2012). Suárez designed a reaction chamber that can record the on-site NO_x concentration, ambient temperature, UV intensity, and relative humidity (Suárez et al. 2014). However, in the existing dynamic and static indoor tests, the flow and distribution of gas and exhaust gas on the road surface are inconsistent. The simulation of the road environment, road material specimen size, etc., is limited. Tests with the final product as the tracking object are also more vulnerable to the interference of similar products generated by non-photocatalytic degradation. Therefore, it is necessary to carry out systematic research on the testing methods of the photocatalytic activity of road materials.

4 ROAD PERFORMANCE OF PHOTOCATALYTIC COMPOSITE MATERIALS

The comprehensive road performance of photocatalytic composite materials includes road skid resistance, durability, mechanical properties, construction performance, road photocatalytic activity, compatibility, and so on.

Hassan MM et al. confirmed that nano TiO_2 could improve the high-temperature stability of mucilage (Hassan et al. 2011). Yang Qun et al. used a black light lamp to accelerate the aging of asphalt mortar containing nano TiO_2, which demonstrated that such a method could delay the process of light aging of asphalt (Yang 2011). Bocci analyzed the mechanical properties of the heterogeneous system formed by emulsified asphalt photocatalysts and measured the modulus and fatigue dynamic properties of the corresponding porous asphalt concrete (Bocci et al. 2014).

The author's research team conducted the skid resistance of the asphalt mixture surface in the pendulum apparatus. The pendulum value of the asphalt mixture surface coated with nano TiO_2 photocatalyst material was 56.3, while the pendulum value of the asphalt mixture surface without nano TiO_2 photocatalyst material was 73.6. Both are greater than the value of 54 required by the specification and meet the design requirements. The test results showed that the anti-skid performance of permeable asphalt mixture coated with nano TiO_2 photocatalyst material changes, but all met the requirements of road use. Li Yi found that when the coating amount of nano TiO_2 photocatalyst material was 20g/900cm^2, 30g/900cm^2, and 40g/900cm^2, the degradation efficiency was 51%, 76.6%, and 78.8%, respectively. It can be seen that the degradation efficiency gradually increased with the increase of the coating amount. When it reached a certain coating amount, the degradation efficiency showed a stable trend (Li et al. 2013). Osborn analyzed the short-term durability of the catalytic performance of the road surface after TiO_2 dispersion spraying treatment through circulation and flushing. The results show that asphalt pavement exhibits photocatalytic degradation effects within 10$\sim$16 months of TiO_2 coating (Osborn et al. 2014). Bocci used a dynamic method to measure the NO_x reduction performance of drill core samples coated with nano TiO_2 for 1, 17, 46, 88, 218, and 527 days. It was found that the performance decreases with the increase in the severity of traffic volume and weather fluctuation (Bocci et al. 2016). Etxeberria et al. found that even if part of the TiO_2 is covered by dust, the nano TiO_2 sprayed on the surface still has a good NO_x removal rate, and the performance can be recovered by washing with clean water, but the impregnation of gasoline and other materials will greatly degrade the catalytic effect on the surface. Alkaline detergent or n-hexane solvent permanently degrades part of the catalytic performance of the surface (Etxeberria et al. 2017). Wei Peng compared the two addition methods of doping a nano TiO_2 photocatalyst and coating a nano TiO_2 photocatalyst. The test showed that the latter had a greater impact on the skid resistance of roads, so the coating amount of the nano TiO_2 photocatalyst should be controlled. The activity of the nano TiO_2 photocatalyst is mainly determined by temperature. With increasing temperature, its degradation efficiency will be significantly reduced. At high temperatures, the adhesion of the nano TiO_2 binder will also be significantly reduced. The nano TiO_2 photocatalyst penetrating into the asphalt mixture will peel off, leading to a significant reduction in the permeability of the asphalt mixture (Wei 2008).

Relevant tests show that the void fraction of the asphalt mixture has little effect on the degradation efficiency of the nano TiO_2 photocatalyst. The high and low temperature performance, water stability, and other road performance aspects of the mixture have little influence and exhibit a certain degree of improvement. A Nano-TiO_2 coating will cause a considerable decrease in the skid resistance of asphalt pavement but has little impact on durability. It must be noted that although a relatively high initial degradation efficiency can be obtained by spraying nano TiO_2 aqueous dispersion suspensions or thin coating dispersion systems on flexible pavement, the loading strength of the catalyst is low, and its durability is poor. With its rapid stripping and wear, the degradation effect will rapidly decline in a short period of time. When nano-TiO_2 is directly mixed into the surface mixture, the amount of photocatalyst used is large, it is not easy to disperse, the content of effective components on the surface of the structure is low, the contact surface with polluting gases and sunlight is small, and the degradation efficiency is relatively low. The absorption of light energy by black or brown binders such as asphalt may also affect the utilization of sunlight by photocatalysts. Therefore, it is of great practical significance to carry out comprehensive road performance evaluation and analysis of photocatalytic asphalt mixtures and pavement structures and to verify the effectiveness of photocatalytic asphalt and its mixtures based on physical engineering.

5 CONCLUSIONS

In summary, TiO_2 is mostly used in road photocatalysts, and there is no effective solution to the secondary pollution caused by the photocatalytic oxidation product NO_3^-. There are few studies on the photocatalytic reaction kinetics and adsorption model of photocatalytic components loaded on asphalt mixtures. Current research on photocatalytic asphalt mixtures has been reported. Still, most of this research focuses on the incorporation of TiO_2, whose photocatalytic efficiency can only be excited by ultraviolet light. In contrast, ultraviolet light only accounts for approximately 5% of the total solar radiation. The photocatalytic energy is limited, and there is no effective solution to the secondary pollution caused by the photocatalytic oxidation product NO_3^-. In addition, most of the research on modified nano-TiO_2 and its loading method belongs to the field of material science. Although the results are rich, the lack of motivation for the transformation of related achievements and technologies to the field of road engineering leads to slow progress in its application and promotion.

Moreover, there are gaps and differences between the road engineering equipment level and material science, environmental engineering, etc., and actual projects often have time limit requirements. Therefore, there are certain requirements for the convenience and maneuverability of photocatalytic activity characterization methods in the comparison stage of road photocatalysts. In addition, the mechanism of photocatalytic component loading asphalt mixtures is not clear; in particular, the photocatalytic reaction kinetics and adsorption model are less studied, and applications in practical engineering are still rare.

In view of the fact that asphalt pavement has become the mainstream development direction at present, the following is suggested:

(1) According to the characteristics of the natural and traffic environment where the road works are located, the preparation and application of asphalt pavement materials with the function of photocatalysis and adsorption synergistic degradation of pollutants should be taken as the research subject, road photocatalytic composite materials should be developed, the factors and mechanisms affecting the photocatalytic performance should be analyzed. Methods for characterizing photocatalytic properties suitable for road photocatalytic materials and functional structures should be proposed.
(2) Furthermore, the design method of asphalt mixtures with photocatalyst loaded wearing in the surface and near surface pores should be explored, corresponding preparation technologies and processing schemes should be formed. The changing trend of fine, microstructure, and macro road performance of composite photocatalyst asphalt mixtures under different processing conditions should be analyzed. In addition, applications should be demonstrated to verify the actual road performance.
(3) Moreover, based on the whole life cycle model, the economic and environmental benefits of photocatalytic asphalt concrete roads in the stages of construction, operation, and maintenance can be analyzed.
(4) Considering the exposure of road structures to outdoor light, it is recommended to develop and prepare photocatalytic materials with a wide light gamut and carry out research on photocatalytic degradation under the action of visible light.

ACKNOWLEDGMENT

The authors wish to acknowledge the financial support of the Science and Technology Project of Zhejiang Provincial Department of Transport (Grant No. 202225), Natural Science Foundation of Zhejiang Province (Grant No. LY20E080002), Science and Technology Project of Ningbo Transportation Bureau (Grant No. 202007).

REFERENCES

Asadi S., Hassan M. M., Kevern J. T. *et al.* Development of Photocatalytic Pervious Concrete Pavement for Air and Storm Water Improvements. *Transp Res Rec.* (2290), 161–167 (2012).

Bai S., Chu H., Xiang X. *et al.*: Fabricating of Fe_2O_3/$BiVO_4$ Heterojunction Based Photoanode Modified With NiFe-LDH Nanosheets For Efficient Solar Water Splitting. *Chem. Eng. J.* –350, 148–156 (2018).

Ballari M. M. and Brouwers H. J. H.: Full Scale Demonstration of Air Purifying Pavement. *J. Hazard. Mater.* 254(6), 406 (2013).

Bocci E., Riderelli L., Fava G. *et al.*: Durability of NO Oxidation Effectiveness of Pavement Surfaces Treated With Photocatalytic Titanium Dioxide. *Arab J Sci Eng.* 41(12), 4827–4833 (2016).

Bocci M., Gianluca C. and Colagrande S. Experimental Investigation of the Dynamic Behavior of Asphalt Concrete Treated With Photocatalytic Mortars. In: *Sustainability, Eco-efficiency and Conservation in Transportation Infrastructure Asset Management.* pp,95–102, Pisa, (2014).

Cao X., Yang X., Li H. *et al.*: Investigation of Ce-TiO_2 Photocatalyst and Its Application in Asphalt-based Specimens for NO Degradation. *Constr Build Mater.* 824–832 (2017).

Chen D. M., Jiang Z. Y., Geng J. Q. *et al.*: Carton and Nitrogen Co-doped TiO2 With Enhanced Visible-light Photocatalytic Activity. *Ind. Eng. Chem. Res.* 46(9), 2741–2746 (2007).

Dylla H., Hassan M. M. and Osborn D. Field Evaluation of the Ability of Photocatalytic Concrete Pavements to Remove Nitrogen Oxides. *Transp Res Rec.* (2290), 154–160 (2012).

Etxeberria M., Guo M., Mauryramirez A. *et al.* Influence of Dust and Oil Accumulation on the Effectiveness of Photocatalytic Concrete Surfaces. *J Environ Eng.* 143(9), 04017040-1-04017040-10 (2017).

Han Xiangchun, Bai Haiying, Guan Qiang, Li Honggang, and Du Xuejing. Design on the Testing System for Degrading Automobile Emission With Photocatalytic Material TiO_2. *J. Northeast For. Univ.* 33(5), 89–91 (2005).

Hassan M. M. and Asadi S.: Laboratory Evaluation of the Environmental Performance of Photocatalytic Titanium Dioxide Warm-mix Asphalt Pavements. *J. Mater. Civ. Eng.* 24 (5), 599 (2012).

Hassan M. M., Mohammad L. N., Cooper S. B. *et al.*: Evaluation of Nano-titanium Dioxide Additive on Asphalt Binder Aging Properties. *Transp Res Rec.* (2207), 11–15 (2011).

Hu C., Ma J., Jiang H. *et al.*: Evaluation of Nano-TiO_2 Modified Waterborne Epoxy Resin as Fog Seal and Exhaust Degradation Material in Asphalt Pavement. *J Test Eval.* 45(1), 260–267 (2016).

Jo W., Moru S. and Tonda S. A Green Approach to the Fabrication of a TiO_2/NiAl-LDH Core-Shell Hybrid Photocatalyst for Efficient and Selective Solar-powered Reduction of CO_2 Into Value-added Fuels. *J. Mater. Chem. A.* 8(16), 8020–8032 (2020).

Leng Z., Yu H. Novel Method of Coating Titanium Dioxide Onto Asphalt Mixture Based on the Breath Figure Process for the Air-purifying Purpose. *J. Mater. Civ. Eng.* 28(5) (2016).

Li Jianfei, Liu Liping, Sun Lijun: Research on the Efficiency of Decomposition of Hydrocarbon in Exhaust Gas From Vehicles by Nanometer Titanium Dioxide. *HE.* 35(2), 151–155 (2010).

Li Yi, Zang Wenjie, Tao Jianqiang and Zhang Chenxu. Study on Preparation and Optimum Dosage of the New Pavement Coating Material for Automobile Exhaust Degradation.VE (2013).

Liu W., Wang S. Y. and Zhang J. Photocatalytic Degradation of Vehicle Exhausts on Asphalt Pavement by TiO_2/rubber Composite Structure. *Constr Build Mater.* 81(6), 224 (2015).

Marwa M. H., Louay N. M., Heather D. *et al.* Photocatalytic Warm Mix Asphalt Laboratory Performance *Testing. J. Mater. Civ. Eng.* 24 (2014).

Osborn D., Hassan M M., Asadi S. *et al.*: Durability Quantification of TiO_2 Surface Coating on Concrete and Asphalt Pavements. *J. Mater. Civ. Eng.* 26(2), 331–337 (2014).

Osborn D., Hassan M. M. and Dylla H. Quantification of Reduction of Nitrogen Oxides by Nitrate Accumulation on Titanium Dioxide Photocatalytic Concrete Pavement. *Transp Res Rec.* (2290), 147–153 (2012).

Paredes S., Valenzuela M., Fetter G. *et al.* TiO_2/MgAl Layered Double Hydroxides Mechanical Mixtures as Efficient Photocatalysts in Phenol Degradation. *J. Phys. D: Appl. Phys.* 72(8), 914–919 (2011).

Qian Guoping., Zhu Junwen and Zhou Dayao: Comparative Experimental Study on the Effect of Nano-TiO_2 Coating on the Degradation of Automobile Exhaust in Asphalt pavement and Slurry Seal. High. *Auto. Appl.* (03),88–90 (2016).

Singh D., Hu C., Valentin J. *et al.*: Study on Exhaust Degradation Material For Asphalt Pavement. In: *Innovative and Sustainable Solutions in Asphalt Pavements.* pp,38–44. ASCE, Shandong (2016).

Suárez S., Portela R., Hernández-Alonso M.D *et al.*: Development of a Versatile Experimental Setup for the Evaluation of the Photocatalytic Properties of Construction Materials Under Realistic Outdoor Conditions. *ESPR.* 21(19), 11208–11217 (2014).

Suh M., Shen Y., Chan C. *et al.*: Titanium Dioxide-layered Double Hydroxide Composite Material for Adsorption-photocatalysis of Water Pollutants. *Langmuir.* 35(26), 8699–8708 (2019).

Todorova N., Giannakopoulou T. Karapati S. *et al.* Composite TiO_2/clays Materials for Photocatalytic NO_x Oxidation. *Appl. Surf. Sci.* 319, 113–120 (2014).

Wang L., Gao X, Cheng Y Q *et al.*: TiO_2@MgAl-layered Double Hydroxide With Enhanced Photocatalytic Activity Towards Degradation of Gaseous Toluene. *J. Photochem. Photobiol. A.* 369, 44–53 (2019).

Wei Peng. Study on Asphalt Mixture Pavement With Degradable Automobile Exhaust. In: Harbin Institute of Technology. Harbin. (2008).

Yang Qun, Ye Qing, Liu Yi: Dispersion and Re-aging Properties of Asphalt Modified With TiO_2. J. *Tongji Univ. Nat. Sci.* (02), 263–265 (2011).

Constructional Engineering and Ecological Environment – Chih-Huang Weng (Ed)
© 2024 The Author(s), ISBN 978-1-032-53198-4

Study on sustainable development path of agricultural science and technology ecological environment

Lin Qiu & Weihua Yang*
Dalian Polytechnic University, School of Management, Dalian, China

ABSTRACT: The ecological environment of agricultural science and technology is an important scientific and technological environment belonging to the primary industry among many benign elements of modern science and technology promoting social development. The construction of agricultural science and technology ecological environment is helpful to strengthen the strength and complement the weakness and help social and economic development by means of science and technology, especially for cities with "incomplete" agricultural development. Exploring the multi-line development of agriculture, forestry, fishery, and tourism through the analysis of the current situation, existing advantages and weaknesses of agricultural science, technology, and agricultural ecological environment, and on the basis of clarifying the principles for the construction of agricultural science, technology, and agricultural ecological environment, the paper puts forward the path and countermeasures for the construction from the three aspects of innovation subject, innovation resources, and innovation environment, which can effectively improve the innovation ability of agricultural science and technology, In turn, it will help ensure the efficient and stable development of agricultural development while properly preparing for various difficult situations.

1 INTRODUCTION

With the transformation of China's development focus and demand, the agricultural ecological environment has formed the characteristics and characteristics of different means, tasks, ideas, and ideas in different industries and fields. As socialism with Chinese characteristics enters a new era, the major social contradictions in our country have changed. However, on the whole, major contradictions still exist in the problems of urban and rural areas and rural development. The No. 1 central document of the Central Committee of the CPC in 2022 is consistent with the basic connotation highlighted every year (Arumapperuma 2016). We will continue to make solving the "three rural issues" the top priority of the work of the whole party and comprehensively promote rural revitalization as a key strategy to achieve the great rejuvenation of the Chinese nation. We will adhere to and strengthen the Party's overall leadership over the "three rural" work, accelerate the modernization of agriculture and rural areas, and let farmers live better life (Ashlee-Ann 2018). The agricultural scientific and technological agricultural ecosystem is based on the overall idea of "scientific layout, optimization of resources, innovation mechanism, and enhancement of capabilities," with the core of improving the sustainable innovation capability and efficiency of science and technology, with the means of integrating resources and innovation mechanisms, and with the construction of scientific and technological agricultural ecosystem as the path to solving this problem.

*Corresponding Author: 964524748@qq.com

DOI: 10.1201/9781003410843-35

2 RESEARCH DESIGN

2.1 *Build index system*

This study takes sustainable development as the premise and sets evaluation indicators by sorting out a large number of research results related to the topic. The positive and negative impacts of each evaluation index of the agroecological environment were determined to form a complete evaluation index system, as shown in Table 1.

Table 1. Evaluation index system of agricultural ecological environment.

Evaluation object	Evaluating indicator	Positive and negative
Agroecological environment	Forest coverage	+
	Per capita agricultural water consumption	+
	Agricultural disaster area	−
	The application amount of chemical fertilizer on farmland	−
	Effective irrigation rate of farmland	+
	Per capita cultivated area	+

2.2 *Data sources*

The data of this study is from China Rural Statistical Yearbook, and the data from 2011 to 2021 is selected as the research support. In consideration of the differences in the positive and negative, the order of magnitude, and the dimension of different evaluation indicators, in order to ensure the reliability of the evaluation and analysis results, this study adopts standardized processing for these data (Janssen 2019). Two treatment methods are designed according to the positive and negative indicators. Assume that the collaborative evaluation index before processing is X_{ij}, which is used to describe the single evaluation index value of item j in the year i. The evaluation index of the agricultural ecological environment is X_i, which is used to describe the single evaluation index value of the ith year. The agricultural economic evaluation index is X_j, which describes the single evaluation index value of item j. The collaborative evaluation index after pretreatment is X'_{ij}, which is used to describe the standardized value of the single evaluation index of item j in the year i. The negative indicator data is processed as follows.

$$X'_{ij} = (\max X_{ij} - X_j) / (\max X_j - \min X_j) \tag{1}$$

The positive index data is processed as follows.

$$X'_{ij} = (X_{ij} - \min X_j) / (\max X_j - \min X_j) \tag{2}$$

3 MEASUREMENT RESULTS

3.1 *Build model*

When building the collaborative development model, this study first discusses the calculation method of the development score of the agricultural ecological environment and agricultural economy and then builds the model on this basis.

The development evaluation score of the agricultural ecological environment is calculated by using the weighted function. The formula is as follows.

$$U_1 = n\Sigma i = 1\lambda_i X_{ij}' \quad (3)$$

The weighted function is also used to calculate the development evaluation score of the agricultural economy. The formula is as follows.

$$U_1 = n\Sigma i = 1\lambda_i X_{ij}' \quad (4)$$

Next, build a coupling measure model of collaborative development, and use the physical theory to create a coupling measure model under the interaction of two evaluation objects as follows.

$$P = 2 \times \left[(U_1 \cdot U_2)/(U_1 + U_2)^2\right]^{1/2} \quad (5)$$

We use formula (5) to calculate the collaborative development coupling measure of the two evaluation objects to evaluate the degree of collaborative development coupling. The formula for calculating the comprehensive index of coordinated development is as follows.

$$T = aU_1 + bU_2 \quad (6)$$

In the formula, both parameter a and parameter b are undetermined coefficients, which are used to describe the contribution of the two evaluation indicators to collaborative development. A is the contribution of the agricultural ecological environment to the coordinated development, and b is the contribution of the agricultural economy to the coordinated development. In this study, the contribution of the two evaluation objects is the same, so a=b=0.5.

3.2 *Calculation results*

According to the statistical results, from 2011 to 2021, the degree of coordination and coupling between the agricultural ecological environment and the agricultural economy in China will gradually increase, and the degree of coordination will also increase (Hall 2021). However, the problem of economic backwardness is still very serious, and in recent years, the

Table 2. Measurement results of sustainable development of agricultural ecological environment.

Particular year	Comprehensive order parameter U_1	Comprehensive order parameter U_2	Degree of collaboration
2011	0.2069	0.0961	Min
2012	0.1538	0.1488	Min
2013	0.3301	0.2031	Mid
2014	0.4260	0.2019	Mid
2015	0.4659	0.3498	Mid
2016	0.5408	0.4368	Mid
2017	0.6861	0.5069	Mid
2018	0.6999	0.5661	Mid
2019	0.7288	0.6115	Max
2020	0.7738	0.7231	Max
2021	0.8039	0.7996	Max

economy is still lagging behind. From the perspective of development history, the agricultural ecological environment evaluation is higher and shows an upward trend of development, while the agricultural economy develops slowly (Bai 2022; Zhang 2022). We can try to find ways to promote the development of the agricultural economy while stabilizing the development of the agricultural ecological environment.

4 COUNTERMEASURES AND SUGGESTIONS

4.1 *Promote intra-industry exchanges and cooperation*

Universities and research institutes are the main force of the knowledge innovation system, the main producers, disseminators, and users of knowledge, and the important intellectual support for the innovation drive. A knowledge innovation system is the foundation of a technology innovation system and a regional innovation system. Therefore, without a strong knowledge innovation system, the innovative construction of an agricultural ecological environment is impossible. The theory of industrial, agricultural ecological environment points out that the aggregation of innovative species in spatial geography or industrial chain is conducive to achieving collaborative cooperation. Open cycle, symbiotic evolution between relevant innovation groups that are organically linked on the industrial chain and value chain, forming innovation clusters, promoting intra-industry exchanges and cooperation, and producing scale agglomeration effect. Therefore, as the main body of innovation, universities, and research institutes are not in a single competitive relationship with enterprises and other intermediaries as the sub-main body but the competition between the industrial chain and the industrial, agricultural ecological environment (Shen 2021). The author believes that the construction of agricultural industry and agricultural ecological environment should emphasize the relevant subjects in the industrial chain and value chain. The leading enterprises in the industrial core and the horizontally and vertically related small and medium-sized enterprises in the industrial extension should form an organic community of common interests through intra-industry cooperation and division of labor. Through the interconnection of relevant R&D and innovation activities, the collaborative innovation, R&D, and application of the three communities in the innovation chain can give play to the advantages of large-scale and specialized division of labor and improve the overall innovation ability of the industry.

4.2 *Comprehensive improvement of the ecological environment*

The construction of agricultural science, technology, and ecological environment requires multiple factors to work together, and funds, policies, and talents are indispensable. First, governments at all levels should regard the special funds for agricultural scientific and technological innovation as one of the important fixed expenditures. Every year, a certain amount of funds is used for the development of the agricultural scientific and technological service industry, and the investment will increase year by year on the premise that the overall economy is favorable. At the same time, with government investment as a seed fund, we will strive for more social funds through investment promotion, loans, subsidies, and other ways, and set up a special fund for agricultural scientific and technological innovation, which is dedicated to the research, development, promotion, and application of new agricultural technologies (Lin 2021). Second, take policy as the starting point, vigorously develop private agricultural science and technology enterprises, gradually cultivate a number of non-governmental science and technology organizations with financial and scientific strength, and encourage enterprises, agricultural science, and technology innovators and individuals to participate in agricultural science and technology investment in financial or intellectual ways. Support and guide enterprises to spend a certain proportion of their operating income on new technology research and development every year. We will accelerate the development of private capital markets and encourage private capital to invest in agricultural science and technology industries. At the same time, strive for bank loans, gradually strengthen loans for agricultural science and technology services through the establishment of a credit system, promote the support of agricultural banks, rural credit cooperatives, and other financial institutions for agricultural science and technology innovation enterprises, and focus on "high-tech, high value-added" agricultural enterprises for financing services (Luo 2021). Third, attach importance to the key role of talents in promoting agricultural scientific and technological innovation and development, fully tap talents, and use talents to promote the application and promotion of modern science and technology and management methods. To pay attention to these "successors," we should shape them on the basis of theoretical expertise, practice, and innovation ability, so as to lay a foundation for training agricultural science and technology talents with agricultural theory and practice ability (Li 2019). These paths and initiatives put forward a common requirement, that is, "flexibility," which should vary from time to time, from event to event, and from mode to mode.

5 CONCLUSION

As the primary industry, the ecological environment of agricultural science and technology can be effectively constructed from the three aspects of innovation subject, innovation resource, and innovation environment after combining the current situation and problems so as to enhance the sustainable development of the ecological environment of agricultural science and technology and achieve high efficiency and stability.

REFERENCES

Arumapperuma F: Agricultural Innovation System in Australia. *Journal of Business System*. 4(15), 67–75 (2016).

Ashlee-Ann F: Exploring an Agricultural Innovation Ecosystems Approach for Niche Design and Development in Sustainability Transitions. *Agricultural Systems*. 164:116–121 (2018).

Bai Erhu F. Coal Mining Method With Near-zero Impact on the Ecological Environment. *Bulletin of Engineering Geology and the Environment*. 76–81(2022).

Hall A F. Challenges to Strengthening Agricultural Innovation Systems. *UNU- MERIT*:102–105(2021).

Janssen W F: *Enhancing Agricultural Innovation: How to Go Beyond the Strengthening of Research Systems*. World Bank:37–38 (2019).

Li Yulin, F: Research on Agricultural Science and Technology Information Service Model and Its Problems Based on Big Data. *Proceedings of the 16th International Conference on Innovation and Management*. 197–202 (2019).

Lin Hongzhen, F: Research on Cooperative Protection Mechanism of Ecological Environment. *IOP Conference Series: Earth and Environmental Science*, 772(2021).

Meifang Luo, F: Rural Finance, Agricultural Science and Technology Input and Urban-Rural Economic Integration. *Journal of Social Science and Humanities*:22–27(2021).

Shen Shiming, F.: Agricultural Ecological Environment Protection Based on the Concept of Sustainable Development. *Plant Science*:65–71(2021).

Zhang Ming, F: Ecological Environment Evaluation Based on Remote Sensing Ecological Index: A Case Study in East China over the Past 20 Years. *Sustainability*. 14:31–37(2022).

Constructional Engineering and Ecological Environment – Chih-Huang Weng (Ed)
© 2024 The Author(s), ISBN 978-1-032-53198-4

Study on construction technology of typical rural houses in Baoan Town, Shaanxi, China

Zhihong Li*, Xianming Huang, Duo Yuan* & Yingying Ma
Architectural Design and Research Institute of Tsinghua University Co., Ltd, Beijing, China

ABSTRACT: This paper focuses on the construction characteristics of a rural house in Baoan Town, Shaanxi Province of China, sorts out the construction mode and technology system of a rural house in Baoan Town, and refines construction technology applied to a rural house in Baoan Town and providing technical support for the development of green and livable new housing construction in Baoan Town.

1 GENERAL INSTRUCTIONS

In order to thoroughly implement the spirit of the Fifth Plenary Session of the 19th Central Committee of the Communist Party of China and the requirement of construction beautiful and livable rural areas in the 13th Five-Year Plan, accelerate the modernization of rural area and housing construction, the state and local governments have issued relevant standards and norms for rural housing to improve the construction level and quality of rural housing.

In 2009, "Demonstration Technology Scheme of Building Energy Efficiency for The Rural Reconstruction in Shaanxi" was issued by the Shaanxi Provincial Department of Housing and Urban and Rural Construction, Provincial Development, and Reform Commission, Provincial Department of Finance, and Provincial Department of Civil Affairs. This scheme proposed an energy-saving path for rural reconstruction. "Regulations on Rural Planning and Construction of Shaanxi Province" were promulgated and implemented in the same year. It required that the relevant departments should strengthen the guidance on the design and construction of rural housing and promote the application of new materials, new technologies, and new processes according to the characteristics of houses in different regions of Shaanxi, such as Guanzhong area, Northern Shaanxi area, Southern Shaanxi area on the basis of safe, applicable, economic, aesthetic and healthy. The regulation puts forward to advocate and promote rural housing with sloping roofs in suitable areas. The new, renovated, or expanded house with more than two layers should be designed and constructed by the units with relevant design qualifications or adopt the general design and construction method provided freely by the construction administrative departments above the county level. Also, the technical guide and fund subsidies are provided for the villagers who use clean energy such as biogas, solar energy, wind energy, and hydropower in the construction and transformation of rural housing (People's Government of Shaanxi Province 2008).

In 2010, the Ministry of Housing and Urban-Rural Development and the Ministry of Science and Technology carried out the declaration, review, and selection of livable housing technology and energy-saving technology for the reconstruction of the existing house in town

*Corresponding Authors: 595571833@qq.com and 4586727@qq.com

 DOI: 10.1201/9781003410843-36

and village. They issued the “Livable Housing Technical Promotion Catalogue in Villages and Towns” and “Energy Efficiency Reconstruction Technical Promotion Catalogue” to promote the application of new technologies, new processes, and renewable energy application technology for rural housing construction in rural areas.

“Livable Housing Technical Promotion Catalogue in Villages and Towns” mainly involves infrastructure construction, planning and design, thermal insulation technology, waterproof technology, structural system, assembly technology, renewable energy application technology, and so on (Ministry of Housing and Urban-Rural Development and the Ministry of Science and Technology 2010). “Energy Efficiency Reconstruction Technical Promotion Catalogue” comes down to renewable energy applications, heating and air conditioning systems, measurement, architectural heat insulation, waterproofing, and other aspects of the technical application (Ministry of Housing and Urban-Rural Development and the Ministry of Science and Technology 2010).

In 2016, the Shaanxi Provincial Department of Housing and Urban and Rural Construction issued the “Rural Characteristics Residence Design Atlas for Shaanxi Province.” Based on the guidelines for applicable, economic, green, and aesthetic, the atlas promotes the application of new technology, new materials, and new processes to meet the requirements of villagers’ life and production from some aspects, including perfect function, environmental protection, appropriate scale, reasonable cost, convenience for construction and helps to present new rural style (Shaanxi Provincial Department of Housing and Urban and Rural Construction 2016).

In 2020, the Shaanxi Provincial Department of Agriculture and Rural Affairs issued “Interpretation on The Rural Homestead Approval Management in Shaanxi Province.” This interpretation makes clear the area standard for rural homesteads, no more than 133 square meters per household in the plain area, no more than 200 square meters per household for the plain field between mountains or rivers, and no more than 267 square meters per household in the mountains and hills (Shaanxi Provincial Department of Agriculture and Rural Affairs).

In 2021, “Guidance on Accelerating the Modernization of Rural Housing and Village Construction” was jointly produced by the Ministry of Housing and Urban-Rural Development, the Ministry of Agriculture and Rural Affairs, and the National Administration for Rural Revitalization. The report calls for refinement design and construction for rural housing, “Making an elaborately spatial layout, realizing the separation of the bedroom, living room and dining room gradually,” “encouraging the use of local materials, a safe and reliable new type of construction such as prefabricated steel structures.” Also, the document promotes energy innovation and clean energy utilization, such as solar thermal utilization in rural areas (Ministry of Housing and Urban-Rural Development 2021).

There is also much research on the construction of rural housing in China.

Yuan Ling *et al.* have conducted the research by literature search and analysis of the study about rural housing construction in China. He indicated that the focus of rural house study since 2009 was mainly on the indoor thermal environment, building energy efficiency, layout design, prefabricated, etc. And in the cold regions, the study of rural houses focused on building energy saving (Yuan 2019).

Sun Dapeng *et al.* compared the sustainable rural house with the ordinary rural house. The sustainable construction strategy pays more attention to ecological protection, renewable energy, land-intensive use, the indoor environment, and cultural transmission (Sun 2006).

Zheng Shiju *et al.* conducted a survey on the housing structure in typical rural areas of China. They found that there are mostly two or three-story houses with brick-concrete structures in the developed area, while it is very common to see rammed earth, wood, and stone structures in undeveloped areas (Zheng 2011).

Gao Xiang investigated and analyzed the functional space and indoor environmental quality of the rural house in China. The study found that the performance of thermal insulation, natural ventilation, and daylighting is generally poor, and physical comfort would be improved (Gao 2016).

2 FIELD SURVEY

Baoan Town is located in the western Luonan County of Shaanxi Province shown in Figure 1. It is in the south of the Qinling Mountains. The area of the town is about 107.8 square kilometers, and it is composed of 12 villages, one community, 150 village groups, 6827 households, and 24500 people.

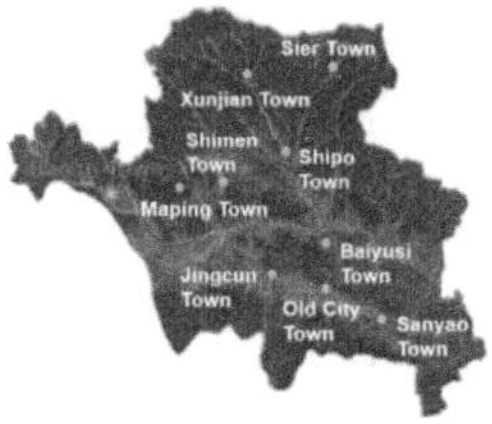

Figure 1. Satellite imagery of Luonan county of Shaanxi.

In order to understand the construction technology of rural housing in Baoan Town, the project team conducted an investigation in Baoan Town, Luonan County, Shaanxi Province. A questionnaire coupled with household interviews, site surveying, and mapping was conducted on housing construction, satisfaction, and target needs in Beidou Village, Cangsheng Community, and Sanyi Village of Baoan Town. The results of the investigation are listed in Table 1.

Table 1. Statistical table of the investigation outcome.

Village	Questionnaire QTY	Sum
Changsheng community	37	97
Sanyi Village	23	
Beidou Village	15	
Heitan Village	22	

2.1 *Household surveys*

The primary industry in Luonan County, Shaanxi Province, is walnut planting and the family farm. And the secondary industry is mainly composed of power generation, flower industry, and food processing. Out-migrant for work and farm work is the main revenue source for the local villagers. The results of household surveys are as follows.

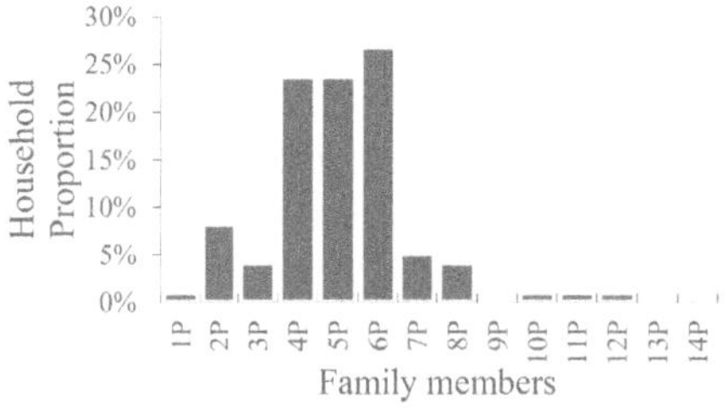

Figure 2. Family structure proportion in household surveys.

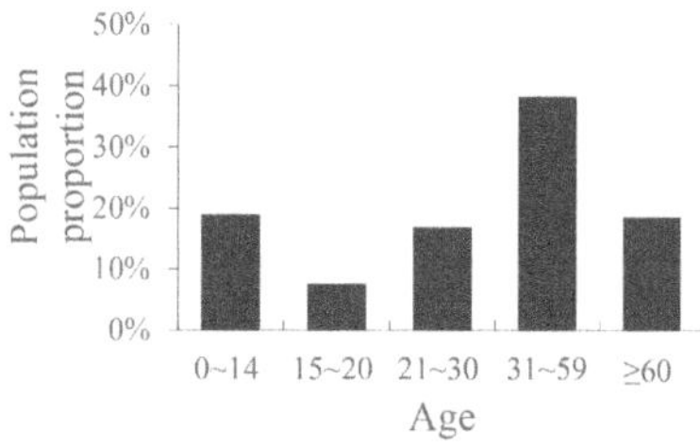

Figure 3. Age structure in household surveys.

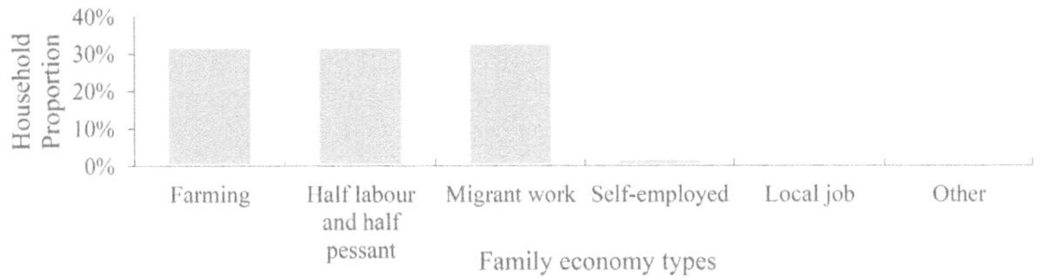

Figure 4. Family economy types in household surveys.

The figures above show that there are mainly multi-person households, and the size of families is normally 4-6 people in the household surveyed. The family age structure characteristics are generally middle-aged and elderly, and the trend of aging phenomena is obvious. The main revenue source for the samples is working outside, working on the farm, and half-time farming, while the number of self-employed and local job households is small. As we researched, most of the people left behind are older people, and villages present the phenomenon of the aging population and vacant villages.

2.2 *Housing types*

There are 97 household samples in total in this survey program. 80% of them are self-built, and the rest are fund-raising rooms. The houses were built in different ages which go back to the 70s and before, 80s, 90s, 2000 years and up to now. Most of the samples were built after 2000 years, and the characteristics of housing are shown in the figures below (Figures 5~7).

As shown in the figures above, most of the samples have brick and concrete structures, accounting for 64%. Behind the brick and concrete structure are rammed earth and timber frame structure which is a type of traditional structure with a unique local flavor. Brick-timber structure, reinforced concrete frame structure, and Reinforced concrete frame structure are approximately the same and account for a small proportion of the samples. The building cost can be attributed mostly to the small ratio of the reinforced concrete frame structure and Reinforced concrete frame structure. The exterior wall materials are mainly clay brick which is cheaper in Baoan town, and the next material type for the exterior wall is rammed earth wall, which is the main component of rammed earth and timber frame structure. Building blocks account for 16% of the samples. The proportion of stone and reinforced concrete materials are lower in the household survey. As to the roof material, Cast-in-situ reinforcement concrete and Timber roof truss and tile account for a large share of roof materials in household surveys (42% and 28%, respectively). The unknown materials not quite clearly narrated by respondent accounts for 20%.

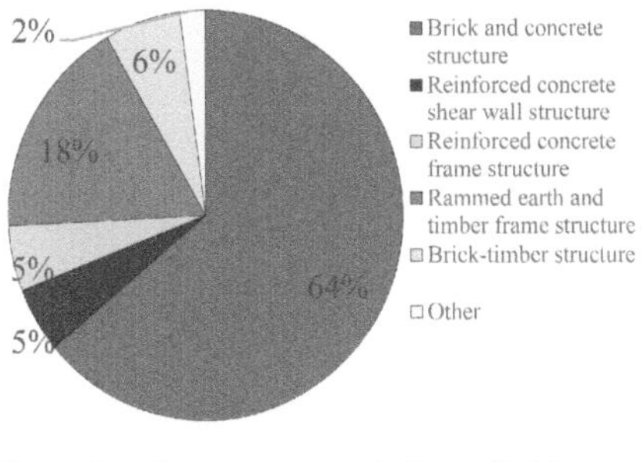

Figure 5. Structure types in household surveys.

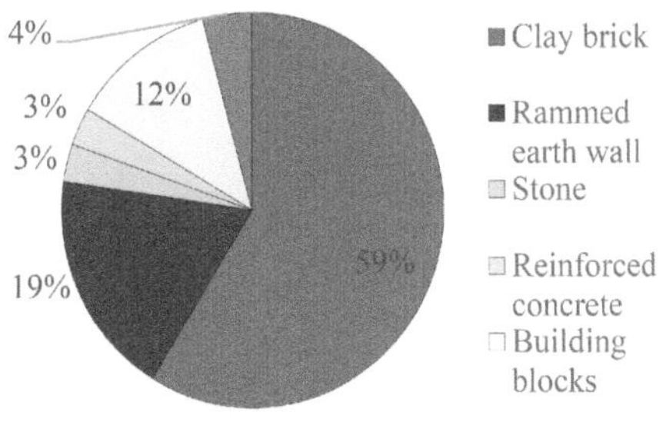

Figure 6. Exterior wall material in the household survey.

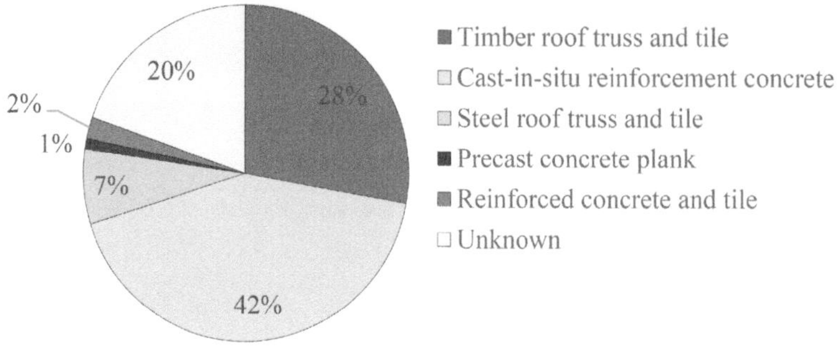

Figure 7. Roof material in the household survey.

2.3 *Residential satisfaction*

Per the data shown in Figures 8~10, 61% of the surveyed households are satisfied with the existing housing, and the satisfaction of building housing built after the 90s is higher. Satisfaction mainly embodies the suitable living environment, size, and quality. The dissatisfaction is observed in the small indoor area, incomplete facilities, and old and shabby. Housing satisfaction is mainly focused on livability, and dissatisfaction is generally a small indoor area for the house built after 2010.

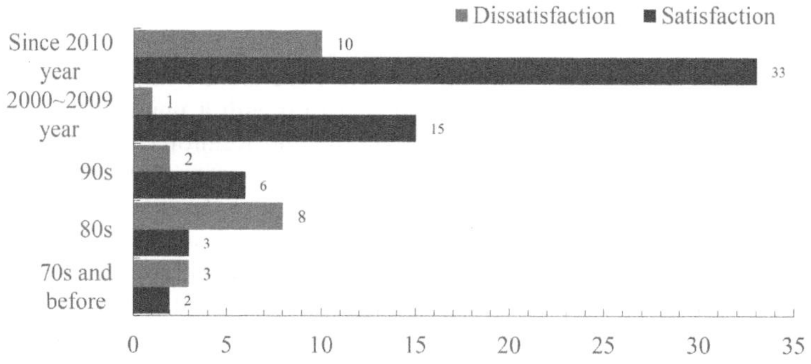

Figure 8. Satisfaction of housing in the household survey.

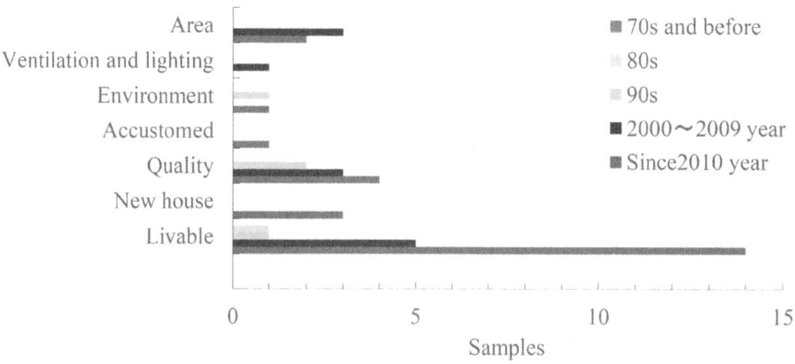

Figure 9. Satisfactory aspects of housing.

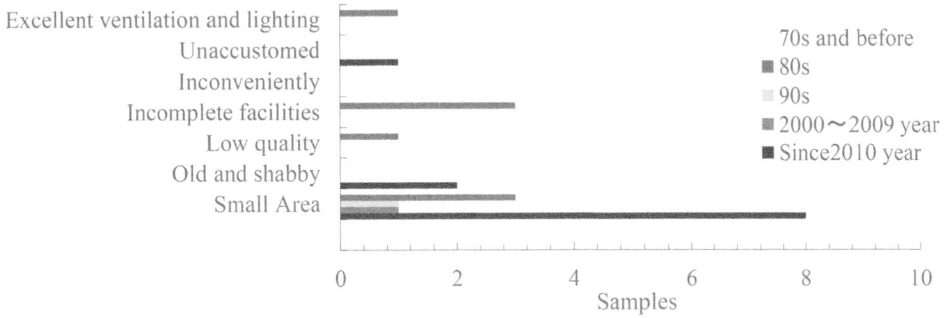

Figure 10. Dissatisfactory aspects of housing.

2.4 *Demand and expectation for housing*

According to the data analysis of Figures 11~14, it is obvious that a single family with single and two storeys is the first choice for the respondents' expectations. 49% of the respondents believe that housing with 100-200 m^2 can meet their needs, and 30% of the respondents express their expectation for a larger house with 200~250 m^2. The durability and cost are the major factors taken into account for construction housing. Mentioned the characteristics of housing demand, the respondents paid more attention to the housing area, housing orientation, and sectorization.

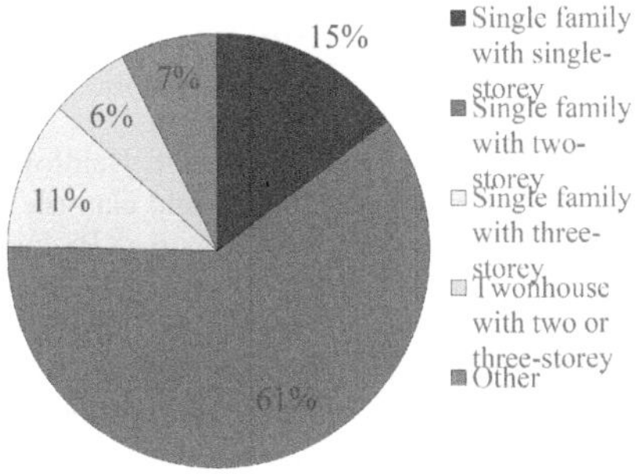

Figure 11. The expectation of housing style in the household survey.

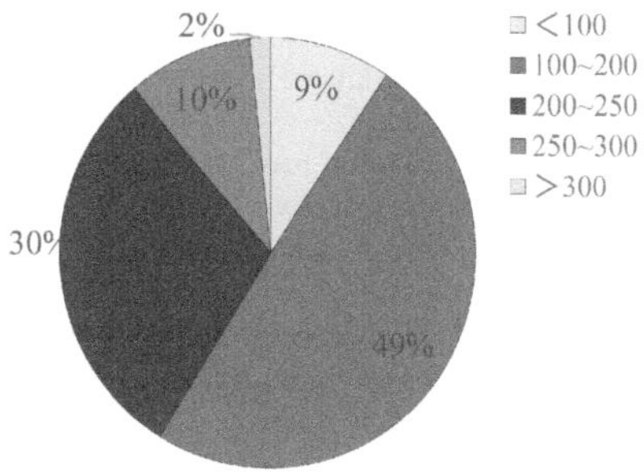

Figure 12. The expectation of housing area in the household survey.

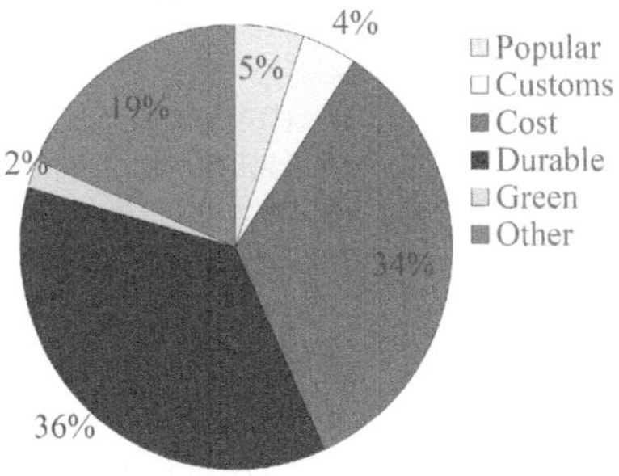

Figure 13. Consideration of housing construction in the household survey.

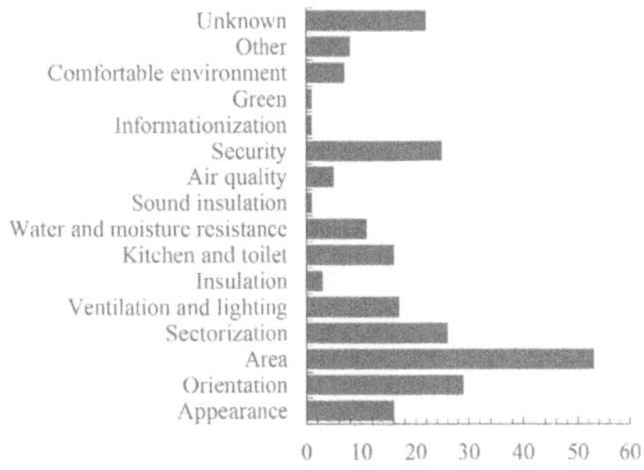

Figure 14. Characteristics of housing demand in the household survey.

3 TECHNICAL CHARACTERISTICS OF HOUSING CONSTRUCTION

In 2016, the Shaanxi Provincial Department of Housing and Urban and Rural Construction issued the "Rural Characteristics Residence Design Atlas for Shaanxi Province." Based on the guidelines for applicable economic, green, and aesthetic, the atlas provides 96 cases of technical measures in detail to guide housing construction for the local villagers. 35 cases are suitable for Baoan Town, and the characteristics of housing construction technologies are concluded in Table 2.

Table 2. Technical list of rural housing construction.

Categories	Technical of construction		
Structure system	Brick and concrete structure	Reinforced concrete structure	Steel frame structure
	Cold-formed thin-walled steel structure	Specially shaped column frame structure	Prefabricated compound wallboard building system structure
Roof system	Flat roof: Reinforced concrete with EPS insulation board and waterproofing layer	Flat roof: Reinforced concrete with hydrophobic pearlite board and damp-proof tile	Pitched roof: Reinforced concrete with hydrophobic pearlite board and tile
	Pitched roof: Reinforced concrete for the base layer with tile	Pitched roof: Reinforced concrete with hydrophobic pearlite board and tile	Pitched roof: Reinforced concrete with cement polystyrene board and tile
	Pitched roof: Reinforced concrete with XPS insulation board and tile	Pitched roof: Reinforced concrete with XPS insulation board, overhead layer, roof board, grey tile	Pitched roof: Steel frame, board, and tile (Attic filled with insulation)
	Pitched roof: Reinforced concrete with EPS insulation board and tile	Pitched roof:Light steel keel filled with rock wool, OSB board, and grey tile	Pitched roof: Reinforced concrete and clay tile (Attic filled with XPS insulation)
	Pitched roof: Reinforced concrete with insulation board and tile	Pitched roof: Reinforced concrete with insulation board and planting soil	Pitched roof: Reinforced concrete and tile (Attic filled with hydrophobic pearlite board insulation)
	Pitched roof: Timber roof truss with timber purlin and tile	/	/
Exterior wall system	Perforated clay brick	Cement pressure plate, Gypsum plasterboard, Light-Gauge steel joist, Rock wool board	Perforated clay brick, The insulation mortar
	Shale hollow brick	Clay hollow brick	Hollow brick
	Fired perforated brick, Glazed hollow beads, thermal insulation mortar	Concrete wall with insulation	Hollow brick,The insulation layer
	Concrete slab 60mm, Air space, Concrete slab 90mm	Prefabricated compound wallboard	Aerated concrete block
	Shale hollow brick, Inorganic insulation mortar	Autoclaved aerated concrete block	Clay hollow brick,EPS insulation board

(*continued*)

Table 2. Continued

Categories	Technical of construction		
	Shale solid brick,Glazed hollow beads thermal insulation mortar	Hollow brick,EPS insulation board	Perforated clay brick, Glazed hollow beads, thermal insulation mortar
Partitioning system	Perforated clay brick	Shale hollow brick	Clay hollow brick
	Fired perforated brick	Shale solid brick	Perforated brick
	Aerated concrete block	Concrete wall	Light-Gauge steel joist, Rock wool board
	Prefabricated compound wallboard	/	/
Window system	Plastics-steel hollow glass window(5+9a+5)	Plastics-steel hollow glass window(5+9a+5Low-E)	Plastics-steel hollow glass window(6+12a+6)
	Plastics-steel hollow glass window(6+6a+6)	Plastics-steel hollow glass window(6+9a+6)	Aluminum alloy hollow glass window
Door system	Solid wood door	Aluminum alloy door	Plastic-steel door
	Wood composite door	Iron door	Steel door
Renewable energy utilization system	Solar water-heating system	The solar photovoltaic power generation system	/

Per the local building regulations, investigation results, and the research listed in the Table above, the thermal insulation of the roof system for rural housing in the Baoan Town area is generally achieved in two ways, laying the thermal insulation layer under tiles of the roof deck and adopting an attic with an insulation layer on the floor. The attic is generally considered a buffer layer to delay variation in temperature. Normally, there are louver windows in the attic for ventilation. The cave-dwelling buildings, traditional houses in Baoan Town, are still occupied in many places, and the loess roof is about 1.5-2.0 meters thick. The thermal insulation performance of the roof is excellent, but the waterproofing performance should be emphasized.

For the wall of the housing in Baoan Town, Solid clay brick is a typical material, and the thickness of the exterior wall is normally 370 mm. This material is suitably applied to the wall of brick and concrete structures for rural housing, and the cost of the wall is about ¥18-¥20 every square meter. Fired perforated brick, Fired hollow brick, and Aerated concrete block is also widely available in the local area for better thermal performance, although the cost is almost doubled. Adopting an insulation layer on the exterior wall is another common measure to improve the indoor thermal environment. The most common insulation materials are EPS insulation board and glazed hollow beads thermal insulation mortar in the local area. The thickness of the insulation layer is generally 30 mm, and the heat transfer coefficient of the exterior wall decreases by 50%~60%. The improvement of the wall's thermal performance is significant, and the cost is about ¥50 for every square meter of the exterior wall.

In older housing, wood window, iron window, and aluminum alloy window are commonly installed on the exterior wall, and most of them are single glass layers. Plastic-steel windows with double-glazing units are popular in new construction and existing building reconstruction. The heat transfer coefficient of it decreases by 50%~60%, and the cost increases by about ¥150~¥180 per square meter compared with wood windows and iron windows with single glass layers.

Solar photovoltaic power generation and solar hot water systems are the most suitable renewable energy technology for rural housing in Baoan Town.

Assembled rural housing is a building form advocated by the local government at present. This innovative construction method can greatly improve the quality of rural housing construction and the energy-saving level. The common forms of prefabricated building structures in rural areas include steel structures, wood structures, and concrete structures.

4 CONCLUSION

This paper fully embodies the demand of villagers for good building quality and a comfortable internal environment by investigating and reviewing the local construction technology of rural housing in Baoan Town. The construction technology in rural areas should be continuously updated to meet the needs of villagers. There is already some new construction technology and materials present in the rural area of Baoan Town. For example, prefabricated construction as a new construction mode has great development potential and a clear advantage in rural areas. We believe it will be the major pattern for rural buildings in the future.

ACKNOWLEDGMENT

This paper is supported by the National Key R&D Program of China during the 13th Five Plan Period-"Research on the construction technology system of green and livable rural residential buildings" (2018YFD1100205).

REFERENCES

Dapeng, Sun. (2006). Elementary Study on the Case of Sustainable Rural Housing Construction System. *Huazhong Architecture.* (12): 25–27.

Ling, Yuan. (2019). Literature Review of Construction Technologies for Green and Livable Rural Housing. *Community Design.* (006): 108–112.

Ministry of Housing and Urban-rural Development (2021). *Guidance on Accelerating the Modernization of Rural Housing and Village Construction.* https://www.mohurd.gov.cn/gongkai/fdzdgknr/tzgg/202106/20210621_250525.html.

Ministry of Housing and Urban-rural Development and the Ministry of Science and Technology (2010). *Livable Housing Technical Promotion Catalogue in Villages and Towns.* http://www.gov.cn/gzdt/att/att/site1/20100603/001e3741a2cc0d71a07601.pdf.

Ministry of Housing and Urban-rural Development and the Ministry of Science and Technology (2010). *Livable Housing Technical Promotion Catalogue in Villages and Towns.* http://www.gov.cn/gzdt/att/att/site1/20100603/001e3741a2cc0d71a07a02.pdf.

People's Government of Shaanxi Province (2008). *Regulations on Rural Planning and Construction of Shaanxi Province.* http://www.shaanxi.gov.cn/zfxxgk/zfgb/2006/d7q_4273/200806/t20080626_1639095.html.

Shaanxi Provincial Department of Agriculture and Rural Affairs (2020). *Interpretation on the Rural Homestead Approval Management in Shaanxi Province.* http://nynct.shaanxi.gov.cn/www/snyncth/20200618/9721777.html.

Shaanxi Provincial Department of Housing and Urban and Rural Construction (2016). Rural Characteristics Residence Design Atlas for Shaanxi Province. https://js.shaanxi.gov.cn/minjuftuji/index. aspx.

Shiju, Zheng. (2011). *Investigation and Analysis of the Present Situation of the Existing Rural House Structure.* (4). 1229–1234.

Xiang, Gao. (2016). *Design Strategies of Rural Houses in Cold Area II (A) Based on the Open Building Theory.* Harbin Institute of Technology. 25–26.

Constructional Engineering and Ecological Environment – Chih-Huang Weng (Ed)
© 2024 The Author(s), ISBN 978-1-032-53198-4

Study on thermal acceptability of people exercising in National Fitness Centers in hot summer and cold winter areas in winter

Hongyu Jin & Hai Zhu*
China Construction Yipin Investment & Development Co., Ltd, Wuhan, Hubei, China

ABSTRACT: The indoor thermal environment of sports buildings is closely related to the outdoor physical environment, which has a strong regional character. In order to study the thermal acceptable range of sports crowd exercising in hot summer and cold winter areas, this paper conducts a field survey of Xiangyang National Fitness Center in winter climate. Through the subjective questionnaire of the sports crowd and the analysis of indoor and outdoor thermal environment data, it is concluded that the thermal acceptable operating temperature range and upper limit of the sports crowd in Xiangyang National Fitness Center in winter under the five conditions of sitting, walking at 2 km/h, walking at 4 km/h, walking at 6 km/h and running at 8 km/h are 15.6°C–22.7°C, 14.5°C–19.9°C, 13.6°C–18.7°C, 13.1°C–19.3°C and 10.9°C. This study helps to better understand the thermal perception of people with different intensities of sports in the real environment of hot summer and cold winter areas in winter and provides a basis for the passive design of natural ventilated indoor sports space in the Xiangyang area.

1 INTRODUCTION

The indoor sports thermal environment of the National Fitness Center has a great impact on the exercise effect and duration of the sports crowd. The existing design strategy of the National Fitness Center pays more attention to the size of the sports space and the number of sports types that can be accommodated, while ignoring the environmental quality of indoor sports (Liu 2017). The role of the National Fitness Center is to help the sports crowd shield itself the outdoor uncontrollable weather conditions. Most of the National Fitness Center's regulation of the indoor thermal environment completely depends on the air conditioner (Zhang 2013). However, the control of the indoor sports thermal environment in the National Fitness Center should not be completely isolated from the outside world. A healthy sports thermal environment should be an organic unity of regional climate, sports crowd, and sports space. In order to effectively improve the indoor thermal environment quality of sports buildings in hot summer and cold winter areas, the indoor and outdoor thermal environment data and the subjective questionnaire survey of the sports crowd of Xiangyang National Fitness Center in winter were collected and analyzed through field survey. The thermal acceptable range of sports crowd in Xiangyang National Fitness Center Basketball Training Hall in winter was obtained. It provides guidance and suggestions for the design and reconstruction of stadiums, gymnasiums, and other sports spaces in hot summer and cold winter areas in the future.

*Corresponding Author: 24586375@qq.com

DOI: 10.1201/9781003410843-37

1.1 *Building overview*

The field survey site of hot summer and cold winter areas in winter was located in Xiangyang National Fitness Center in Xiangyang City, Hubei Province, as shown in Figure 1. The building of the Xiangyang National Fitness Center covers an area of 41,000 m^2, 21,000 m^2 above the ground, and 20,000 m^2 underground. There are three floors above the ground and one floor below the ground. The indoor sports space is about 10 meters high. The first floor is a basketball hall and a table tennis hall, and the third floor is a tennis hall and a badminton hall. Considering the convenience of the venue, the operability of the field survey, and other factors, the basketball hall of the National Fitness Center was finally selected as the sports space for the field survey.

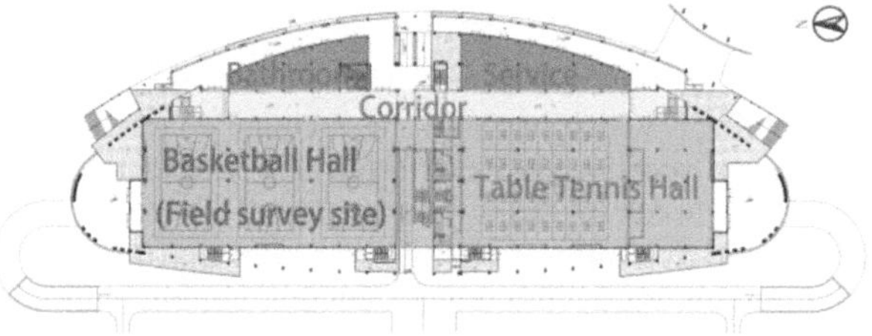

Figure 1. Site selection of field survey.

2 RESEARCH METHODS

The time of field survey is from January 20 to March 5 in 2021, which is about 6 weeks in total. The daily measurement is carried out from 12:00 p.m. to 5:00 p.m. During the field survey period, except that the National Fitness Center was closed one week before and after the Spring Festival, which was not accessed, the weather was cloudy to sunny for the rest of the time. On rainy days, data processing and analysis are carried out to ensure the stability and accuracy of the data. During the field survey in winter, the basketball hall of the National Fitness Center ensured that no active regulating equipment was functioning, such as heating and air conditioner, and the indoor space was in a natural ventilation state.

2.1 *Subjects and clothes*

Due to the strict local epidemic control measures and the need to purchase unified winter sports clothes for the subjects, 10 subjects were recruited during the field survey, including 5 men and 5 women. All the subjects had lived in the Xiangyang area for more than 10 years, adapting to the environment of hot summer and cold winter areas. 48 valid subjective questionnaires were collected, and 240 groups of valid field data were obtained. The basic information on the winter subjects is shown in Table 1.

Table 1. Subjects' general information.

Gender	Age	Height (cm)	Weight (kg)	BMI[a] (kg/m^2)
Men	22.2±4.1[b]	177.2±6.1	74.2±7.1	23.8±3.6
Women	18±3.9	163.6±6.8	50.2±6.9	18.8±2.6
Total	20.1±4.4	170.4±9.4	62.2±14.3	21.3±4.0

[a]Body Mass Index (BMI)= Mass /Height^2 (kg/m^2)
[b]Standard deviation

Before the formal field survey, it was found through three days of Preliminary survey that, for winter sports crowd, the amount of clothing (corresponding to the clothing insulation) had a greater impact on the thermal perception during sports. In order to avoid the influence of too much or too little clothing on the measured thermal perception during the measurement and also to avoid physical discomfort symptoms such as cold and cough. In the winter field survey, the subjects purchased the same winter sports suit produced by the same manufacturer, including a sports sweater, a sports coat, and a pair of sports cotton trousers, to strictly control the clothing insulation during the field survey, as shown in Figure 2.

Figure 2. Subjects wearing uniform sportswear in winter field survey.

The subjects were required to keep their underwear, underpants, socks, and sports shoes but were required to change into uniform sports clothes 20 minutes before the beginning of the field survey. During the field survey, they were not allowed to change clothes. The average clothing insulation of the subjects in the winter field survey is 1.15 Clo (American Society of Heating, Refrigerating and Air Conditioning Engineers 2017).

2.2 *Measuring instruments*

According to the thermal comfort equation of Professor Fanger (1970), the impact of the thermal environment on sports people is mainly divided into two parts: the objective physical parameters of the thermal environment and the subjective thermal perception evaluation of sports people. The objective physical parameters of the thermal environment mainly include air temperature, relative humidity, air velocity, and radiation temperature. The subjective thermal perception of sports people includes the metabolic rate of the human body and the clothing insulation. In the process of measurement, the metabolic rate of the human body is kept stable by controlling the movement speed on the treadmill, and the clothing insulation is controlled by the uniform winter sports clothing. The objective physical parameters in the field survey and research in this paper take the operating temperature as the evaluation standard of the indoor thermal environment of the National Fitness Center and use the calculation formula provided by ISO Standard 7726-2002 (Geneva 2006) to calculate the operating temperature, as shown in Formula (1) and Formula (2):

$$T_{op} = A \times T_a + (1 - A)T_r \tag{1}$$

$$T_r = \frac{(T_g + 273)^4 + (1.10 \times 10^8 V^{0.6})(T_g - T_a)}{\varepsilon D^{0.4} - 273} \tag{2}$$

Where, T_{op} is the operating temperature, in°C; T_a is the air temperature, in°C; T_r is the average radiation temperature, in°C; A is a constant. When the air velocity is less than 0.2m/s, the value is 0.5; T_g is the black ball temperature, in°C; V is the air velocity, in m/s; D is the radius of the black ball. The standard black ball is used in the field survey, and the value is 0.15 m; ε is the absorption rate of the black ball, taken as 0.95.

It can be seen from Formula (1) and Formula (2) that the operating temperature of an indoor thermal environment requires the measurement of physical quantities such as black ball temperature, air temperature, indoor air flow rate, and relative humidity. According to the requirements of ISO7726 on the measuring range and accuracy of measuring instruments, Testo480 multi-function testing instrument was finally selected as the data acquisition instrument for measuring the indoor thermal environment of the National Fitness Center. The Testo480 multi-function detector mainly consists of five parts: Testo480 host machine, the Thermal radiation black-bulb sensor, hot-wire wind speed sensor, the indoor air quality sensor, and the fixed support, as shown in Figure 3. The wearable motion detector is used to monitor and record the heart rate of the subjects in different states at the same time.

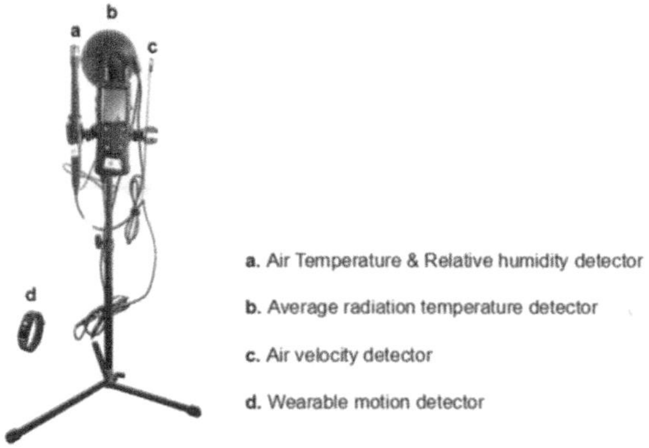

Figure 3. Objective physical parameters measuring instruments.

2.3 *Subjective survey questions*

In previous studies, Gagge (1970), Fanger (1970), and other scholars believed that "thermal comfort" meant that the human body was in a "neutral" state of "neither cold nor hot," that is, when the human body was in a neutral temperature condition, the thermal feeling was thermal comfort. Later, some scholars pointed out that thermal comfort is different from thermal sensation. When the human body is at a neutral temperature, it may not be able to achieve thermal comfort (Zhao 2000). Moreover, the PMV (Predicted Mean Vote) index is no longer applicable to the evaluation of human thermal sensation under high metabolic intensity (Fletcher et al 2020). Due to the separation of thermal comfort and thermal sensation, the subjective questionnaire used in the field survey focused on thermal acceptability. In this study, paper questionnaires were used to collect and sort out the subjective feelings of the subjects. The filling of the questionnaire was carried out synchronously with the collection of environmental parameters. The questionnaire was filled in once every 3 minutes of each exercise state. Each exercise state was maintained for 15 minutes. The thermal acceptability was measured following the same pattern, from 0.01(slightly acceptable) to +4 (clearly acceptable), representing satisfaction, and from 0.01 (slightly unacceptable) to -4 (clearly unacceptable), representing thermal dissatisfaction. The scale of the subject's subjective questionnaire on thermal sensation and thermal acceptability is shown in Figure 4.:

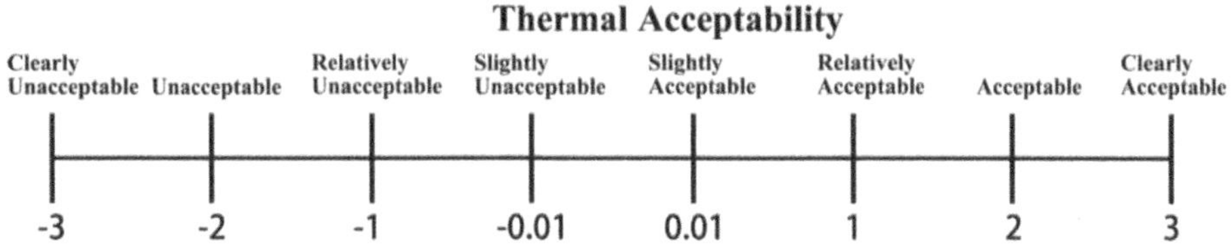

Figure 4. Rating scales used in the field survey.

2.4 *Survey plan*

Five exercise modes were set in the winter field survey. According to the 2011 Compendium of Physical Activities published by the American College of Sports Medicine (Orth et al. 2011), the exercise modes and metabolic rates measured in winter are shown in Table 2.

The whole measurement process takes 135 minutes, as shown in Figure 5. Subjects arrived at the site 15 to 20 minutes before the start of the field survey. After arriving at the site, the subjects put on the winter uniform sportswear prepared by the measuring personnel, bound the skin temperature measurement sensor, and began to fill in the subjective questionnaire for sitting. After the survey began, the exercise and the rest of the subjects were carried out at the test site. The subjective questionnaire is a paper questionnaire, which is filled in by the subjects themselves during the sitting and rest periods, and the subjective questionnaire of the subjects during exercise is filled in by the recorder after inquiry. A subjective questionnaire and heart rate reading monitoring were filled in every 3 minutes, which did not affect the subject's exercise status.

Table 2. Exercise modes and metabolic rates in winter field survey (Orth et al. 2011; Zhai et al 2020).

	Metabolic rate	
Exercise Mode	W/m^2	METS
Sitting	53.8	1.0
2 km/h walking	107.6	2.0
4 km/h walking	161.4	3.0
6 km/h walking	242.1	4.5
8 km/h running	446.5	8.3

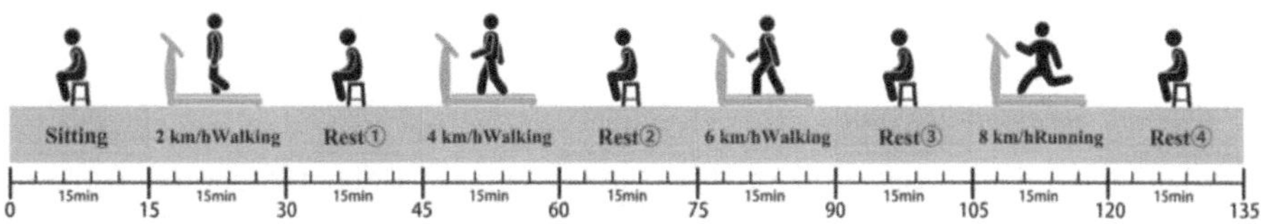

Figure 5. Experimental procedures.

3 DATA ANALYSIS

The field survey was designed to be repeatable. In this study, SPSS v22.0 was used to keep T_{op} in 0.5°C steps. Statistical analysis results were performed through Excel, and all the differences were accepted as significant at a 0.05 level. The observed subjective thermal responses for each individual were averaged, and mean values plus standard deviations (SD) were reported in the paper.

3.1 *Thermal acceptability analysis*

In this study, the thermal acceptability of subjects in sitting and four exercise states was collected by filling in a subjective questionnaire. This paper first collects the voting values of the thermal acceptability of the subjects at different times through the field survey and counts the number of positive and negative values of the thermal acceptability, where positive values such as 0.01, 1, 2, and 3 represent acceptable values, negative values such as -0.01,–1,–2 and–3 represent unacceptable values, and then counts the percentage of the number of unacceptable negative votes in the total number of votes in this state, Finally, the percentage of dissatisfaction (PD) of subjects in this state was obtained. Furthermore, take 0.5°C as an interval, take the average operating temperature in this interval as the independent variable, and the percentage of thermal dissatisfaction as the dependent variable, and establish a binary quadratic equation data model. Finally, draw the linear regression curve between the percentage of thermal dissatisfaction and the operating temperature under five exercise conditions, as shown in Figure 6.

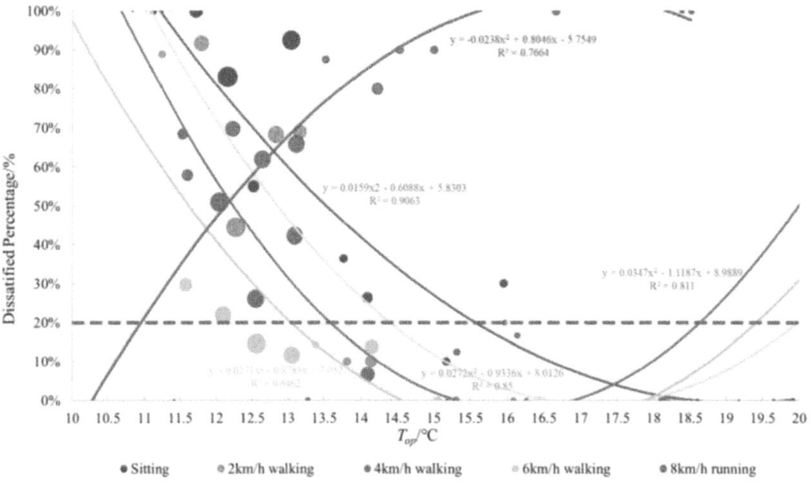

Figure 6. Relationship between the operating temperature and thermal dissatisfaction.

3.2 *Thermal acceptability range*

By predicting the linear regression between the percentage of dissatisfaction and the operating temperature, the regression formula between the thermal unacceptability and the operating temperature and the determination coefficient R^2 are obtained. As shown in Table 3, the determination coefficient R^2 of the regression relationship between thermal dissatisfaction and the operating temperature under the five states is greater than 0.30, indicating that the fitting degree of each regression equation is good, and the linear correlation is obvious.

According to the good linear regression correlation between thermal dissatisfaction and operating temperature, the regression equation obtained has certain predictability. At the same time, ASHRAE 55 (American Society of Heating, Refrigerating and Air Conditioning

Table 3. Relationship between predicted thermal dissatisfaction and operating temperature.

Activities	Equation	R^2	Thermal acceptability range/upper limit	Operating Temperature range
Sitting	$PPD = 0.0159T_{op}^2 - 0.6088T_{op} + 5.8303$	0.91	15.6°C–22.7°C	10.8°C–19.9°C
2 km/h walking	$PPD = 0.0272T_{op}^2 - 0.9336T_{op} + 8.0126$	0.85	14.5°C–19.9°C	
4 km/h walking	$PPD = 0.0347T_{op}^2 - 1.1187T_{op} + 8.9889$	0.81	13.6°C–18.7°C	
6 km/h walking	$PPD = 0.0271T_{op}^2 - 0.8783T_{op} + 7.052$	0.65	13.1°C–19.3°C	
8 km/h running	$PPD = -0.0238T_{op}^2 + 0.80464T_{op} - 5.7549$	0.77	10.9°C	

Engineers 2017) specifies the range of thermal acceptability. Only the thermal environment that is acceptable to at least 80% of the population can be considered acceptable. At this time, the temperature range is the temperature range when less than 20% of the subjects vote unsatisfied with their thermal environment. The predicted percentage of dissatisfaction (PPD) $\leq$ 20% (i.e., thermal satisfaction>80%) is considered the acceptable temperature range for the test population. Substitute PPD=20% into equations of Table 3 to obtain the range or upper limit of thermal acceptability under five conditions, as shown in Table 3.

When the thermal dissatisfaction is 20%, the intersection with the five state curves is marked with red dotted lines in Figure 6. In the national fitness center with natural ventilation in winter, the thermal acceptable range of the sports crowd in the state of sitting is 15.6°C–22.7°C; walking at 2 km/h slow walking and low metabolic intensity is 14.5°C–19.9°C; walking at 4 km/h and moderate metabolic intensity is 13.6°C–18.7°C; walking at 6 km/h and high metabolic intensity is 13.1°C–19.3°C. The thermal acceptability range of 8 km/h running under extremely high metabolic intensity is 10.9°C. The value solved by the equation, in the indoor environment with a low temperature in winter, the lower limit value of the thermal acceptable temperature range of sitting, 2 km/h walking, 4 km/h walking, and 6 km/h walking and the upper limit value of 8 km/h running were calculated from the actual operating temperature measured. In contrast, the upper limit temperature of the thermal acceptable temperature range of sitting, 2 km/h walking, 4 km/h walking, and 6 km/h walking were predicted through the regression curve.

4 CONCLUSION

Based on the analysis of the questionnaire voting data of the thermal acceptability of indoor sports crowds and the data of indoor and outdoor thermal environment physical parameters measured from January to March in winter, the following conclusions are drawn:

(1) In the Xiangyang National Fitness Center in the hot summer and cold winter area, the thermal acceptable operating temperature range and upper limit under the five conditions of sitting, walking at 2 km/h, walking at 4 km/h, walking at 6 km/h and running at 8 km/h are 15.6°C–22.7°C, 14.5°C–1.9°C, 13.6°C–18.7°C, 13.1°C–19.3°C and 10.9°C respectively.

(2) In the winter, when the indoor environment temperature is low, starting from sitting, the lower limit of the thermal acceptable operating temperature range decreases with the increase of exercise intensity, indicating that in a lower temperature environment, the human body can promote heat production by increasing exercise intensity and exercise duration, thereby improving the thermal adaptability of the human body to the low-temperature environment.

(3) In order to achieve the most healthy exercise state, extend the time of exercise thermal comfort, and obtain the best exercise effect, it is recommended that the exercise intensity be kept at medium. High intensity (such as 4 km/h moderate speed walking and 6 km/h fast walking) when the sports crowd is exercising in the low-temperature conditions in winter in the National Fitness Center in hot summer and cold winter areas, and try to maintain the balance between heat production and heat dissipation of the body at the same time, exercise time under extremely high metabolic intensity (such as 8 km/h running) shall be reasonably controlled to avoid discomfort such as catching a cold after exercise.

REFERENCES

Ashrae Standard 55: *Thermal Environmental Conditions for Human Occupancy, American Society of Heating, Refrigerating and Air Conditioning Engineers*, Atlanta, Georgia, 2017.

Fanger P. *Thermal Comfort: Analysis and Applications in Environmental Engineering*. Copenhagen: Danish Technical Press. 1970.

Fanger P.O. *Thermal Comfort. Copenhagen*. Danish Technical Press, 1970.

Fletcher M. J., Glew D. W., Hardy A. *et al.* A Modified Approach to Metabolic Rate Determination for Thermal Comfort Prediction During High Metabolic Rate Activities. *Building and Environment*, 2020, 185:107302.

Gagge A.P. *Introduction to Thermal Comfort*. INSERM. 1977.

ISO 7726: *Ergonomics of the Thermal Environment-Instruments for Measuring Physical Quantities International Standards Organization*, Geneva, 2006.

Liu Jiawang. *The Research of the National Fitness Center Activity Space Based on Health Needs*. Hefei University of Technology, 2017.

Orth B.E. *et al.* 2011 Compendium of Physical Activities. *Medicine & Science in Sports & Exercise*. 2011. 43 (8): p. 1575–1581.

Zhai Y., Zhao S., Gao Y., *et al.* Preferred Temperatures With and Without Air Movement During Moderate Exercise. *Energy and Buildings*. 2020, 207(5):109565.

Zhang Wei. *Research on Synergism between National Fitness Center and Natural Ventilation in Humid Tropical Area*. South China University of Technology, 2013.

Zhao Rongyi. Discussion on "Thermal Comfort." *Journal of HV&AC*, 2000 (03): Page 25–26.

Constructional Engineering and Ecological Environment – Chih-Huang Weng (Ed)
© 2024 The Author(s), ISBN 978-1-032-53198-4

Prediction of landslide deformation trend of a substation access road based on time-series InSAR

Yang Tang
State Grid Sichuan Electric Power Company, Chengdu, China

Like Huang
Ganzi Power Supply Company, State Grid Sichuan Electric Power Corporation, Kangding, China

Songmei Lv*
Chengdu University of Technology, Chengdu, China

Jun Xiang, Xianping Zhou & Lihao Yin
Ganzi Power Supply Company, State Grid Sichuan Electric Power Corporation, Kangding, China

Xili Yang
Chengdu University of Technology, Chengdu, China

ABSTRACT: In this paper, the application of the time-series InSAR analysis method in landslide deformation trend prediction is discussed by taking the landslide approach road of a substation in Danba as the experimental area. By comparing the cumulative displacement monitoring map of landslides, the large deformation area is obtained, and it is taken as the key research object for zoning monitoring. According to the comprehensive field investigation and time series InSAR monitoring, there were no obvious macro deformation signs in the trailing edge I area and the downstream side II area. At the same time, there were obvious macro deformation signs and more complex deformation phenomena in the upstream side approach road section III area. There were many deformations in the micro monitoring. In March 2017, both the deformation amount and deformation rate increased. Progressive deformation occurred in the slope, and the deformation tended to intensify. The maximum cumulative deformation in this area exceeded 150mm. Then the landslide is in slow creep.

1 INTRODUCTION

A landslide is an important type of geological disaster, which not only harms human life, but also causes great harm to the environment, resources, and property. Geological disasters are one of the factors that influence our economy and casualties greatly every year. Our country is also a country that carries out geological disasters frequently. At present, landslide research is in the development stage from qualitative to quantitative, from static to dynamic. However, the traditional technical means are limited, so the development of remote sensing technology is accelerating. The early identification, monitoring, warning, and risk assessment of landslide disasters are benefiting from new remote sensing technologies, especially InSAR, etc. (Ferretti et al. 2000; Su et al. 2017; Xu et al. 2017; Yin et al. 2017; Zhang et al.

*Corresponding Author: 1965347983@qq.com

DOI: 10.1201/9781003410843-38

2018) Compared with the traditional surface deformation monitoring method, synthetic aperture radar interferometry (InterferometericSyntheticApertureRadar, InSAR) has many advantages, such as monitoring range, large density, higher accuracy, especially not being restricted by the weather. As a new technical means for landslide research, it has been a new surface deformation monitoring technology in recent years. In 2000, Ferretti, an Italian scientist, proposed Permanent Scatter InSAR (PS InSAR) for the first time. The basic idea of PermanentScatterinsar (PS InSAR) is to compose N+1 SAR images covering the same area into a time series. Based on the image quality, time, and space baseline, one of the images was selected as the common main image. The remaining N images were registered and resold to the main image space to generate the time series differential interferogram. The points with high coherence were selected to effectively separate the phase components of other errors by establishing and solving the deformation inversion model, so as to obtain the rate and time series of surface deformation (Li et al. 2021; Liao et al. 2017; Shi et al. 2016; Strozzi et al. 2018; Zhang et al. 2016).

In this paper, the landslide on the approach road of a substation in Danba was taken as the research object. Based on 27 views of sentry rail descent data and 69 views of sentry rail ascent data, the surface deformation rate in the study area was inverted by InSAR technology, and the overall state of the slope was found to be in slow creep through comparison. The results can not only provide basic data support for the investigation, monitoring, warning, and prevention of regional geological disasters, but also serve for the planning and construction of Danba and the surrounding towns. They also have important theoretical and engineering significance for the deformation trend analysis and risk assessment of a single landslide.

2 THE GEOLOGICAL BACKGROUND SETTING OF THE LARGE LANDSLIDE

The landslide on the access road of a substation in Danba is an ancient landslide with a clear boundary. The rear edge is bounded by the steep and gentle junction of bedrock slope and secondary platform, the front edge is bounded by the riverbed of Dadu River, and the upstream and downstream sides are bounded by gullies (Figure 1). The landslide is "bell-shaped" on the plane, with a narrow rear edge and wide front edge. The elevation of the rear edge of the landslide is about 2270 m, the elevation from the front edge to the riverbed bottom is about 1820 m, and the elevation difference between the front edge and the rear edge are about 450 m, and the main sliding direction is 40°. The longitudinal length of the landslide is about 1010 m, the width along the river is 428~734 m, the average width is 470 m, and the plane area is 42.76×104 $m^{2\cdot}$ The overall slope of the landslide is steep, with a

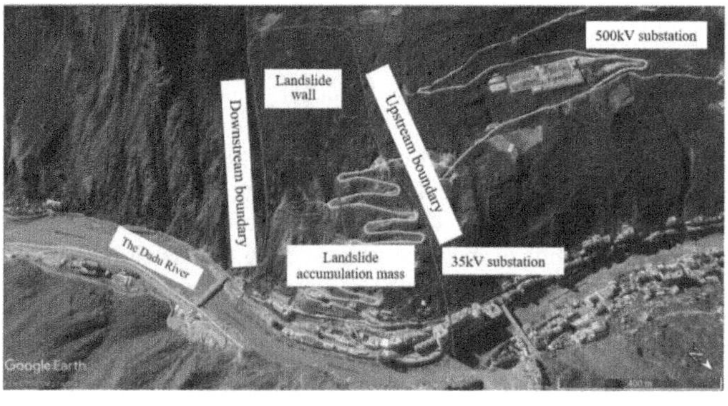

Figure 1. Overview of Landslide(self-drawn).

natural slope of 30~40°. The average thickness of the landslide is about 30 m, and the total volume is about 1283×104 m^3, which is a large rock landslide.

3 DATA AND METHODS TO MONITOR THE LANDSLIDE

Figure 2 is a geometric schematic diagram of the stereo measurement principle of interferometric SAR image in RTI mode. It is assumed that the distance from the main image intake point S1 to the ground monitoring point is r, and the distance from the image intake point S2 to the ground point with the same name is r+dr. H represents the vertical height from S1 to the ground point during the first imaging, and h is the ground elevation. The spatial distance between S1 and S2 is called spatial baseline B. The Angle between the baseline and distance direction is α, and θ is the incidence Angle of the radar wave. Then, the elevation h of the ground point target can be calculated by the Formulas (1) and (2) through the geometric relationship of stereo measurement.

$$(\mathrm{r} + dr)^2 = r^2 + B^2 - 2rb \sin(\theta - \alpha) \tag{1}$$

$$\mathrm{h} = H - r \cos \alpha \tag{2}$$

The phase of S1 and S2 relative to the target P can be obtained by the relationship between phase and wavelength during wave propagation, which can be obtained by formulas 3 and 4, where λ is the radar wavelength.

$$\phi_1 = -\frac{4\pi}{\lambda} r \tag{3}$$

$$\phi_2 = -\frac{4\pi}{\lambda}(r + dr) \tag{4}$$

$$\varphi = \phi_2 - \phi_1 = -\frac{4\pi}{\lambda} dr \tag{5}$$

Formula 5 is the interference phase of the ground target, then:

$$dr = \phi_2 - \phi_1 = -\frac{\lambda}{4\pi}\varphi \tag{6}$$

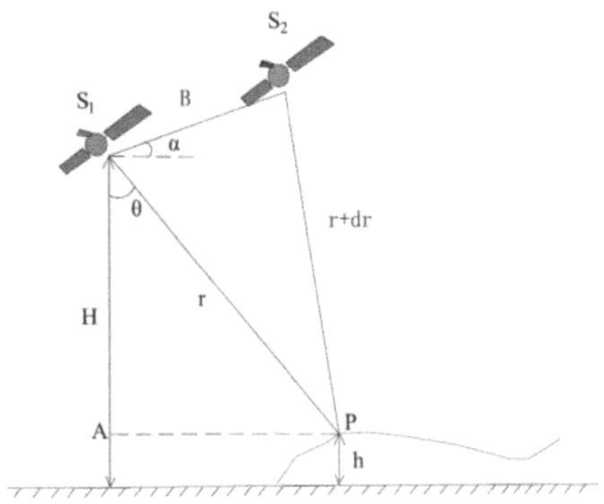

Figure 2. Schematic of interferometric SAR image stereo measurement principle (self-drawn).

Under the radar system, r»B and r»dr Are true, so the real lines S1 and S2 to the ground target point can be regarded as parallel. In △S1S2P, the law of cosine can be obtained:

$$(r + dr)^2 = r^2 + B^2 - 2rB\sin(\theta - \alpha) \tag{7}$$

Hence:

$$dr \approx -B\sin(\theta - \alpha) \tag{8}$$

Combining formulas 6 and 8, the incidence Angle θ is derived as:

$$\theta = arc\sin\left(\frac{\lambda\varphi}{4\pi B}\right) + \alpha \tag{9}$$

By using the orbit lowering data of 27 sentries in time period 1 (2014/10/26~2017/03/08) and the orbit raising data of 69 sentries in time period 2 (2017/03/13~2020/12/22), and using the PS InSAR technology, a series of calculations and processing are carried out through images to obtain the ground deformation data of the landslide on the access road to Danba Substation and the surrounding areas, and the average annual ground deformation rate in the LOS direction of the landslide is obtained (Figures 3 and 4).

It can be seen from Figure 3 that from October 26, 2014, to March 8, 2017, the deformation rate in the middle of the landslide was relatively high, mainly showing red and orange coherent points, with a deformation rate of -7~-37 mm/year. There were few coherent points in some parts of the middle of the landslide, which led to incoherence; According to the distribution map of deformation rate, the deformation of the landslide is concentrated on the access road in the middle and front of the landslide, and the maximum deformation rate is – 37 mm/year. The results of the annual average deformation rate from March 2017 to December 2020 (Figure 4) show that the deformation of the slope mainly occurs in the middle and front of the landslide, especially in the middle of the slope, where orange or red coherent points are dominant. The deformation rate is significantly higher than that of the surrounding areas, with a deformation rate of -17~-53 mm/year.

It can be seen from Figure 5 that obvious deformation occurred in local areas of the slope from October 2014 to March 2017, of which the deformation in Area I was small, the

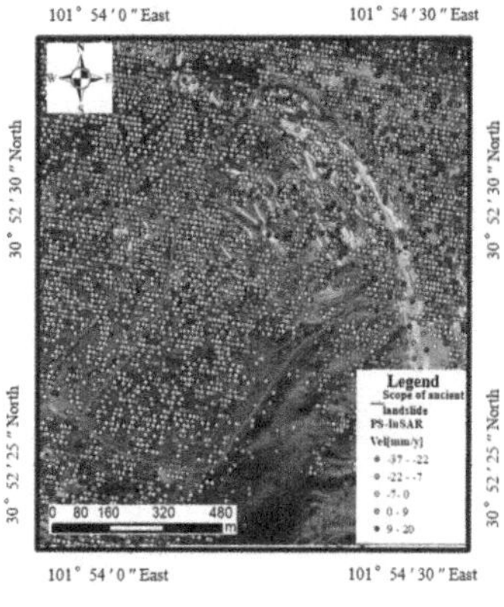

Figure 3. Distribution of landslide deformation rate from October 26, 2014, to March 08, 2017 (self-drawn).

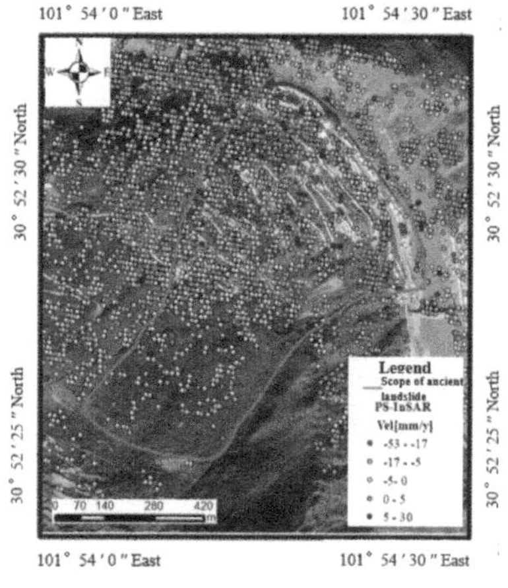

Figure 4. Distribution of landslide deformation rate from March 13, 2017, to December 22, 2020 (self-drawn).

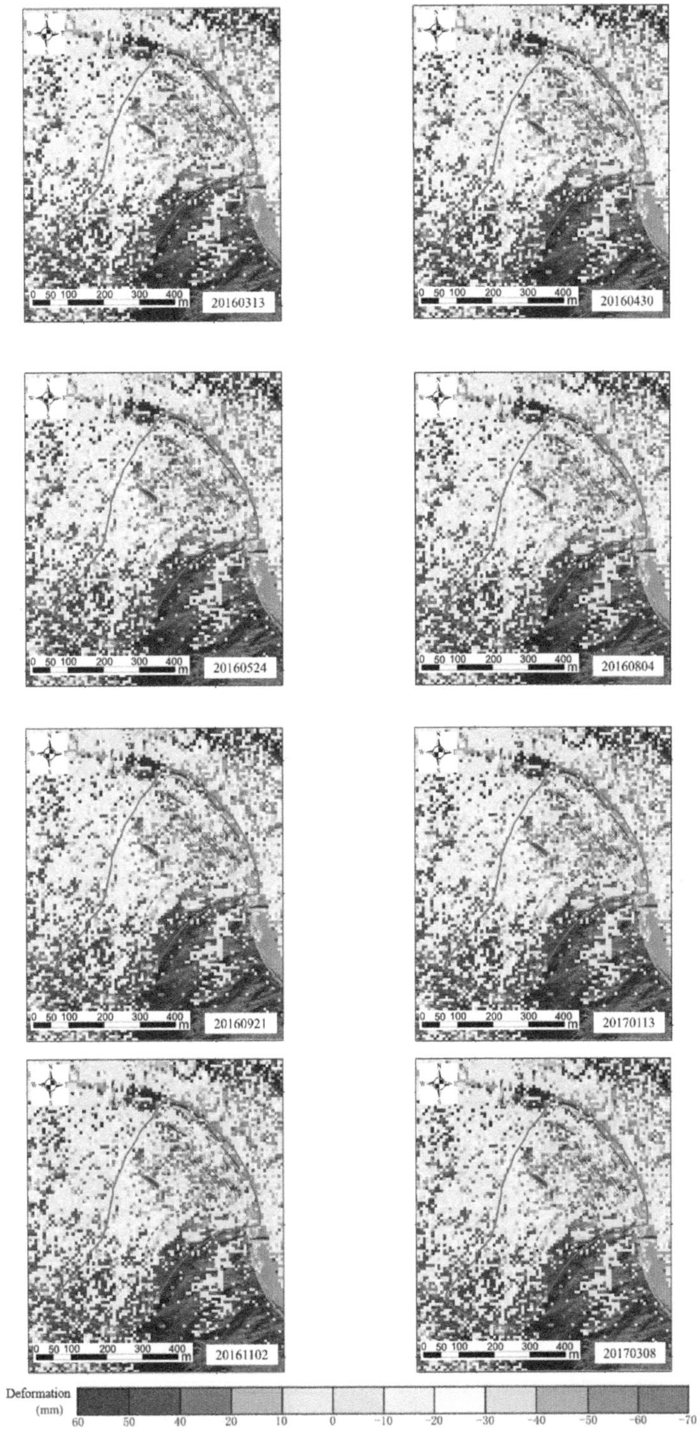

Figure 5. Monitoring map of cumulative displacement of landslide based on PS InSAR (2014/10/26~2017/03/08) (self-drawn).

deformation occurred in the middle and rear of Area II, and large local deformation occurred in the slopes on both sides of the ③~⑦ turnback of the access road in Area III in the middle of the landslide, with a maximum deformation of up to 70 mm; The side of the access road has large deformation, including the west side of ③turn back, ⑥turn back, and ②~⑦ turn back in the middle of the slope.

4 LANDSLIDE DEFORMATION MONITORING BASED ON THE PS INSAR METHOD

According to the field survey data and InSAR time series monitoring results, the landslide can be divided into three large deformation areas. Of this area, I is the rear edge of the landslide. From the image and the current time series monitoring results, there is no obvious deformation; Area II is the loess ridge landslide control area, which has experienced integral sliding in history, but the control measures taken in the later period are mainly the micro deformation of the structure surface at present; Zone III is located in the middle and front part of the landslide, and there is deformation in both macro and micro aspects, and the deformation is complex. Based on this, 23 monitoring points are arranged at different parts of the landslide along each profile.

Based on the comprehensive field investigation and time series InSAR monitoring, it is known that the landslide is in the process of slow creep as a whole. There is no obvious deformation sign in the rear edge area I and the downstream area II, which are relatively stable. The macroscopic deformation sign in the upstream access road area III is obvious, and the deformation phenomenon is more complex. There are also many deformations in the microscopic monitoring. From 2014 to 2017, the displacement of each monitoring point in Zone III was small, but from March 2017, its deformation amount and deformation rate have increased, and the slope has undergone progressive deformation, with a trend of

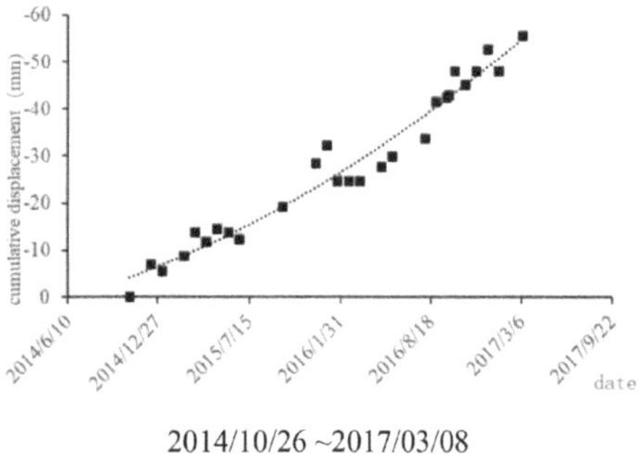

2014/10/26 ~2017/03/08

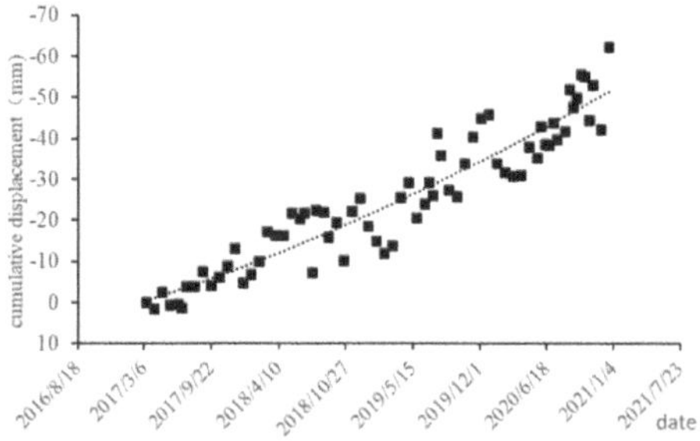

2017/03/13 ~2020/12/22

Figure 6. Cumulative Displacement Time Chart of Monitoring Points in Zone III (self-drawn).

increasing deformation. In space, the terrain at the rear of Zone III is relatively gentle, with small deformation, while the terrain at the front edge is gentle and has no obvious deformation; There is a downward trend along the slope in this area. It is very likely that instability will occur first in this area, and then the rock and soil mass at the rear edge will be pulled to instability.

5 CONCLUSION

The landslide on the access road of a substation in Danba is a large rock landslide with a plane area of 42.76×10^4 m^2; The average thickness of the landslide is about 30 m, and the total volume is about 1283×10^4 m^3.

Based on the comprehensive field investigation and time series InSAR monitoring, there is no obvious macroscopic deformation sign in the trailing edge area I and the downstream area II, which are relatively stable. The macroscopic deformation sign in the upstream access road area III is obvious, and the deformation phenomenon is more complex. The slope has undergone progressive deformation, and there is a trend of increasing deformation. The maximum cumulative deformation in this area exceeds 150 mm. The deformation of the rear and front edges of Zone III is smaller than that of the middle and front parts. The cumulative deformation of the front edge is between 36 mm and 65 mm, and the deformation of the rear is between 30 mm and 80 mm. On the whole, the landslide is in slow creep.

ACKNOWLEDGMENT

The authors would like to thank State Grid Sichuan Electric Power Company for funding this paper: (SGSCGD00JSJS2200361).

REFERENCES

Ferretti A., Prati C. and Rocca F. Nonlinear Subsidence Rate Estimation Using Permanent Scatterers in Differential SAR Interferometry. *IEEE Transactions on Geoscience & Remote Sensing*, 38, 2202–2212 (2000).

Li X.E., Zhou L., Su F.Z. *et al.* Application of InSAR Technology in Landslide Hazard: Progress and Prospects. *National Remote Sensing Bulletin*. 25, 614–629 (2021).

Liao M.S., Zhang L., Shi X.G. *et al. Method and Practice of Landslide Deformation Monitoring by Radar Remote Sensing*. Beijing: Science Press. (2017).

Shi X., Liao M., Li M. *et al.* Wide Area Landslide Deformation Mapping with Multi-path ALOS PALSAR Data Stacks: A Casc Study of Three Gorges Area, China. *Remote Sensing*. 8, 136 (2016).

Strozzi T., K1imes J., Frey H. *et al.* Satel1ite SAR Interferometry for the Improved Assessment of the State of Activity of Landslides: A Case Study From the Cordilleras of Peru. *Remote Sensing of Environment*. 217, 111–125 (2018).

Su L., Hu K., Zhang W. *et al.* Characteristics and Triggering Mechanism of Xinmo Landslide on June 24, 2017 in Sichuan, China. *Journal of Moun-lain Science*. 14, 1689–1700 (2017).

Xu Q., Li W.L., Dong X.J. *et al.* The Xin-mocun Landslide on June 24, 2017, in Diexi Mao-xian, Sichuan; Characteristics and Failure Mechanism. *Chinese Journal of Rock Mechanics and Engineering*. 11, 2612–2628 (2017).

Yin Y.P., Wang W.P., Zhang N. *et al.* Long Runout Geological Disaster Initiated by the Ridge-Top Rockslide in a Strong Earthquake Area: A Case Study of the Xinmo Landslide in Maoxian County, Sichuan Province. *Geology in China*. 44, 827–841 (2017).

Zhang L., Liao M.S., Dong J. *et al. Early Detection of Landslide Hazards in Mountainous Areas of West China Using Time Series SAR Interferometry—A Case Study of Danba. Sichuan*, Geomatics and Information Science of Wuhan University. 43, 2039–2049 (2018).

Zhang Y., Meng X., Chen G. *et al.* Detection of Geohazards in the Bailong River Basin Using Synthetic Aperture Radar Interferometry. *Landslides*. 13, 1273–1284 (2016).

Constructional Engineering and Ecological Environment – Chih-Huang Weng (Ed)

Planning methods for anti-epidemic residential quarters after the epidemic

Huajin Zhou*
PingYang Natural Resources and Planning Bureau, Wenzhou, China

ABSTRACT: Residential quarters, in addition to providing residential functions, bear more social significance. Especially during the epidemic period, a small residential community is a small social collective. If the city is compared to a living body, then the community is the cells that make up the living body. In the era of the epidemic, the residential area has become a container for people during the city's lockdown period, a distribution center for materials, information, and management. It can be said that if the epidemic situation in the community is well-managed. The built community lacks necessary supporting facilities, such as pharmacies, fitness venues, material distribution centers, and express processing facilities, and even the lack of planning and implementation of the routes for the epidemic personnel to walk. The untimely input and disorderly management have caused residents to complain more, physically starve, and bear the burden of the lockdown and restricted freedom in their hearts. The article uses the crawler technology, uses the data housekeeper to obtain the most urgently needed help information of the residents of the community, and generates the word cloud map through the word segmentation tool, from which the most urgent needs of the residents are the lack of materials. From the perspective of people's needs, plan and think about how to do community planning in the post-epidemic era and think about making planning contributions to people's most urgent needs. That is, in community support, epidemic prevention route setting, community green environment creation, apartment space design, etc. Changes to adapt to major event emergencies such as epidemics and even wars.

1 CONCEPT

In this paper, the community refers to the community with the property management company or the community owner's committee as the main body of management, and its upper-level managers are the community neighborhood committees and sub-district offices in the traditional sense.

2 RELATED RESEARCH AND INNOVATION

From the analysis of the situation since the outbreak of the epidemic in Shanghai, the authority of the neighborhood committee is limited, and it can only play the role of uploading and issuing. Many things cannot play a decisive role, resulting in slow information flow and poor anti-epidemic effects. This requires community residents to voluntarily unite to save themselves. Such self-rescue actions require social circles of acquaintances. Generally, each community has its own WeChat owner group and other communities. Daily

*Corresponding Author: 523401843@qq.com

 DOI: 10.1201/9781003410843-39

interaction and communication make the residents in the community naturally become a closed community. In a society of acquaintances, once an event such as the closure and control of the community occurs, it becomes possible for the residents of the community to manage themselves, and this requires guidance and training before the outbreak. It is difficult for a clever woman to cook without rice. The people in the acquaintance society and the community environment are available, but the supporting facilities in the community are not complete. The most important thing is the lack of material reserves, which leads to overwhelmed self-care. In the face of the epidemic, each community is vulnerable, and the epidemic is not in the community. The organization spreads.

The current planning of residential areas is based on the "Planning and Design Standards for Urban Residential Areas" GB50180-2018, and its relevant regulations do not involve major public health events such as epidemic prevention. Its setting specifications and configuration requirements only serve daily life. Living needs, and it is to meet the concept of a certain life circle. During the epidemic period, many service stores in the life circle were also forced to close their business. What residents need is more to solve the problem of the last mile, not life. The problem of insufficient circle matching (Xu et al. 2020). Community planning is closely related to health events. In his famous book The City of Tomorrow, Le Corbusier wrote: "Hygiene and Moral Health are determined by the planning and arrangement of cities. Without sanitation and moral health, social cells will shrink (Le 1987)."

In the early 20th century, pollution in industrial revolution cities led to infectious diseases, which introduced public health into urban planning. After the 1840s, chronic diseases appeared in people's field of vision, and the community was an important spatial unit that affected the healthy environment and people's behavior. The practice of a "Healthy Community" emerged and continued to innovate (Barton et al. 2003). In 1986, WHO proposed a series of action plans to promote community health (WHO 1986). In 2007, British scholar Mala Rao discovered the "disorders related to urban architectural space" (Rao et al. 2007). The SARS and new crown epidemics in recent years have once again warned people of the importance of public health construction, and the community is an important unit of public space. Scholars have made a lot of explorations on how to do a good job in community planning. Focus on theoretical aspects, such as proposing community planning guidelines and architectural design guidelines (Yu et al. 2020). There are also suggestions for prevention and control of the spatial structure of the community (Qin 2021). In recent years, scholars have returned to study the community itself, focusing on community development, resilient communities, and green spaces. Domestic research has focused on the construction of evaluation criteria for healthy communities (Zhu 2019) and method research (Chen et al. 2020; Wang 2015). However, less thinking is made from the perspective of human needs, and less research is done using big data methods. The innovation of this paper is that people-oriented, with the needs of people as the first starting point, to study the areas that need to be improved in epidemic prevention community planning. The technology uses data stewards to analyze data and refine people's needs as the basis for planning and thinking about epidemic prevention communities.

3 RESEARCH METHODOLOGY AND DATA ANALYSIS

The article collected 1,348 original Weibo data from April 30 to May 31, 2022, through the data steward, using the keyword "help in Shanghai epidemic situation," and used the word segmentation tool of the data steward to sort out the collected data. And analysis to generate a word cloud map and a social network relationship map, respectively.

It is found from the word cloud map that the words "we," "community," "nothing," "materials," "nucleic acid," "isolation," "positive," "hospital," "resident," "video," "elderly person," etc. to represent the word with the most frequent occurrences of the participle.

Among them, "we" appeared 742 times, "community" appeared 836 times, "no" appeared 758 times, "materials" appeared 594 times, "nucleic acid" appeared 668 times, and "quarantine" appeared 492 times. "Positive" appeared 514 times, "hospital" appeared 416 times, "resident" appeared 394 times, "video" appeared 346 times, and "elderly" appeared 274 times (see Table 1). The content of help and concerns related to the word "community" is "help the elderly in the community," "the community environment is very bad," "enter the community casually," "unblock the community without forgetting," "and the community is positive without transfer," "Community management is chaotic," etc.; Taking "materials" as an example, more helpful information is "material shortage," "materials not received," "vegetable materials not distributed," "where are the materials going" and so on (see Table 2).

Table 1. Part of the data of the word selection result is compiled by the author from the data steward [Self-painted].

Tag words	Village	Nothing	We	Nucleic acid	Material	Quarantine	Hospital	Resident	Video	The elderly	
Word frequency	836	758	742	668	594	492	416	394	346	274	... ...
Document frequency	504	574	452	424	384	308	326	272	338	176	

Table 2. Part of the data in the word selection matrix is compiled by the author from the data steward [Self-painted].

Serial Number	Text	Village	Material
23590	#Baoshan materials # # Sitang second village # # Shanghai epidemic help # the elderly have no food, the epidemic has been isolated at home for a month, director Zhou of the neighborhood committee never comes out, the management of the community is in a mess, and the security materials in the neighborhood committee are rotten by more than 100 boxes, @ xuankexuan @ STV newsroom @ Baoshan, Shanghai @ Baoshan, Shanghai @ Xinhua	1	2
23600	Shanghai rescue Yangpu District Changbai Xincun Street Neijiang road 476 Lane Community!!! 1. There is a shortage of materials in the community. Since the closure of Puxi on April 1, only five materials (plus a bag of rice) distributed by the neighborhood committee have been received. None of them has received real and real meat (the so-called meat is replaced by three cans of luncheon meat). The fifth time the materials were delivered, the ducks in empty packages were still spoiled and contained colonies. Data disclosure	2	3
23899	#Shanghai epidemic # # Shanghai epidemic help # that is to say, if you don't have food, you can apply for materials. People in the community can help you. If you don't have money to pay the rent, you can borrow it from family and friends. If you don't have money, you can borrow it. There are many ways and means, such as JD white slips and credit cards. How can you send a collection code to expect strangers to help? How can it be so kind? People's anti-epidemic help is a matter of life. Do you beg on this line?	1	1

(continued)

Table 2. Continued

Serial Number	Text	Village	Material
24099	After the static management started in Shanghai on March 28, 2022, all materials distributed by a Pudong community are shown in the figure. The distribution times are April 5, 11, 18, and 27. For residents who do not stock up on food, if they can't grab food, can't use their mobile phones, or have no source of income at home, can they eat enough for a month? For residents who can't cook without pots and pans, or even without a stove, can these be eaten raw@ Released in Shanghai	1	1
24657	#Shanghai epidemic # # Shanghai epidemic help # # Shanghai supplies # # food grabbing anxiety under the Shanghai epidemic # # Shanghai # # things around Shanghai # just because there is a confirmed diagnosis on the floor, all the residents on the floor of my parents' house have been sealed. Even the takeout has not been delivered. Do you want to starve people to death at home? No nucleic acid has been done in the last week, only antigen has been sent. Did the lax epidemic prevention in the community lead to the absence of the original building?	1	1
24987	#Shanghai epidemic # # Shanghai epidemic help # # Shanghai supplies # # food grabbing anxiety under the Shanghai epidemic # # Shanghai # # things around Shanghai # just because there is a confirmed diagnosis on the floor, all the residents on the floor of my parents' house have been sealed. Even the takeout has not been delivered. Do you want to starve people to death at home? No nucleic acid has been done in the last week, and only antigen has been sent. Did the lax epidemic prevention in the community lead to the absence of the original building? In the last 10 days, 30 netizens have contributed: the oldest is 55 years old, and the youngest is only seven months old. Materials are unexpectedly robbed. It is said that the shelter originally only accepts 600 people and has accepted 1200 people today without notice. When we had a fever a week ago and wanted to save our lives, no one paid attention to us. Now we have turned overcast, but we have to be forcibly dragged to the shelter hospital by the community. The big white in the shelter asked why we brought a seven-month-old baby here. I also want to ask why I came	1	1
25557	#Shanghai epidemic help # be a person who listens to the arrangement, that is, pinduoduo's food is not allowed in. Meituan blocked the community yesterday, and it can't grab the food every morning. There are various problems with the things sent by the community. The group purchase is limited to 5 per day, and the leader must distribute them himself and provide protective articles. (now, except for the leader who has made money, few people in the community have done it, and they have to paste it upside down.) the community said that there was no news after the statistics of the direct docking of materials, Jd.com placed an order at the beginning of this month, but it hasn't done so yet	4	1
… …			

After the co-word matching analysis of the data housekeeper, the result of the co-word matrix is as follows, and the results of some words of the data housekeeper are intercepted.

Finally, a visual social network relationship diagram has been formed. Click on the social network relationship diagram. The research found that by clicking on the network relationship

Table 3. Part of the data of the co-word matrix is compiled by the author from the data steward [Self-painted].

	Nothing	Village	We	Positive	Nucleic Acid	Material	Maybe	Quarantine	Hospital	Resident
Nothing	574	152	132	110	116	108	88	72	84	76
Village	152	504	154	108	100	130	54	72	42	166
We	132	154	452	78	100	90	50	70	50	78
Positive	110	108	78	360	154	32	28	94	42	64
Nucleic Acid	116	100	100	154	424	22	24	104	48	56
Material	108	130	90	32	22	384	40	28	20	80
Maybe	88	54	50	28	24	40	282	34	40	18
Quarantine	72	72	70	94	104	28	34	308	38	42
Hospital	84	42	50	42	48	20	40	38	326	16
Resident	76	166	78	64	56	80	18	42	16	272

Figure 1. Word clouds figure[Self-painted].

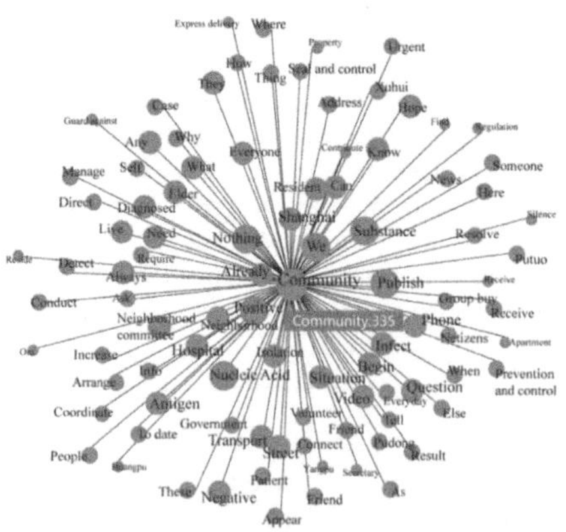

Figure 2. "Community" social network diagram [Self-painted].

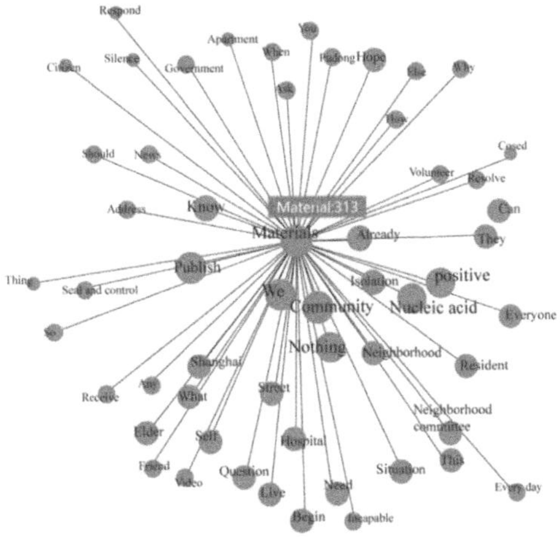

Figure 3. "Material" social network diagram [Self-painted].

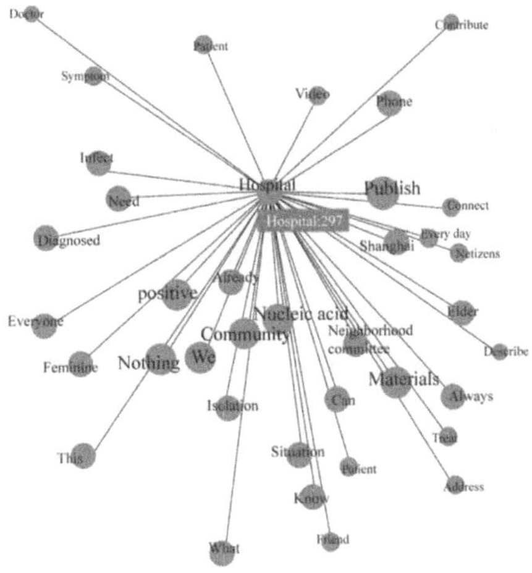

Figure 4. Social network diagram of "hospital"[Self-painted].

diagram of "community," "materials," and "hospital," the main ones related to the community are "no," "materials," "nucleic acid," and "positive" show that the residents of the community who were closed during the period of silence throughout the city had a strong demand for supplies and various problems such as difficulties in seeking medical treatment travel difficulties for residents, and discontinuation of medicines for the elderly with chronic diseases, coupled with disorderly nucleic acid testing queues The methods, positive news reports and video reposts have made the residents of the community even more uneasy.

4 ANTI-EPIDEMIC COMMUNITY PLANNING

Before the outbreak of the epidemic, community planning pursued the maximization of economic value and was guided by maximizing the building area. High-density, high-volume ratio, and high-rise buildings became a common construction model. The architectural design description content of modern community planning mainly includes architectural design, structural design, special articles on green building energy conservation, water supply, electrical, drainage (HVAC, fire protection, civil air defense), and other professional special design, as well as accessibility, sponge city design, etc., but there is no epidemic prevention design involved in the description. One is that there is no standard and specification for epidemic prevention buildings as a basis; the other is traditional design thinking, which leads to the content of epidemic prevention design in the architectural design text of the community, even if there are some sporadic descriptions of epidemic prevention design, it is extremely simple and scientifically reasonable. Discuss. The epidemic prevention community is a systematic project. No matter the design of the physical space or the construction of a small community, many things need to be configured, which are massive, and it is impossible to start. However, people live in the community, and people's needs are clear. Thinking from the perspective of people's needs can greatly reduce the difficulty of community planning research. From the above data analysis, it is known that the following aspects are urgent and needed in the future community planning and setting, that is, planning from the aspects of community support, intelligent contactless traffic, community plant configuration, and epidemic prevention apartment building design.

5 REFLECTION ON COMMUNITY PLANNING

5.1 *Living facilities*

The modern community is mainly based on a single residential function, and the existing supporting facilities are set up to meet the basic starting point of life. There is a lack of corresponding complex functional supporting facilities at your fingertips. Although the latest urban residential area planning and design standards stipulate solutions for supporting facilities in five-minute, ten-minute, and fifteen-minute living circles, the requirements are "relatively centralized layout according to their service radius," and such facilities serve the minimum number of residential units. To be in a residential area with more than 300 units is in conflict with the requirements of epidemic prevention, home isolation, and non-essential not leaving home. It is also different from the community referred to in this article. In the post-epidemic era, living facilities are required to be dispersed and easily accessible to residents. What is most needed is to solve the last mile problem. The regulations on cultural and sports facilities in the "Zhejiang Provincial Residential District Public Cultural Facilities Supporting Construction Standards" issued by Zhejiang Province are worthy of reference and study. The coefficient is required for matching construction, which can solve the problem of supporting facilities when the scale of the community in the following areas, such as the county seat, is small and the number of residential units is less than 300. However, from the perspective of this epidemic, there are problems such as difficulty in shopping for food, difficulty seeing a doctor, difficulty buying medicine, and difficulty picking up express delivery.

About 20% of the community health service centers in the streets and towns of Shanghai do not have wards, and about 50% do not meet the set standards; during this epidemic, among the first batch of 110 fever clinics opened in the city, only some community health service centers in the suburbs were used as supplements medical institutions. In addition, there is still a 30% gap in the scale of family doctors in the city, and the first consultation rate of community health centers is far lower than that of developed countries. Therefore, in the post-epidemic era, the supporting facilities of the community also include fresh vegetable shops, small clinics, pharmacies, express delivery, and disinfection mail rooms, etc. (Zhang et al. 2020) At present, Zhejiang's future community construction pilot program, especially the proposal of "three modernizations and nine scenarios," has solved some of the pain points of residents' needs in epidemic prevention communities. However, it is not comprehensive enough, and there is still a lack of configuration in terms of purchasing materials, purchasing medicines at pharmacies, and seeing a doctor for minor diseases. Therefore, in terms of community planning and design standards, increase the living facilities oriented to people's needs. For example, when building new communities, encourage the addition of necessary supporting facilities such as micro supermarkets, small pharmacies, and small clinics, and give a certain price when the land is sold. Preferential concessions, or when formulating land transfer conditions, it is stipulated that certain necessary supporting facilities must be set up as the preconditions and basis for community construction. In old communities, some idle properties and other buildings are rebuilt to add movable partitions to isolate living areas. For supporting facilities space, certain public spaces for supporting facilities can also be set up in some green grasslands to meet the living needs of residents during the lockdown period and achieve the purpose of combining epidemic control.

5.2 *Emergency access*

Residents are connected to the community from the moment they enter the community, both spatially and temporally. Residents need a complete monitoring and disinfection system from access control to the entry door. The first is the division of lines. The planned community generally has two entrances and exits. During the epidemic period, a designated channel for epidemic prevention personnel and a channel for healthy personnel. Automatic temperature test doors are set at the two passage openings of the community, among which a spray disinfection space and a purple light disinfection space are added in the A channel, and the B channel only has a temperature measurement device. If the key susceptible people enter the community through the A channel, if the general non-risk population enters the community through the B channel. If people with abnormal temperatures occur, they will seek medical treatment in time, and the clinic is also divided into zones. The first area is the medical area for healthy people, and the second area is the medical area for risk groups. Receive an initial diagnosis at a community clinic for continued home isolation or out-of-community isolation (treatment). Judging from the current progress of the epidemic, the symptoms are mild, like a fever and a cold, and there is an asymptomatic infection. In this case, the community clinic can completely solve the problem and relieve the shortage of social medical resources so that the resources of large general hospitals can be reserved for people with serious diseases. In addition to the access control system at the entrance of the community, fully automatic spray disinfecting and purple light disinfecting systems are further installed at the basement elevator entrance, lobby entrance, and resident entrance, especially at the high-rise elevator entrance, the light curtain will automatically appear to disinfect the whole body. At the same time, with the development of 5G, in these access control places, face recognition is carried out without contact to reduce the spread of viruses. In addition to setting up a sterilization system at the entrance of the residents, especially at the entrance, in the design of the residential unit, increase the full level of the entrance to increase the space for epidemic prevention, coat disinfection area, and disinfection and hand washing area. In addition to the joint efforts of architects, the formulation of new residential design standards is also a necessary prerequisite for such a design.

5.3 *Green plant configuration*

Attention is given to the requirements of green space construction and plant configuration. Many traditional Chinese medicines are plant components, which is enough to prove that plants have natural antibacterial effects. The WHO also pointed out that traditional Chinese medicine can effectively treat the new crown (Li 2022). Studies have shown that certain green plants can release volatile organic compounds (VOCs) (Volatile Organic Compounds), which are named "phytoncides." For example, the volatiles of lemon eucalyptus can effectively inactivate various pathogens such as tuberculosis and encephalococcus. Cinnamon oil volatilized from cinnamon, cypress, thick plum, cedar, geranium, and other plants can play a role in sterilization and disinfection. Each acre of pine and Berlin can volatilize more than two kilograms of bactericides every day. Phytocides can significantly increase the activity of NK cells and significantly enhance the expression of perforin, GrA, and GRN.NK cells are important immune cells in the body that play a role in anti-tumor, anti-viral infection, and immune regulation. Phytocides can also enhance cellular immune activity. Therefore, deploying effective plants in the community is beneficial to the human body no matter in the stage of epidemic prevention or the stage of daily healthy life (Zhang 2020). The sterilization properties of different plant configurations are different. Studies have shown that arbor-shrubbing has the obvious bacteriostatic effect, and the bacteriostatic effect of osmanthus x heather + camellia x sea tungsten + Phnom Penh boxwood community is stronger than other arbor-shrub configurations. Therefore, cultivating a certain plant configuration landscape architect is a future direction. In addition, it is necessary to draft and formulate the community greening plant configuration standard and the recommended list of plant species. In addition, it is also recommended method to plant some edible fruit trees inside the community, such as planting pear trees, apple trees, etc., so as to have fruits all year round. It's also a nice innovation, and it's a good choice when something big happens, like public health or war (Luo & Li 2010).

Certain plant landscaping also has a positive impact on people's psychology. Whether it is the landscaping of Western plant application forms or the oriental plant artistic conception landscaping method, it can give people the enjoyment of physical and mental beauty. At the same time, international studies on the built environment and health have shown that there is a certain connection between green space and disease. Therefore, adding three-dimensional greening and sky garden space is friendly to future buildings and living environments. In addition to the greening of the community, roof greening, balcony greening, and vertical space greening of buildings are friendly to private spaces. Planting plants that release bactericides is beneficial to human health and can also purify the air.

5.4 *Living space environment*

Compared with the air in nature and the large environment of the community, in terms of the small environment, the design of residential units and the design of HVAC is more important links because from the perspective of a human lifetime, more than half of the time is real. The sleep is spent at home. Since the outbreak of the epidemic, there have been frequent reports of infections from unknown sources of transmission on different floors of a building, and the WHO survey results of the Amoy Garden incident show that the traditional model of room design, HVAC and toilet (bathroom) exhaust systems have not been affected by the epidemic. When it comes, there is a significant risk of transmission. Therefore, changing and considering the cross-border design of health and safety has become a demand. First, reduce the design of patios to reduce vertical airflow. Second, avoid the use of circulating central air conditioners, adopt fresh air ventilation systems and add disinfection control technology, and use window screens to filter virus materials. Finally, Japan's overall bathroom technology, and toilets. The integrated technology of sanitary and epidemic prevention and health and safety performance assurance, such as the independent horizontal exhaust system of the bathroom and the single-family parallel and same-floor systemized and maintenance-replaceable building infill drainage items, is worth learning and referencing (Liu 2020).

Community planning is a huge system. Changes are not only made in physical space but also in soft environments, such as design specifications and design methods. Focus on thinking about how to change the traditional residential building design model. Update the residential community specification, increase the design of epidemic prevention content, and add epidemic prevention chapters and healthy community content in residential community design so as to ensure that the residential community can still maintain a certain resilience in the face of epidemics and even wars, so as to make the community more suitable for people to live in healthily Ability.

6 CONCLUSIONS

The construction of the city is built according to the needs of the people, and the people living in the community are mainly the people of the city. The progress of urban planning is inseparable from the occurrence of public health incidents and countermeasures. The planning and development of residential quarters are also related to public health affairs. A healthy living environment for people requires healthy community planning. The epidemic prevention community is built with the needs of people as the starting point. Through the data analysis of the data steward, it is found that planning a healthy epidemic prevention community requires the construction of living physical spaces such as pharmacies, fresh supermarkets, and small clinics to ensure life and requires contactless access. Achieving the smart system for entering the home requires a scientific green planting community-building environment. The new apartment building design scheme with an advanced fresh air ventilation system and toilet and bathroom drainage system is another aspect that needs to be planned. Half of the time is spent in sleep in the bedroom.

REFERENCES

Barton H, Grant M & Guise R. (2003). *Shaping Neighborhoods for Health, Sustainability, and Vitality. M.* London: Routledge.

Chenyu Zhu. (2019). *Establishment and Feasibility of Healthy Community Evaluation Index System.* D. Shenyang: Shenyang Jianzhu University.

Chun Chen, Xi Chen & Zhirong Luo. (2020). Research on the Impact of Community-built Environment on Respiratory Health. *J. Planners.*36(9):71–76.

Dongwei Liu. (2020). Focusing on Sanitation, Epidemic Prevention, and the Safety and Health of Residential Buildings. *J. China Construction News.* 25 (2):008.

Fan Zhang, Minqing Zhang & Suqian Guo. (2020). Planning Methods for Shanghai Communities Confronting Major Public Health Risk Affair. *J. Shanghai Urban Planning.*02.

Ji Li. WHO. (2022). *Traditional Chinese Medicine Can Effectively Treat Covid-19.* EB / OL. https://m.gmw.cn/baijia/2022-04/05/1302883353.html.

Jianyang Qin. (2021). Study on the Influence of City Community Spatial Structure on CoVID-19 Prevention and Control, *Thesis of Professional Master's Degree.*06.

Le C. (1987). *The City of Tomorrow and its Planning.* M. New York: Dover.

Leiqing Xu, Zhen Zhang & Mengqi Li. (2020). *Guidelines for Community Planning of Epidemic Prevention Based on Risk Management.* Beijing Planning and Construction. 04.

Rao M, Prasad S & Adshead F. (2007). The Built Environment and Health. *J. The Lancet.* 370 (9593): 1111–1113.

WHO. *Ottawa Charter for Health Promotion* [M].1986.

Yi Wang. (2015). Community Planning Guided by Healthy City. *J. Planner.* 31(10):101–105.

Ying Luo & Xiaochu Li. (2010). Study on the Bacteriostatic Function of Plant Community Configuration Mode in Residential Green Space. *J. Forestry Science and Technology Development.* 24 (2): 61–64.

Yu Li, Ying Liang, Yu Li & Liang Y. (2020). Community Planning for Epidemic Prevention. *J. Architectural Skills.*05.

Zhe Zhang. (2020). Impact and Reflection of Urban Green Space on Public Health in the Context of the Epidemic. *J. Urban and Rural Construction.* 15: 42–44.

Constructional Engineering and Ecological Environment – Chih-Huang Weng (Ed)

Suspended sediment transport mechanism in the West shoal of Pearl river Estuary in the dry season

Yulong Xiong*, Wanhua Yuan, Meixuan Li & Shaobo Wang
Bureau of Hydrology and Water Resources, Pearl River Water Resources Commission of Ministry of Water Resources, Guangzhou, China

ABSTRACT: The west shoal of Lingdingyang bay is a vital sediment channel and trap, which is also a natural barrier to protect the Lingdingyang deepwater channel. According to the measured data and hydrodynamic simulation results, the dynamic characteristics, sediment transport, and deposition characteristics were discussed to explore the transport mechanism of suspended matter in the study area in the dry season, which can provide an essential theoretical basis for waterway regulation and engineering construction in this area. The results show that: 1. In the dry season, West shoal is a flood-dominated region, and vertical circulation is more evident on the south side of point 4# and the Lingding waterway on the eastern side of West shoal, where the residual current of the bottom and the middle layer is from sea to land, and the surface is from land to sea. 2. Flood and ebb channels strengthen the asymmetry of ebb and flood current, thus affecting the suspended sediment transport and deposition. 3. West shoal has a dynamic environment with bidirectional trapping sediment. In the ebb period, river flow diffuses in the beach surface and is filtered by the estuarine front. The two processes quicken sediment deposition. In the flood period, the sediment is transported from sea to land, and the fresh and saltwater mixing environment is conducive to suspended sediment deposition. 4. According to the suspended sediment transport mechanism, horizontal flow sediment transport is primary, followed by tidal trapping and vertical circulation transport. 5. Human activity has a more significant impact on suspended sediment transport. After 2000, large-scale port construction, channel dredging, and other water engineering increased the suspended sediment concentration transported by the tidal current to the West shoal and deposited. Silt strength in 2000 was 2.65 times than before.

1 INTRODUCTION

The estuary is a transition zone from the river to the ocean, influenced by river and ocean dynamics. The vulnerability of the estuary is becoming increasingly evident under the influence of the high-intensity impact of human activities. In recent years, due to the effect of reservoir sand interception and river sand extraction in the middle and upper reaches of the estuary, the amount of incoming sand in the estuary has decreased sharply (Chen 2002; Liu *et al.* 2008). Human interference, such as the construction of ports and waterways in the estuary section, has increased the content of suspended sand. The transport process of estuarine suspended sand includes flocculation and maximum turbidity zone (Eisma 1986), estuarine circulation (Dyer 1996; Schubel 1968) and tidal action (Jay & Smith 1990; Shi 2010), geomorphological processes, etc. Human activities also become an essential factor.

*Corresponding Author: 874123818@qq.com

 DOI: 10.1201/9781003410843-40

Estuaries are essential sediment traps (Dyer 1986). In tidal estuaries, the tide is the primary driver of sediment transport, upstream at flood tide and downstream at ebb tide (Prandle 2004). Subtidal, intertidal, and supratidal zones are formed by adding fine-grained sediment through multiple mechanisms. The maximum turbidity zone is an important process of sediment trapping, and the main processes include flocculation, resuspension, flocculation, and deposition (Franz *et al.* 2012). Maximum deposition occurs during the slack flow phase when the flow velocity is intensified, and the bottom shear is smaller afterward. Maximum erosion occurs when the flow velocity is higher between ebb and flood (van Rijin 2012; Wu *et al.* 2012).

2 STUDY AREA

West shoal of Lingdingyang is an important buffer zone between the East Three Gates and Lingdingyang, a transport channel for water and sand from Hongqi Gate, Hengmen, and Jiaomen, as well as an essential barrier to protect the deep water channel of Lingdingyang and a barometer of human activities such as dredging and sand mining in the upstream estuary and remediation of the deep water channel of Lingdingyang downstream.

The west shoal of Lingdingyang is at the front edge of the Pearl River Delta, bearing the water and sand of Jiaomen, Hongqimen, and Hengmen on the top and the Lingding channel on the east. The west shoal can be divided into three parts, including the shoal on the south side of Jibao Sand, the West shoal, and the shoal on the northwest side of Qiyao Island (Figure 1).

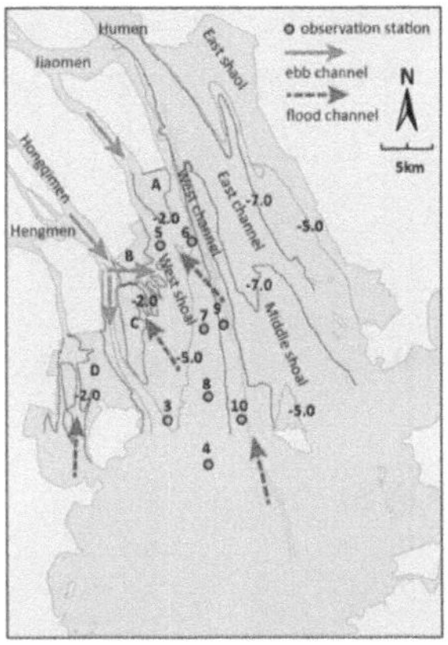

Figure 1. Topography of Lingdingyang and location of observation points.

3 METHODS

3.1 *Field observation data*

In this study, the observation data of November 2005 from the Lower Yangtze River Hydrological and Water Resources Survey Bureau of the Hydrological Bureau of the Yangtze River Water Resources Commission were used. The observation period was from 10:00 on November 3, 2005, to 15:00 on November 4, 2005, and the test items included current, water level, salinity, and suspended sand. The representative hydrological combination was the dry season flood tide (Figure 1).

3.2 *Numerical simulation*

In our study, the hydrodynamic model was set up as a two-dimensional, depth-averaged model. In a depth-averaged 2D configuration, the model is based on the numerical solution of the depth-integrated incompressible Reynolds averaged Navier-Stokes equations. The models consist of continuity, momentum, temperature, salinity, and density equations. A Boussinesq assumption is applied in our model, in which it is assumed that momentum transfer caused by turbulent eddies can be modeled with an eddy viscosity. In our study, a 2D model including the 50m depth range from the Badakou gates to the outer sea was established, and the model validation results were good (Figure 2).

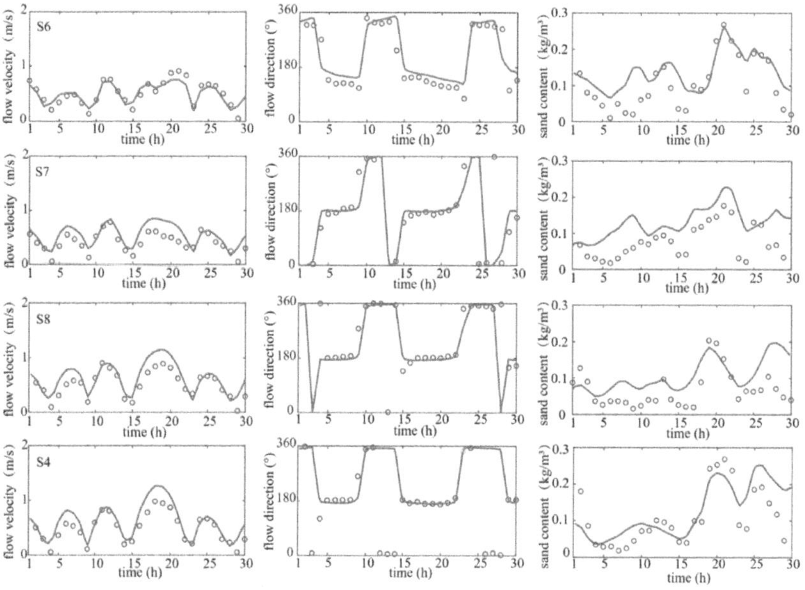

Figure 2. Numerical model validation.

4 RESULTS AND DISCUSSIONS

4.1 *Dynamic characteristics*

As is shown in Figure 1, in the dry season, the seaward side of the shallow 2m water depth is dominated by flood tide dynamics, the landward side is dominated by ebb-tide dynamics, and the shallow 2m water depth is relatively balanced between high and ebb-tide dynamics and becomes the main sediment capture area.

In the dry season, the water coming from the upstream is low, the tidal water in the river area of the delta network is vast, the tidal storage volume is large, the tidal capacity of the river network is large, and the tide on the beach surface is high.

The flow velocity is larger. During the ebb-tide period, because of the faster and smoother convergence of the water in the net river area into Humen through these lateral branches of the Upper Hengli, Lower Hengli, and Eider Island waterways (the horizontal ratio drop is greater than the vertical ratio drop), the advantage of convergence during the ebb-tide period at West shoal is less than that during the flood tide period, and the geomorphological features of the flood tide channels developed on the shallows also indicate the predominance of flood tide dynamics.

In the dry season, West shoal is controlled by the mixed water of the estuary, which is an interface between the outer high-salt shelf water and the freshwater inside the mouth, with the high salinity in the northwest and low salinity in the southeast, and the salinity values range from 2‰ to 24‰. Figure 3 shows the process line of flow velocity, sand content, and salinity change at observation point 7. The maximum flow velocity appears in the rapid rising period of 11~12 hours, the maximum flow velocity in the surface layer is close to 1m/s, and the flow velocity is around 0.5m/s most of the time. The value of bottom salinity fluctuation is small, generally around 20‰. The maximum value of sand content in observation point 7 appears in the fall rest stage.

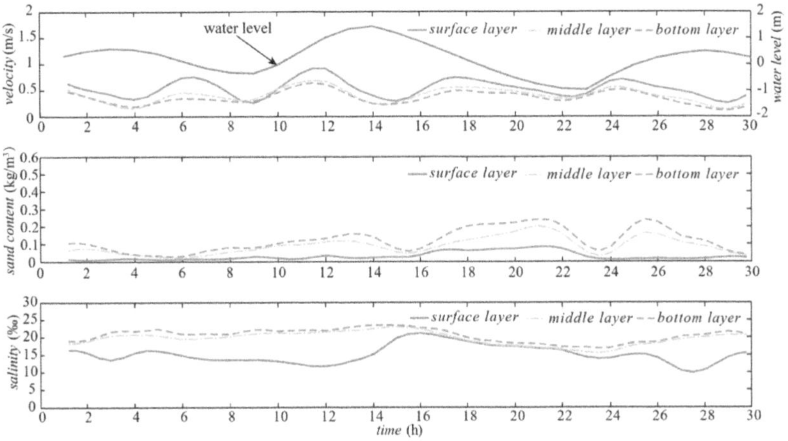

Figure 3. Variation process of flow rate, sand content, and salinity at observation station 7.

Figure 4 shows the distribution of the residual flow vertical structure, the lower end of the shoal 4# and the southeast side of 9# and 10# measurement points due to the influence of

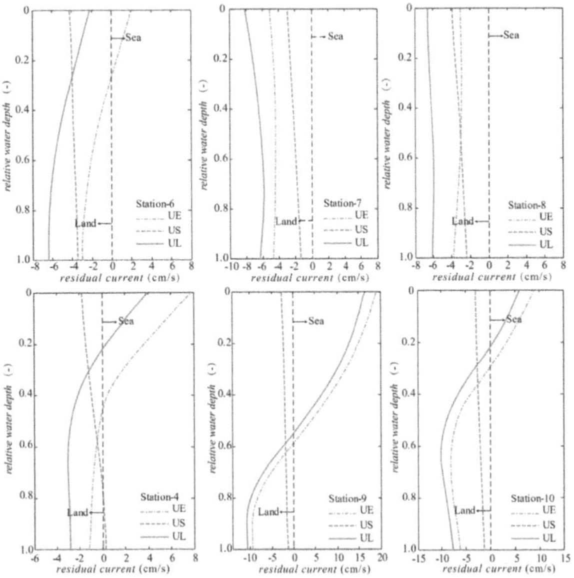

Figure 4. Vertical distribution of residual flow in the dry season (blue line represents Eulerian residual flow, the orange line represents Lagrangian residual flow, and the green line represents residual flow).

the vertical circulation caused by the oblique pressure, the residual flow of the middle and bottom layer to land, the surface residual flow to the sea, the middle and upper part of the shoal residual flow to land.

4.2 *Suspended sand transport*

4.2.1 *Suspended sand transport mechanism(Li & Zhang 1998; Li 1986; Uncles et al. 1985)*

The instantaneous flow velocity can be approximated as the sum of the vertical average term and the deviation term:

$$u(x, z, t) = \bar{u} + u' \tag{1}$$

where $\bar{u}$ and u' can be further decomposed into the sum of the mean tidal term and the tidal deviation term:

$$\begin{cases} \bar{u} = \bar{u}_0 + \bar{u}_t \\ u' = u'_0 + u'_t \\ u(x, z, t) = \bar{u}_0 + \bar{u}_t + u'_0 + u'_t \end{cases} \tag{2}$$

The salinity can be expressed as:

$$c(x, z, t) = \bar{c}_0 + c'_t + c'_0 + c'_t \tag{3}$$

The water depth can be expressed as:

$$h(x, t) = h_0 + h_t \tag{4}$$

Single-width water body transport volume:

$$\begin{cases} \langle Q \rangle = h_0(\bar{u}_E + \bar{u}_s) = h_0\bar{u}_L \\ \langle Q \rangle = \dfrac{1}{T}\displaystyle\int_0^T \int_0^1 uhdzdt = \bar{u}_0 h_0 + \langle \bar{u}_t h_t \rangle \\ \bar{u}_E = \bar{u}_0 \\ \bar{u}_s = \langle \bar{u}_t h_t \rangle / h_0 \\ \bar{u}_L = \langle Q \rangle / h_0 = \bar{u}_E + \bar{u}_s \end{cases} \tag{5}$$

where $\bar{u}_E, \bar{u}_S, \bar{u}_L$ are the Eulerian, Stokes, and Lagrangian residual flow.

Single wide material transport volume:

$$\begin{aligned} \frac{1}{T}\int_0^T \int_0^h ucdz_0 &= \frac{1}{T}\int_0^T \int_0^1 uhcdzdt \\ &= \underset{T_1}{h_0\bar{u}_0\bar{c}_0} + \underset{T_2}{\langle h_t\bar{u}_t \rangle \bar{c}_0} + \underset{T_3}{\langle h_t\bar{c}_t \rangle \bar{u}_0} + \underset{T_4}{\langle h_t\bar{u}_t\bar{c}_t \rangle} + \underset{T_5}{h_0\overline{u'_0c'_0}} + \underset{T_6}{\langle h_t\overline{u'_0c'_t} \rangle} + \underset{T_7}{\langle h_t\overline{u'_tc'_0} \rangle} \\ &\quad + \underset{T_8}{\langle h_t\overline{u'_tc'_t} \rangle} \end{aligned} \tag{6}$$

where T_1 is material transport due to mean flow; T_2 is Stokes drift term; $T_1 + T_2$ is the advection term; T_3 is tidal variation related term of tidal and material content; T_4 is material concentration related term of tidal variation (tidal trapping term, tidal pumping term); T_5, T_6, and T_7 are the shear terms due to time-averaged volume and tidal oscillation shear; T_8 is the tidal oscillation shear term.

The mechanism decomposition method is an excellent way to study the transport of suspended sand in tidal estuaries, which is highly related to tidal motion, and the calculation principle can be found in the related literature (Chen 1993; Li 1986; Uncles *et al.* 1985). The mechanism decomposition method decomposes sand content into 8 terms: T_1 is mean flow induced material transport; T_2 is Stokes drift term; T_3 is tidal variation related term of tide and material content; T_4 is tidal variation related term; T_5 is vertical flow velocity variation and material content related term; T_6 and T_7 are shear terms due to time-averaged volume and tidal oscillation shear; T_8 is vertical tidal oscillation shear effect (Table 1).

Table 1. Decomposition terms of sand content and their percentages (+ means seaward, -means landward).

Points	T_1	T_2	T_3	T_4	T_5	T_6	T_7	T_8	T_9
4	$2.3e^{-3}$	$-1.74e^{-2}$	$-1e^{-4}$	$-1.0e^{-2}$	$-5.5e^{-3}$	$-1e^{-4}$	$1.1e^{-3}$	$-3e^{-4}$	$-2.3e^{-2}$
	7.5	-58	-0.2	-33.7	-18.3	-0.2	3.8	-0.9	-100
8	$-7.3e^{-2}$	$-5.5e^{-3}$	$2.4e^{-3}$	$-4.7e^{-3}$	$2.6e^{-2}$	$5e^{-4}$	$-1.3e^{-3}$	$1e^{-4}$	$-5.5e^{-2}$
	-132.3	-10	4.4	-8.5	47.6	0.9	-2.4	0.2	-100
7	-1.3	$-5.7e^{-3}$	$4e^{-4}$	$-5.8e^{-3}$	$5e^{-4}$	0	$9e^{-4}$	$1e^{-4}$	$-2.2e^{-2}$
	-56.8	-25.6	1.9	-25.8	2.1	0	4	0.3	-100
6	$-5e^{-3}$	$-1.7e^{-2}$	$2e^{-4}$	$-1.4e^{-2}$	$-3.2e^{-3}$	$-1e^{-4}$	$5e^{-4}$	$-2e^{-4}$	$-3.8e^{-2}$
	-13.1	-44.3	0.5	-35.3	-8.4	-0.1	1.3	-0.4	-100

1) Advective sand transport items T_1, T_2

   In the dry season, the West shoal is dominated by tidal action and mostly by upwelling dynamics, and the net transport of material is landward. The net transport of suspended sand advection includes two items, T_1 and T_2. The net transport of sand is caused by the average flow term T_1. Except at point 4#, the material is transported to the sea, and material at the rest of the measurement points is transported to land.

   At point 8#, point 7#, and point 6#, substances transport to the land, the average flow term gradually decreases, and these points after the flow direction land, and the rising tide dynamics strength to reduce the law upstream. T_2 is caused by the tide Stokes drift term; point 8# accounts for a small proportion of only 10%. At points 4#, 6#, and 7#, the proportion is larger, -58.0%, -25.8%, and -44.3%, respectively.

2) Tidal trap sand transport item T_4

   When the phase of the variation of flow velocity and suspended sand content is close to each other, T_4 is close to 0. Points 7# and 6# are located in the middle and upper part of the West shoal, and point 4# is located in the lower part of the beach. T_4 accounts for a larger proportion, which is caused by the large phase difference between the variation of sand content and the variation of flow velocity caused by the convergence of water flow in the upper part of West shoal, as well as the exchange of sediment in the beach channel and the resuspension of sediment in the bottom boundary layer. The lower point 4# is caused by the inconsistency between the change of sand content and the change of flow velocity due to the effect of salt water chiseling, resulting in a larger T_4 item.

3) Vertical net circulation sand transport term T_5

   Point 4# is at the junction of high salinity shelf water and relatively low salinity estuarine mixed water body. The vertical circulation effect is obvious; the T_5 term is negative and occupies a significant proportion. Point 6# is at the junction of fresh water from the south branch of Jiaomen and a relatively high salinity estuarine mixed water; the T_5 term occupies a certain proportion but relatively less than the proportion of point 4#. The T_5 term is positive because the water from the north branch of the confluence section and the west side of Jijiao Men South Branch is merging with the water from the west side of Jijiao Men South Branch, and the velocity of the surface water increases, and the vertical net circulation is seaward.

4) The role of other sand transport items

T_6 and T_7 are the shear terms caused by time-averaged volume and tidal vibration shear, and T_8 is the effect of vertical tidal oscillation shear because the flow velocity is opposite to the vertical gradient direction of the sand content, the effect of shear diffusion term caused by tidal vibration term is not obvious and can be ignored in general.

4.2.2 *Suspended sand components and their variations*

At ebb-tide, the suspended sand from runoff at the West shoal is refined from land to sea; at flood tide, the resuspended sediment from the Lingding channel is transported to the West shoal, and the phenomenon of refinement from sea to land occurs, which reflects that West shoal has a two-way capture effect.

The composition of suspended sand in West shoal is chalky clay, and the composition of chalky sand and clay varies with the tide at each measurement point. Figure 5 shows the Gaussian fit of the suspended sand component, and the four curves represent the four periods of rising, falling, falling, and rising. S3 and S7 downstream of the West shoal show coarsening during the rising resting period, and S10 downstream of the West shoal in the Lingding channel shows coarsening during the rising urgency. S9 and S10 near the channel showed a sudden increase in bottom sand content during the rapid flood tide (Figure 6), and the maximum sand content in the bottom layer reached $0.3 km/m^3$, and the maximum turbidity zone appeared. The coarsening of S3 and S7 in the rising rest was related to the influence of the maximum turbidity zone. Coarse powder sand to very fine sand under the action of rising and falling tide, presenting stationary - fine particles sediment rise, suspension - coarse particles eruption transport at the same time for large-scale bottom-shaped movement - fine particles rise, suspension - stationary transport cycle mode(Franz *et al.* 2012).

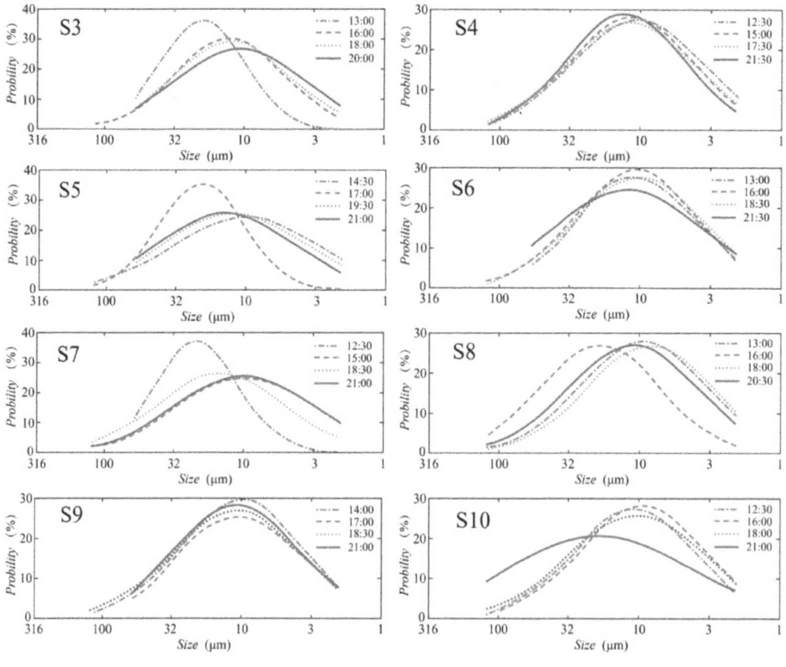

Figure 5. Gaussian fit for the content of each component of suspended sand.

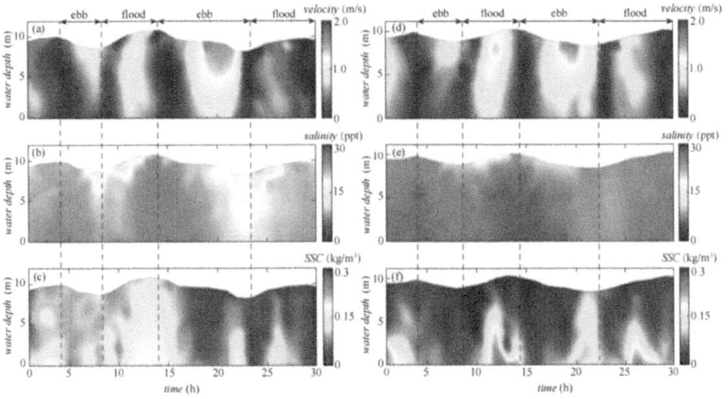

Figure 6. Salinity and suspended sand vertical variation graphs at S9 and S10.

4.2.3 *Numerical simulation results*

In our simulation, the salinity distribution (Figure 7), suspended sand distribution (Figure 8), and siltation distribution (Figure 9) of the Lingdingyang were calculated. In the dry season, West shoal is conducive to sediment deposition: 1) The flow velocity on the beach is small.

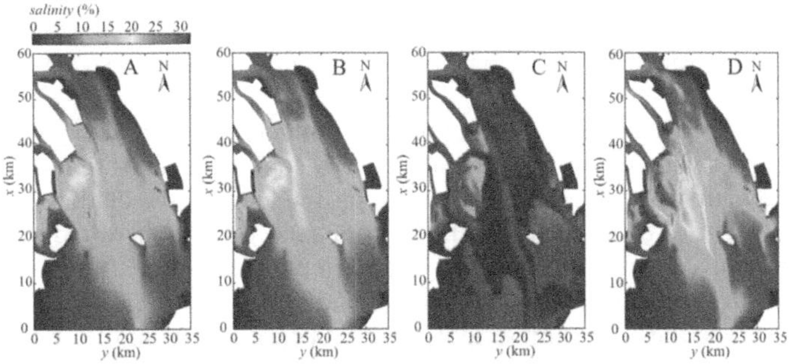

Figure 7. Salinity and flow field diagrams (A: surface layer at rapid ebb tide; B: bottom layer at rapid ebb tide; C: surface layer at rapid flood tide; D: bottom layer at rapid flood tide).

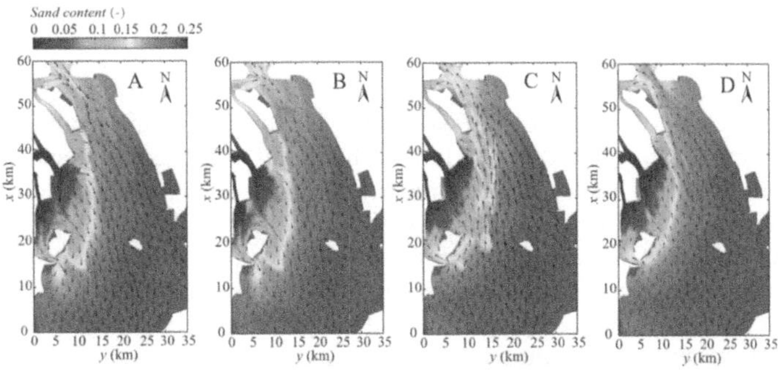

Figure 8. Sand concentration fields (A: surface layer at rapid ebb tide; B: bottom layer at rapid ebb tide; C: surface layer at rapid flood tide; D: bottom layer at rapid flood tide).

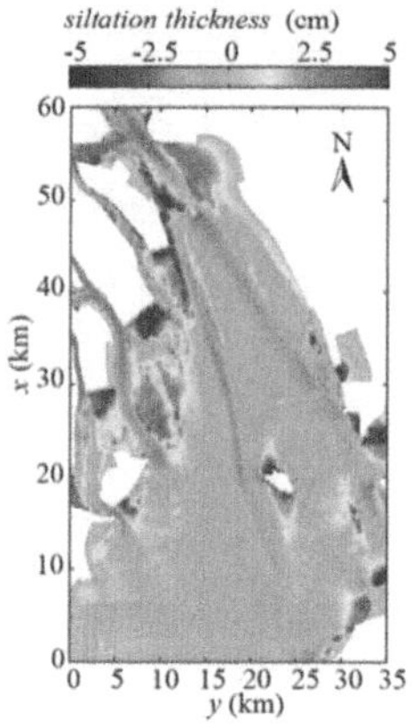

Figure 9. Numerical simulation of siltation distribution (cm/0.5a).

2) The dynamic environment is dominated by flood tide dynamics in the case of a small amount of sand coming from upstream. In the dry season, the flood tide dynamics will transport the sediment deposited in the outer Lingdingyang during the flood period to the West shoal. 3) The salinity environment is favorable to the sediment deposition of suspended sand; neither too large nor too small salinity is utilized for sediment deposition. The salinity has a great influence on the sedimentation speed; when the salinity is less than 20%, the sedimentation speed increases with the increase of salinity; in 4 ~ 6% and 14 ~ 20% increase faster, the rest of the range changes slowly, the high value of the sink speed near 20% (Chen 2002)..

The middle and lower sections of the Lingding channel are in the shape of a trumpet with a wide bottom and a narrow top, and this shape has a converging and strengthening effect on the flood tide dynamics; that is, it will strengthen the asymmetry of the tide. If the burst process of fine particles of sediment in the bottom sand occurs, the lifted sediment will be brought to the West shoal for deposition. The two flood tide channels on the West shoal will also enhance the transport of water and sand to the upstream side. The north branch is an ebb-tide channel, and the water and sand in the confluence section of the ebb-tide period will be discharged along West shoal, but the flat bottom conditions of the beach surface will make the water diffusion flow rate lower, and the coarse powder sand in the suspended sand will be deposited. The deepening of the channel will also restrict the development of West shoal to the channel.

5 CONCLUSION

The large tidal storage volume of the Pearl River Delta network river area, as well as the difference in response time between the longitudinal and transverse waterways of the delta for the ebb and flood tide, makes the difference in the flow velocity of the flood and ebb tide through West shoal, forming a dynamic condition dominated by the flood tide, thus promoting the net transport of suspended sand from the sea to the land. During ebb-tide, the flow velocity decreases due to the diffusion of the water from the north branch and the south branch of Jiaomen on the beach surface, and the natural sedimentation velocity of the powder sand in the suspended sand is accelerated so that the suspended sand appears to be refined from land to sea.

West shoal has dynamic conditions for capturing suspended sand in both directions during the dry season. The shallow beach with a water depth of 2m has the strongest capturing ability and is also the main siltation area. The dynamics environment is dominated by flood tide dynamics; the suspended sand from the sea is transported to land, while the mixed salt and freshwater environment are favorable to the deposition of suspended sand. Sand transport by advection is the main one, followed by sand transport by tidal capture and sand transport by vertical circulation.

REFERENCES

Chen Yao-tai, Modern Sedimentary Velocity and Sedimentary Environments in the Pearl River Mouth, *Acta Scientiarum Naturalium Universitatis Syatseni*, 1992, 31(2).

Chen Zishen, Analysis on Longitudinal Net and Circulations and Material Fluxes in Lingding Estuary, Pearl River and Adjacent Inner Shelf Waters, *Tropical Oceanology*, 1993, 12(4), 47–53.

Cheng He-qin, Song Bo, Xue Yuan-zhong, *et al.*, Mechanics on Transport of Coarser Silt and Very Fine Sand in the Changjiang Estuary- Episodic Re-suspension and Large-scale Bedform Movement, *Journal of Sediment Research*. No. 1, 2002, 20–27.

Cheng Liu, Jueyi Sui, and Zhao-Yin Wang, Sediment Load Reduction in Chinese Rivers, *International Journal of Sediment Research*. 23 (2008) 44–55.

Dyer K.R., 1986. *Coastal and Estuarine Sediment Dynamics*. John Wiley & Sons Ltd., New York. 342 pp.

Eisma D., 1986. Flocculation and Deflocculation in Estuarine Bays, *Netherlands Journal of Sea Research*. 20 (2/3), 183–199.

Franz G., Pinto L., Ascione I., Mateus M., Fernandes R., Leitão P. and Neves R. Modelling of Cohesive Sediment Dynamics in Tidal Estuarine Systems: A Case Study of Tagus Estuary, Portugal, *Estuarine, Coastal and Shelf Science* .151 (2014) 34–44.

Jay D.A. and Smith J.D., 1990. Residual Circulation in Shallow Estuaries: 1. Highly Stratified, Narrow Estuaries. *Journal of Geophysical Research*. 95(C1), 711–731.

Jiaxue Wu, James T. Liu and Xia Wang, Sediment Trapping of Turbidity Maxima in the Changjiang Estuary. *Marine Geology*. 2012,14–25.

Jiyu Chen, and Shenliang Chen, Challenges to China's Estuarine Coast, Marine Geology Letters. 2002,18 (1):1–5.

John Z. Shi, Tidal Resuspension and Transport Processes of Fine Sediment Within the River Plume in the Partially-mixed Changjiang River Estuary, China: A Personal Perspective, *Geomorphology*. 121 (2010) 133–151.

Keith R. Dyer, Estuaries. A Physical Introduction (2nd edition), 1996, 14–17.

Li Jiufa, and Zhang Chen, Sediment Resuspension and Implications for Turbidity Maximum in the Changjiang Estuary. *Marine Geology*. 148 (1998) 117–124.

Li Suqiong, and Li Bisheng, Exploration of Some Problems of Saltwater Wedge Activities in the Grinding Knife Gate, A Collection of Research Papers on a Comprehensive Investigation of the Coastal Zone and Seabed Resources in the Pearl River Estuary, Guangdong Science and Technology Press, 1986, 258–267.

Prandle D., 2004. Sediment Trapping, Turbidity Maxima, and Bathymetric Stability in Macrotidal Estuaries. *Journal of Geophysical Research*. 109 (C08001). doi: 10.1029/2004JC002271.

Ren Jie, Bao Yun and Lin Wei-qiang, Analyses on Water Suspended Sediment Fluxes in Lingdingyang Estuary of Zhujiang River Mouth. *Journal of Tropical Oceanography*. 2001, Vol. 20, No. 3, 35–40.

Schubel J.R. 1968. Turbidity Maximum of the Northern Chesapeake Bay. *Science*. 161, 1013–1015.

Uncles R J. Elliortt R C A and Weston S A. Observed Fluxes of Water, Salt, and Suspended Sediment in a Partly Mixed Estuary. *Estuarine, Coastal and Shelf Science*. 1985,20:147–167.

van Rijn L.C. 2012. Principles of Sedimentation and Erosion Engineering in Rivers, *Estuaries and Coastal Seas. Netherlands* p. 580.

Author index